# Third Madrid Summer School on Turbulence 2017



Journal of Physics: Conference Series Volume 1001

Madrid, Spain
29 May - 30 June 2017

**Editor:**

**Javier Jimenez**

ISBN: 978-1-5108-6168-8
ISSN: 1742-6588

**Printed from e-media with permission by:**

Curran Associates, Inc.
57 Morehouse Lane
Red Hook, NY 12571

proceedings.com

**Some format issues inherent in the e-media version may also appear in this print version.**



Printed by Curran Associates, Inc. (2018)



Institute of Physics
Dirac House, Temple Back
Bristol BS1 6BE UK

Phone: 44 1 17 929 7481
Fax: 44 1 17 920 0979

techtracking@iop.org

**Additional copies of this publication are available from:**

Curran Associates, Inc.
57 Morehouse Lane
Red Hook, NY 12571 USA
Phone: 845-758-0400
Fax: 845-758-2633
Email: curran@proceedings.com
Web: www.proceedings.com

# Table of contents

## Volume 1001

**Third Madrid Summer School on Turbulence**

**29 May to 30 June 2017, Madrid, Spain**

**Accepted papers received: 23 March 2018**
**Published online: 12 April 2018**

### Preface



### Papers

**STRUCTURES**

**THEORETICAL MODELS**



**COMPLEX FLOWS**

IOP Conf. Series: Journal of Physics: Conf. Series **1001** (2018) 011001 doi:10.1088/1742-6596/1001/1/011001

# Third Madrid Summer School on Turbulence

**Editor: Javier Jiménez**

## PREFACE

This third Summer School on Turbulence at the School of Aeronautics of the Universidad Politécnica de Madrid follows two previous ones, in 2013 and 2015. Those previous workshops were funded by the European Research Council Multiflow grant, whose goal was to improve our understanding of the multiscale dynamics of turbulence in fluids. After the initial grant ran its course, funding for the Summer School was transferred to the ERC grant Coturb, directed towards an understanding of turbulence from a structural point of view. The third Madrid Summer School took place from May 29 to June 30, 2017. As for the first two, its goal was to provide a meeting place for theoreticians, experimentalists and simulators, in which to develop and test new ideas on turbulence physics and structure. Around forty, mostly young, participants from twenty one international groups met for five weeks of collaborative work, primarily using the computational data archived in the receiving institution but, in many cases, also contributing their own. Although the format included some invited formal seminars and periodic plenary meetings, most of the work took place in small groups whose composition occasionally changed during the workshop. These proceedings reflect the results of the work of these groups, which, in many cases, later continued in the form of new collaborations.

erc
POLITÉCNICA
'Ingeniamos el futuro'

IOP Conf. Series: Journal of Physics: Conf. Series **1001** (2018) 011001 doi:10.1088/1742-6596/1001/1/011001

## Workshop organiser:

Javier Jiménez
School of Aeronautics, Universidad Politécnica de Madrid
28040 Madrid, Spain

### Acknowledgements

The organizers are grateful to the European Research Council for their funding of this meeting, and to the School of Aeronautics of the Universidad Politécnica de Madrid for their help in hosting it. The meeting would not have been possible without the dedication of those participants who acted as hosts to the visiting groups, facilitating everything from everyday problems to computer access and data usage: José I. Cardesa, Miguel P. Encinar, Oscar Flores, Adrián García-Gutierrez, Ricardo García-Mayoral, Tony Kwon, Adrián Lozano-Durán, Atsushi Sekimoto, Vassilis Theofilis, Eusebio Valero and Alberto Vela-Martín. Very especially, we are indebted to Aurora García Sarrión, without whose administrative help nothing would have worked.

### Picture credits

The cover figure displays the temporal evolution of a compact structure of the wall-normal velocity, as synthesised from its spatio-temporal correlation in a turbulent channel at $Re_\tau = 950$. Several snapshots are overlaid in the same figure, with time increasing with increasing forward tilting. Colours code the structure intensity, which is maximum at the central approximately vertical stage of the evolution. The figure is due to Adrián García-Gutierrez. The picture of the participants was graciously provided by Miguel P. Encinar.

IOP Conf. Series: Journal of Physics: Conf. Series **1001** (2018) 011002 doi:10.1088/1742-6596/1001/1/011002

# Peer review statement

All papers published in this volume of *Journal of Physics: Conference Series* have been peer reviewed through processes administered by the proceedings Editors. Reviews were conducted by expert referees to the professional and scientific standards expected of a proceedings journal published by IOP Publishing.

IOP Conf. Series: Journal of Physics: Conf. Series **1001** (2018) 012001 doi:10.1088/1742-6596/1001/1/012001

# Reynolds stress structures in a self-similar adverse pressure gradient turbulent boundary layer at the verge of separation.

**C.Atkinson[1], A. Sekimoto[1], J. Jiménez[2] and J. Soria[1,3]**

[1] Laboratory for Turbulence Research in Aerospace and Combustion, Department of Mechanical and Aerospace Engineering, Monash University, Victoria 3800, Australia
[2] School of Aeronautics, Universidad Politécnica de Madrid, Pz Cardenal Cisneros 3, 28040 Madrid, Spain
[3] Department of Aeronautical Engineering, King Abdulaziz University, Jeddah 21589, Kingdom of Saudi Arabia

E-mail: callum.atkinson@monash.edu

**Abstract.** Mean Reynolds stress profiles and instantaneous Reynolds stress structures are investigated in a self-similar adverse pressure gradient turbulent boundary layer (APG-TBL) at the verge of separation using data from direct numerical simulations. The use of a self-similar APG-TBL provides a flow domain in which the flow gradually approaches a constant non-dimensional pressure gradient, resulting in a flow in which the relative contribution of each term in the governing equations is independent of streamwise position over a domain larger than two boundary layer thickness. This allows the flow structures to undergo a development that is less dependent on the upstream flow history when compared to more rapidly decelerated boundary layers. This APG-TBL maintains an almost constant shape factor of $H = 2.3$ to 2.35 over a momentum thickness based Reynolds number range of $Re_{\delta_2} = 8420$ to 12400. In the APG-TBL the production of turbulent kinetic energy is still mostly due to the correlation of streamwise and wall-normal fluctuations, $\langle uv \rangle$, however the contribution form the other components of the Reynolds stress tensor are no longer negligible. Statistical properties associated with the scale and location of sweeps and ejections in this APG-TBL are compared with those of a zero pressure gradient turbulent boundary layer developing from the same inlet profile, resulting in momentum thickness based range of $Re_{\delta_2} = 3400$ to 3770. In the APG-TBL the peak in both the mean Reynolds stress and the production of turbulent kinetic energy move from the near wall region out to a point consistent with the displacement thickness height. This is associated with a narrower distribution of the Reynolds stress and a 1.6 times higher relative number of wall-detached negative $uv$ structures. These structures occupy 5 times less of the boundary layer volume and show a similar reduction in their streamwise extent with respect to the boundary layer thickness. A significantly lower percentage of wall-attached structures is observed in the present case when compared with a similar investigation of a rapidly decelerating APG-TBL, suggesting that these wall-attached features could be the remanent from the lower pressure gradient domain upstream.

## 1. Introduction

In a desire to understand the complex dynamics that drive the turbulent boundary layer that develops when a flow passes over a solid surface, we often focus upon the canonical zero pressure gradient turbulent boundary layer (ZPG-TBL). This ZPG-TBL can originate from the leading edge of a thin plate and retain a zero and hence constant non-dimensional pressure gradient as it develops along the plate. While examples of such a flow can be found in nature and industry, the occurrence of constant zero pressure gradient flow pales in comparison to the cases in which the pressure gradient varies in the streamwise direction. The case of an adverse pressure gradient is of particular interest owing to the resulting momentum deficit and if the pressure gradient is strong enough, the separation of the boundary layer, which can significantly affect both the mean and fluctuating forces that are experienced by the surface or in the wake of a body. Even in the absence of separation, the presence of a strong pressure gradient significantly alters the mean and fluctuating velocity profiles, resulting in peak turbulent activity and production moving from the vicinity of the wall to a location close to the displacement thickness [1]. In all boundary layers it is the product of the Reynolds stresses and the mean velocity gradients that drives the production of turbulent kinetic energy, with spatially and temporally coherent regions of intense Reynolds stress subsequently being responsible for the majority of the turbulence production and most of the wall-normal transfer of momentum. It is for this reason that such structures have formed the focus of numerous investigations [2, 3, 4].

Recently, following the quadrant analysis of Wallace et al. [3], Lozano-Durán et al. [5] performed a detailed statistical investigation of the extent, orientation, density and spacing of coherent three-dimensional Reynolds stress structures using data from direct numerical simulation (DNS) of turbulent channel flows at $\mathrm{Re}_\tau = 934$ and 2003. This study found that most of the mean Reynolds stress was contained in wall-attached structures, which tended to be larger than the smaller, more isotropically aligned, wall-detached structures. Far less is known about how these structures behave in regions of adverse pressure gradient, which as the pressure gradient increases result in substantially different wall-normal Reynolds stress profiles [6, 7]. Unlike the ZPG-TBL, an adverse pressure gradient turbulent boundary layer (APG-TBL) is almost always associated with a change in pressure gradient, such as that over a curved surface, meaning the flow at a given point along the surface is generally a function of not only the local pressure gradient, but also the pressure distribution that precedes it. Using a series of large eddy simulations (LES) Bobke et al. [8] compared the turbulent statistics associated with a number of APG-TBLs under the influence of a near constant non-dimensional pressure gradient, with APG-TBL at the same displacement thickness based Reynolds number developing under a non-constant non-dimensional pressure gradient. Results demonstrated that APG-TBL over flat surfaces required a streamwise development distance of approximately 7 boundary layer thickness to overcome the history of their evolution and approach the local turbulent statistics associated with the equivalent constant non-dimensional pressure gradient.

Maciel et al. [9] applied the same methodology as Lozano-Durán et al. [5] to investigate Reynolds stress structures in a DNS of a ZPG-TBL and in DNS of a rapidly decelerating non-self-similar APG-TBL, just upstream of a thin separation bubble. Sweeps and ejections were found to be longer in the ZPG-TBL, ejections extending to streamwise lengths of 5 boundary layer thicknesses $\delta$, yet were both shorter and more inclined with respect to the wall, but occupied a larger proportion of the overall volume in the APG-TBL, with respect to the local boundary layer thickness. Near wall and large scale wall-attached sweeps and ejections were found to be far less numerous in the APG-TBL, consistent with the turbulent activity being lifted away from the wall. In the case of the APG-TBL, statistics were accumulated over a domain spanning five local boundary thickness ($5\delta$), over which the shape factor $H = \delta_1/\delta_2$ demonstrated a substantial variation from H = 1.97 to 3.42. It is unclear what influences this rapid deceleration might have on the identified structures, the extent of which can cover this entire domain.

One way to separate the influence of upstream flow evolution is to consider self-similar APG-TBLs, where the flow evolves over a domain in which the relative contribution from each term in the governing equations remains approximately independent of streamwise position [10, 11, 12]. As a consequence the statistics in these self-similar boundary layers collapse under a local outer velocity $U_e$ and displacement thickness scaling $\delta_1$, with the exception of an exceedingly smaller near wall region which continues to follow a viscous scaling until separation, analogous to the dual inner and outer scaling that exists in a ZPG-TBL. A range of self-similar APG-TBL can also be found in which the non-dimensional pressure gradient, $\beta = \delta_1 P_e/\tau_w$, where $P_e$ is the far-field pressure gradient and $\tau_w$ is the mean wall shear stress, remains constant. These particular cases allow for the exploration of the structure of an APG-TBL that experiences the influence of a constant contribution from the pressure gradient term in the governing equations.

In the present study, DNS of both a ZPG-TBL over a domain of $8.8\delta$ with momentum thickness based $\mathrm{Re}_{\delta_2}$ = 3400-3770 and a self-similar APG-TBL at the verge of separation over a domain of $2.6\delta$ with $\mathrm{Re}_{\delta_2}$ = 8430-12400 [13], are used to investigate the influence of a strong constant non-dimensional pressure gradient on the instantaneous Reynolds stress structures that are present in a flow near the point of separation. To be consistent with previous studies [5, 9], detection of these instantaneous Reynolds stress structures is performed based on a threshold of the instantaneous product of the streamwise $u$ and wall-normal $v$ velocity fluctuations and the use of a three-dimensional clustering algorithm.

## 2. Boundary layer datasets

The ZPG-TBL and APG-TBL data sets were computed by solving the incompressible Navier-Stokes equations in a three-dimensional rectangular volume, with constant kinematic viscosity $\nu$ and a fluid density of unity. The three flow directions are streamwise ($x$), wall-normal ($y$) and spanwise ($z$), with mean velocity components in these directions denoted by $\langle U\rangle$, $\langle V\rangle$ and $\langle W\rangle$, respectively, with velocity fluctuations represented by $u$, $v$, $w$. The desired pressure gradient is set by applying a wall-normal velocity at the upper surface of the simulations, combined with an irrotationality condition. The distribution of wall-normal velocity required to simulate an APG-TBL at the verge of separation was determined based on the potential flow solution in an expanding duct, corrected for the growth of the boundary layer and was chosen to satisfy an expected mean outer streamwise velocity distribution following Mellor & Gibson [11]. An inlet recycling and rescaling was used for both boundary layers. Further details of the boundary condition and recycling can be found in Kitsios et al. [13]. For details of the numerical methodology, parallelism and inlet recycling the reader is referred to Borrell, Sillero & Jiménez [14] and Sillero [15]. Numerical details associated with the data sets are given in table 1. Note the domain of interest ($DoI$) used for the detection of structures in the APG-TBL is slightly larger than the self-similar domain to ensure that the full streamwise extent of the structures within the domain of interest is captured. The domain and detailed evaluation of this self-similar APG-TBL can be found in Kitsios et al. [13].

Throughout this paper the outer velocity $U_e$, displacement thickness $\delta_1$ and momentum thickness $\delta_2$ are defined as by Spalart & Watmuff [16] using the spanwise vorticity based velocity proposed by Lighthill [17]. Under this definition the outer velocity is given by

$$U_e(x) = U_\Omega(x, y_\Omega), \quad \text{where} \tag{1}$$

$$U_\Omega(x,y) = -\int_0^y \langle \Omega_z\rangle (x,\tilde{y})\, \mathrm{d}\tilde{y}, \tag{2}$$

$\langle \Omega_z\rangle$ is the mean spanwise vorticity and $y_\Omega$ is the distance from the wall at which this vorticity has decayed to 0.2% of the mean vorticity at the wall. This definition was used to account for the absence of a uniform free stream velocity in the case of the strong adverse pressure gradient. Following this definition the displacement and momentum length scales are given by

$$\delta_1(x) = \frac{-1}{U_e}\int_0^{y_\Omega} y\,\langle \Omega_z\rangle (x,y)\, \mathrm{d}y, \quad \text{and} \tag{3}$$

$$\delta_2(x) = \frac{-2}{U_e^2}\int_0^{y_\Omega} yU_\Omega\,\langle \Omega_z\rangle (x,y)\, \mathrm{d}y - \delta_1(x), \text{ respectively.} \tag{4}$$

**Table 1.** Numerical details of the ZPG and APG-TBL DNS data sets. Domain sizes are non-dimensionalized by the displacement thickness ($\delta_1$) at the position $x_*$ where $\mathrm{Re}_{\delta_1} = 4800$ and by the displacement thickness at the center of the domain of interest $x_{DoI}$.

| | | ZPG | APG |
|---|---|---|---|
| Nominal pressure gradient | $\beta$ | 0 | 39 |
| Streamwise data points | $N_x$ | 8193 | 8193 |
| Wall-normal data points | $N_y$ | 315 | 1000 |
| Spanwise data poitns | $N_z$ | 1362 | 1362 |
| Streamwise domain size | $L_x/\delta_1(x_*)$ | 480 | 303 |
| Wall-normal domain size | $L_y/\delta_1(x_*)$ | 22.7 | 73.4 |
| Spanwise domain size | $L_z/\delta_1(x_*)$ | 80.1 | 50.7 |
| Streamwise grid spacing | $\Delta x/\delta_1(x_*)$ | 0.0585 | 0.0370 |
| Wall-normal spacing at wall | $\Delta y_{wall}/\delta_1(x_*)$ | $1.53 \times 10^{-3}$ | $9.71 \times 10^{-4}$ |
| Wall-normal spacing free stream | $\Delta y_{\infty}/\delta_1(x_*)$ | 0.0992 | 0.254 |
| Spanwise grid spacing | $\Delta z/\delta_1(x_*)$ | 0.0585 | 0.0370 |
| Displacement thickness ratio | $\delta_{1,DoI}/\delta_{1,DoI,APG}$ | 0.098 | 1 |
| Reynolds number range in DoI | $\mathrm{Re}_{\delta_1}$ | $4460 \to 5140$ | $18450 \to 30200$ |
| Reynolds number range in DoI | $\mathrm{Re}_{\delta_2}$ | $3400 \to 3770$ | $8430 \to 12400$ |
| Streamwise length of DoI | $L_{DoI}/\delta_1(x_{DoI})$ | 54 | 11 |

In the case of the ZPG-TBL this outer velocity and length scales are consistent with classical definitions.

## 3. Flow description and comparison

The substantial differences in the flow between the ZPG-TBL and the APG-TBL at the verge of separation are demonstrated by the mean velocity and the $\langle uv \rangle$ Reynolds stress profiles in figure 1. Profiles are shown for both local outer velocity $U_e$ and displacement thickness $\delta_1$ scaling and inner viscous scaling based on the local friction velocity $u_\tau$ and kinematic viscosity $\nu$, denoted by the $^+$ superscript. Profiles are presented for the beginning, middle and end of the domain that will be used for the structural detection. The ZPG-TBL shows a good collapse under both scalings over this range, while the APG-TBL profiles remain reasonably consistent under the outer scaling. As discussed in section 2 the domain used for the present structural detection in the APG-TBL is larger than the self-similar domain investigated in Kitsios et al. [13], resulting in the small variation in the present profiles when compared to those of Kitsios et al. [13]. Unlike the ZPG-TBL, which possess a fairly broad mean Reynolds stress profile, the peak Reynolds stress in the APG-TBL occurs further from the wall and is confined to a much narrower region of the flow in regards to both inner and outer length scales.

The contribution of the $\langle uv \rangle$ component of the mean Reynolds stress to the production of turbulent kinetic energy is shown in figure 2. In the ZPG-TBL the Reynolds stress results in a single peak production in the vicinity of $y^+ = 14$ viscous units above the wall. The APG-TBL maintains a similar peak that move slightly closer to the wall in terms of both inner and outer units, however this pales in comparison to the peak that develops in the vicinity of $y = \delta_1$. Given the logarithmic axis used for these plots the relative contribution of these peaks is better illustrated using the premultiplied profile in figure 3, which highlight the fairly even contribution to the total kinetic energy production in the ZPG-TBL and the negligible role of the inner production peak in the APG-TBL.

In the ZPG-TBL, the negligible mean streamwise velocity gradients $\frac{\partial\langle U\rangle}{\partial x}$, and negligible mean wall-normal and spanwise velocities $\langle V\rangle$, $\langle W\rangle$, leaves the $\langle uv\rangle$ Reynolds stress component as the only component that makes any significant contribution to the production, given by

$$\mathcal{P} = -\langle u_i u_j\rangle S_{ij}, \quad \text{where} \tag{5}$$

$$S_{ij} = \frac{1}{2}\left(\frac{\partial\langle U_i\rangle}{\partial x_j} + \frac{\partial\langle U_j\rangle}{\partial x_i}\right). \tag{6}$$

This is demonstrated by the comparison of the total production and the $\langle uv\rangle$ associated production shown in figure 3(b). For the ZPG-TBL the difference is indeed negligible, however in the APG-TBL the deceleration of the flow in the streamwise direction and the larger wall-normal velocity component result in a noticeable contribution from $\langle uu\rangle$ and $\langle vv\rangle$ components

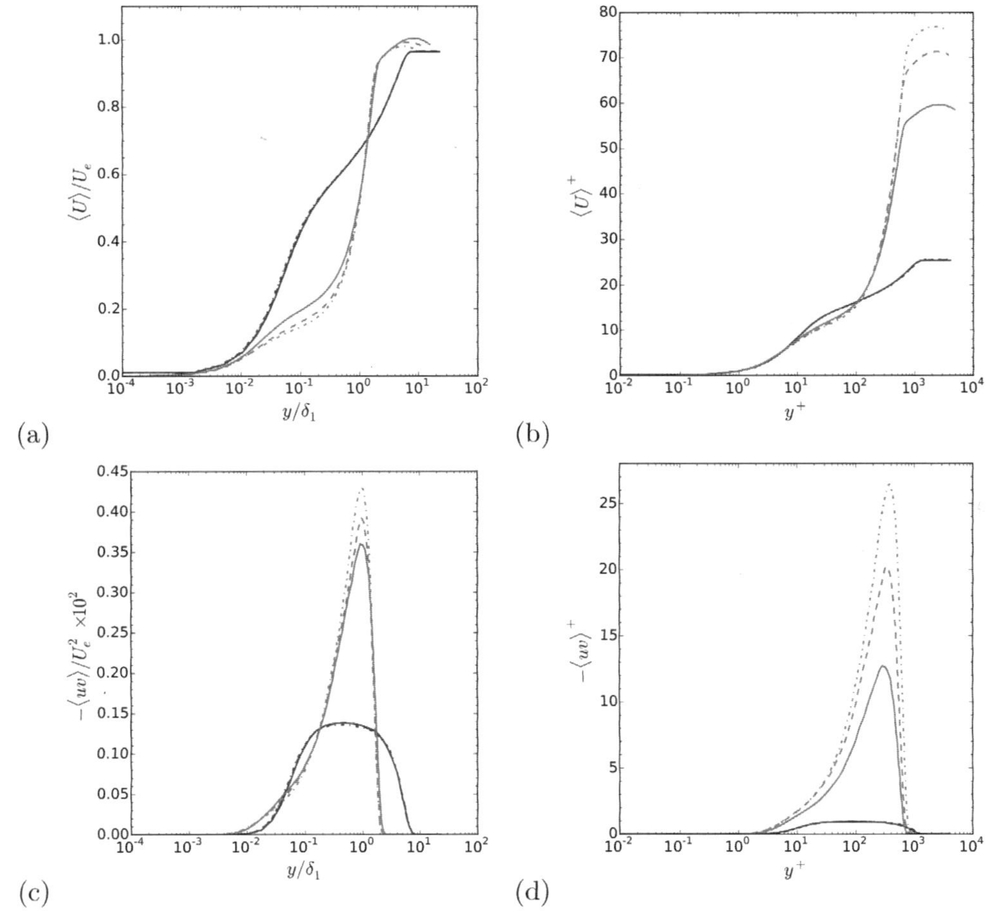


**Figure 1.** Profiles of mean streamwise velocity (a,b) and $\langle uv\rangle$ Reynolds stress (c,d) non-dimensionalized by the local mean outer velocity $U_e$ and displacement thickness $\delta_1$ (a,c) and by viscous units based on local friction velocity $u_\tau$ and kinematic viscosity $\nu$ (b,d). Individual profiles are plotted for the beginning (——), middle (– – –) and end (— · —) of the domain of interest of the ZPG-TBL, and the beginning (——), middle (– – –) and end (— · —) of the domain of interest of the APG-TBL.

IOP Conf. Series: Journal of Physics: Conf. Series **1001** (2018) 012001 doi:10.1088/1742-6596/1001/1/012001

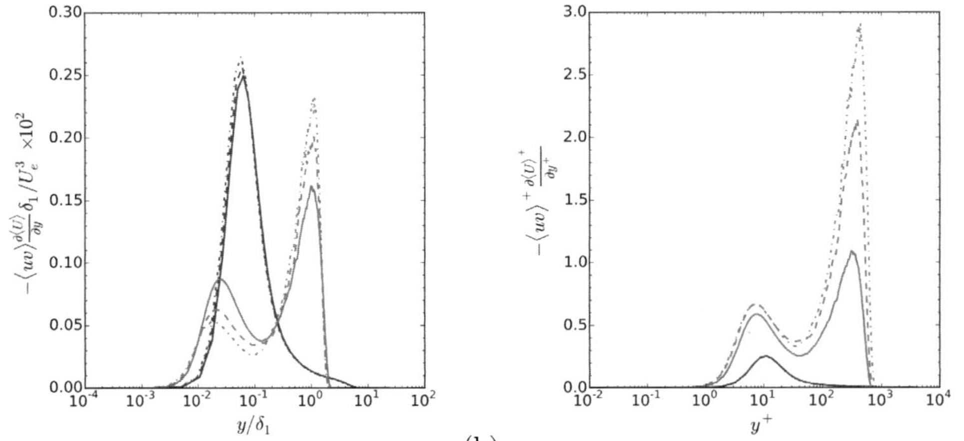


(a) (b)

**Figure 2.** Production of turbulent kinetic energy associated with the $\langle uv \rangle$ Reynolds stress component, non-dimensionalized by the local mean outer velocity $U_e$ and displacement thickness $\delta_1$ (a) and by viscous units based on local friction velocity $u_\tau$ and kinematic viscosity $\nu$ (b). Individual profiles are plotted for the beginning (——), middle (- - -) and end (— · —) of the domain of interest of the ZPG-TBL, and the beginning ( ), middle ( ) and end ( · ) of the domain of interest of the APG-TBL.

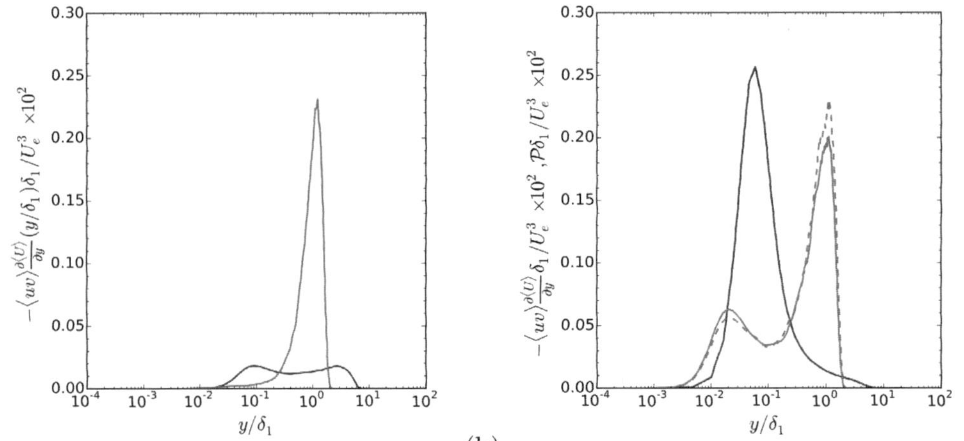


(a) (b)

**Figure 3.** Production of turbulent kinetic energy associated with the $\langle uv \rangle$ Reynolds stress component non-dimensionalized by the local mean outer velocity $U_e$ and displacement thickness $\delta_1$ premultiplied by the height above the wall (a) and relative to the total production associated with all terms of the Reynolds stress tensor $\mathcal{P} = -\langle u_i u_j \rangle S_{ij}$ where $S_{ij}$ is the mean strain rate tensor (b). Profiles are plotted for the middle (——) of the domain of interest of the ZPG-TBL, and the middle (——) of the domain of interest of the APG-TBL for the contribution of the $\langle uv \rangle$ component. The profiles for the corresponding total production are given by (- - - -) and (- - - -).

that reduces the total production at the inner peak and increase the peak in the outer region. Nevertheless, the majority of the contribution is still associated with $\langle uv \rangle$.

The degree to which the streamwise and wall-normal components remain correlated is shown in figure 4. In both boundary layers these fluctuations remains negatively correlated with a peak correlation occurring at around $y^+ = 10$ in the ZPG-TBL, moving to around $y^+ = 4$ in the APG-TBL. In the APG-TBL this correlation decreases slightly through the buffer layer before increasing again through the outer layer, consistent with the influence of near wall streaks and large scale low-speed and high-speed structures in the outer layer. Despite this dip the correlation remains in the vicinity of 0.4 for most of the ZPG-TBL. The APG-TBL involves a much lower correlation of around 0.2 in the near wall region, rising to a correlation of 0.4 at the outer production peak. A similar trend was observed for the rapidly decelerating APG [6], indicating that in both cases the adverse pressure gradient reduced the orderly structure of the flow, particularly in the near wall region, while a high correlation remains in the outer region.

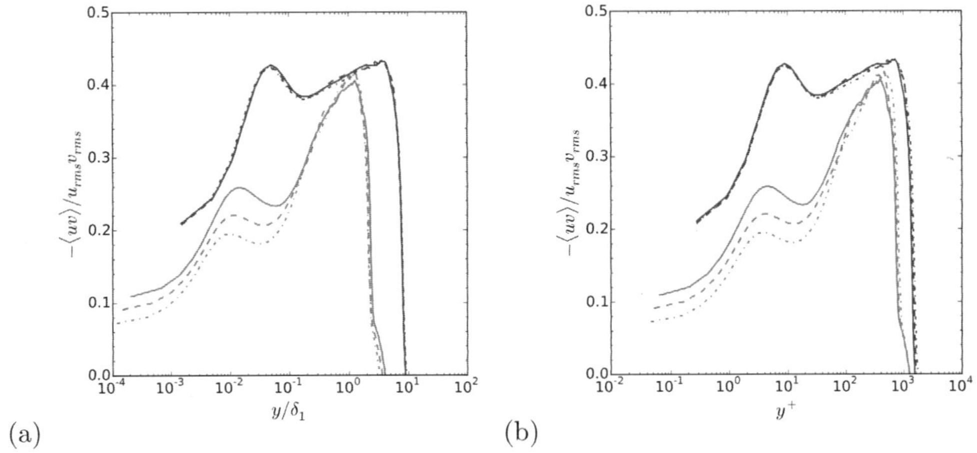


**Figure 4.** Correlation of streamwise $u$ and wall-normal $v$ velocity fluctuations non-dimensionalized by the local displacement thickness $\delta_1$ (a) and by viscous units based on local friction velocity $u_\tau$ and kinematic viscosity $\nu$ (b). Individual profiles are plotted for the beginning (——), middle (– – –) and end (— · —) of the domain of interest of the ZPG-TBL, and the beginning (——), middle (– – –) and end (— · —) of the domain of interest of the APG-TBL.

## 4. Structure detection methodology

From section 3 it is clear that despite significant differences in the Reynolds stress profiles, the individual structures that result in the mean $\langle uv \rangle$ Reynolds stress, continue to play a significantly role in APG-TBLs. The structures that provide the greatest contribution can be defined as spatially coherent regions in which the product of the local fluctuating streamwise and wall-normal velocity components, $uv$, is large with respect to the background fluctuations at that location. Following Lozano-Durán et al. [5] these structures can be identified by first satisfying the local condition

$$\frac{|uv(x,y,z)|}{u_{rms}(x,y)v_{rms}(x,y)} > H^*, \tag{7}$$

where $u_{rms}$ and $v_{rms}$ represent the root mean square velocity fluctuations and $H^*$ is the hyperbolic hole size associated with a percolation analysis that is performed to determine the value of $H^*$ that provides an appropriate balance between the volume of the largest detected structure with respect to the volume of the domain of interest and the number of individual detected structures. The effect of varying $H^*$ is demonstrated in figure 5 with a sudden drop in the largest volume $V_{largest}$ indicating the level at which the largest structure begins to be broken into seperate structures. In the ZPG-TBL this crisis occurs for $H^* = 0.9$ compared to $H^* = 1.75$ for the APG-TBL. In order to be consistent with previous studies [5, 9] a decision was made to use $H^* = 1.75$ for both flows.

Individual structures are computed by clustering adjacent cells in the dataset where equation 7 is satisfied, cells share a common face, and all cells fall into the same $uv$ quadrant. Quadrants are defined following Wallace et al. [3] where $Q_1$ represents the quadrant where $u > 0$, $v > 0$, $Q_2$ where $u < 0$, $v > 0$, $Q_3$ where $u < 0$, $v < 0$ and $Q_4$ where $u > 0$, $v < 0$. $Q_2$ structures are commonly referred to as *ejections* due to their role in transporting low momentum flow away from the wall, while $Q_4$ structure are referred to as *sweeps*, which transport high momentum flow towards the wall. It is these sweeps and ejections that provide the negative $uv$ correlation associated with the mean Reynolds stress and the production of turbulent kinetic energy. We will refer to this combination of $Q_2$ and $Q_4$ structures as $Q_-$.

Following Maciel et al. [9] individual structures were rejected using the criteria

$$V_{struct} < 3\Delta x^3. \tag{8}$$

To prevent possible biasing of the statistics, structure that cross the boundaries of the domain of interest were also rejected. In the case of the ZPG-TBL structures were detected using 50 statistically independent snapshots of the domain of interest and 39 independent snapshots for the APG-TBL.

## 5. Results

The number of identified structures in each quadrant $N_Q$ relative to the total number of identified structures $N_{total}$ and the percentage of the total volume occupied by the combined volume of different quadrant structures $V_Q$ are presented in table 2. The total volume $V_{total}$ was specified as the volume from the wall up to the mean boundary layer thickness. The ZPG-TBL shows a propensity for $Q_2$ ejections followed by $Q_4$ sweeps, combining to make up 60% of the detected structures. $Q_2$ and $Q_4$ events also occupy a much larger proportion of the volume indicating that

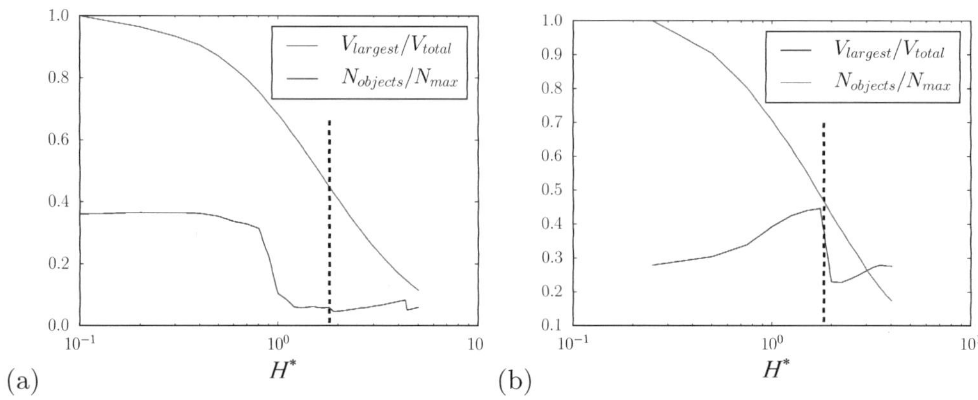


**Figure 5.** Percolation analysis for the identification of $uv$ structures in the ZPG-TBL (a) and the APG-TBL (b) showing the effect of chosen hyperbolic hole size $H^*$ on the volume ratio of the largest structure $V_{largest}$ and on the number of detected structures $N_{objects}$ relative to the maximum number of detected structures $N_{max}$.

they are not only more numerous but also larger than both the $Q_1$ and $Q_3$ structures. In the case of this strong APG-TBL the relative number of structures in each quadrant are far more even with the $Q_-$ structures now only making up just over 50% of the detected structures. Similar difference were observed by Maciel et al. [9] between the relative number of $Q_-$ structures in ZPG-TBL and rapidly decelerating non-self-similar APG-TBL, however the individual number of sweeps and ejects were not reported. The relative number of different quadrant structures in the present ZPG-TBL are similar to the those reported in DNS of a fully developed channel flow [5].

Unlike the previous APG-TBL study, which reports an increase in the relative volume that is occupied by $Q_-$ structures from 8.4% in the ZPG-TBL to 10.6% equivalent in the rapidly decelerating APG-TBL, the present study shows a significant decrease in the volume occupied by all the different quadrant structures from 33% in the ZPG-TBL to 6.3% in the strong self-similar APG-TBL. This is consistent with the narrower peak in the Reynolds stress away from the wall in the APG-TBL, compared to the more even distribution across the ZPG-TBL. It is not clear why the volume occupied by $Q_-$ structures in the present ZPG-TBL is so much higher than that reported by Maciel et al. [9], however it maybe associated with smaller domain used in their analysis of the ZPG-TBL, with their PDFs of the streamwise extent of detected structure indicating that a number of structure reach the bounds of their domain of interest.

The wall-normal extent of the $Q_-$, $Q_2$ and $Q_4$ structures can be examined by considering the joint probability density function (PDF) of the nearest and furthest point from the wall for the structures. In the ZPG-TBL the peak in the near wall region indicates that the majority of structures extend down to the wall, with heights up to the boundary layer thickness. The clustering along the diagonal line represents the extent of structures that begin at a fixed height $y_{min}$ above the wall. The averaged values of the lower boundary of each structure $y_{min}$, upper bound $y_{max}$ and centroid $y_{cent.}$ are presented in table 4. PDFs for the APG-TBL demonstrate a considerably larger percentage of structures that begin above the wall, referred to here as a wall-detached structures. The relative number and volume of wall-attached structures, where $y_{min} < 0.2\delta_1$, and near wall structures, where $y_{cent.} < 0.2\delta_1$, are given in table 3. The percentage of attached structures is approximately 4 times higher for the ZPG-TBL with 3 times as many near wall structures. Maciel et al. [9] report a slightly higher percentage of wall-attached $Q_-$ and near wall structures in their ZPG-TBL but also report a percentage of attached structures in their rapidly decelerating APG-TBL roughly 3 times higher than that observed in the present self-similar case.

In the present APG-TBL a larger relative volume is occupied by a smaller fraction of attached structure in the APG-TBL, when compared to the ZPG-TBL, indicating that the structures

**Table 2.** Number of structures identified in each quadrant $N_Q$ for the ZPG-TBL and the APG-TBL relative to the total number $N_{total}$ of structures identified and the combined volume occupied by structures of each quadrant $V_Q$ relative to the total volume $V_{total}$.

| | | $Q_1$ | $Q_2$ | $Q_3$ | $Q_4$ | $Q_-$ |
|---|---|---|---|---|---|---|
| ZPG-TBL | $N_Q/N_{total}$ | 0.202 | 0.327 | 0.195 | 0.276 | 0.603 |
| | $V_Q/V_{total}$ | 0.051 | 0.197 | 0.048 | 0.136 | 0.333 |
| APG-TBL | $N_Q/N_{total}$ | 0.253 | 0.257 | 0.237 | 0.253 | 0.510 |
| | $V_Q/V_{total}$ | 0.010 | 0.039 | 0.010 | 0.024 | 0.063 |

IOP Conf. Series: Journal of Physics: Conf. Series **1001** (2018) 012001 doi:10.1088/1742-6596/1001/1/012001

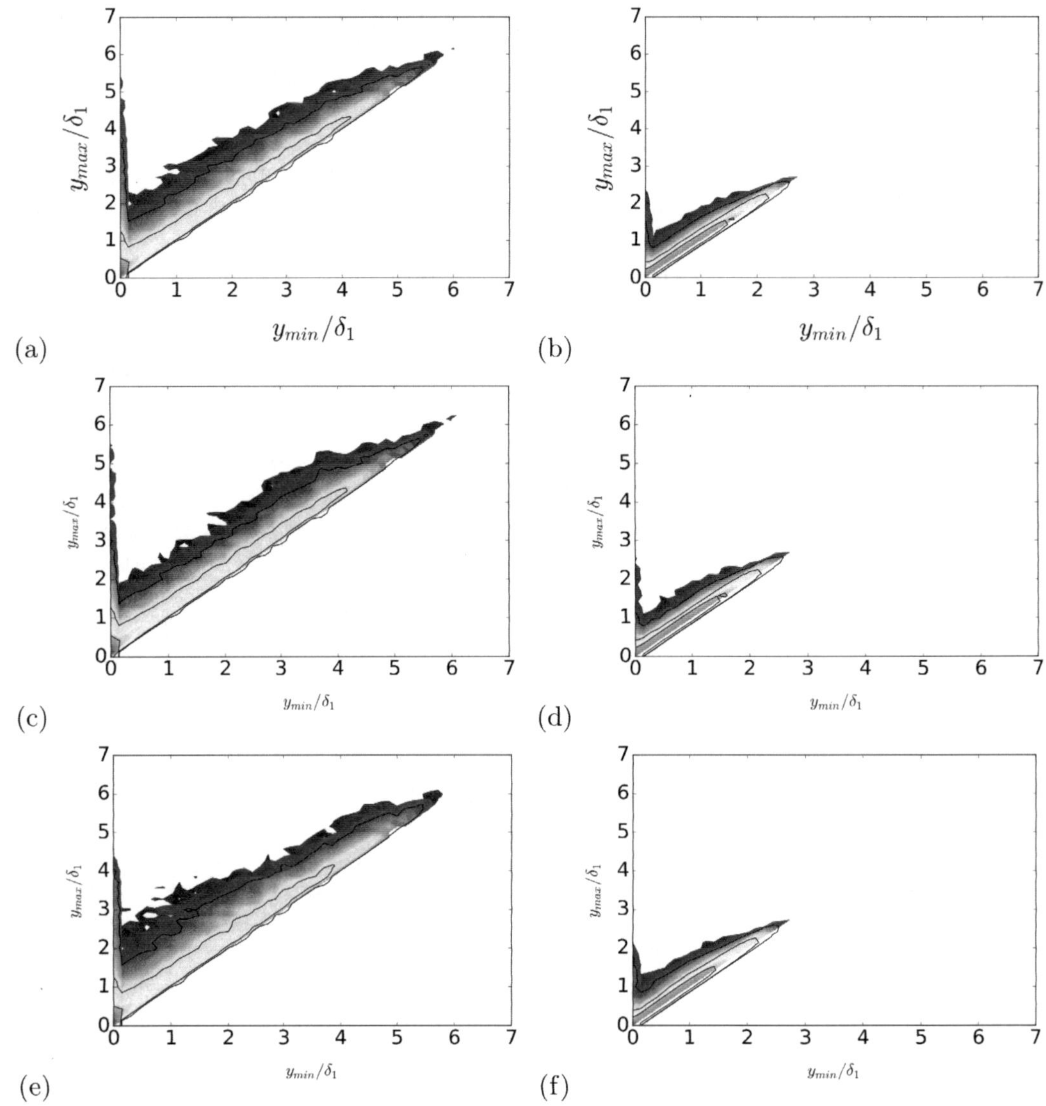


**Figure 6.** Joint PDFs of the wall-normal extent of the $Q_-$ structures in the ZPG-TBL (a) and APG-TBL (b), the $Q_2$ structures in the ZPG-TBL (c) and APG-TBL (d), and the $Q_4$ structure in the ZPG-TBL (e) and APG-TBL (f). Contour levels are shown for 1, 0.1, 0.01, and 0.001.

that remain must be significantly larger. In both boundary layers the $Q_4$ structures show a greater probability of being wall-attached and for a given $y_{min}$ tend to have a larger $y_{max}$ value, indicating that the detached $Q_4$ structures tend to be taller than their $Q_2$ counterparts.

Figure 7 shows the joint PDFs associated with the streamwise $L_x$ and wall-normal $L_y$ extent of the $Q_-$ structures and those of the streamwise and spanwise extent $L_z$. The dashed-lines represent the correlation between the spatial extent of the structures, indicating that for both boundary layers the height of the $Q_-$ structures tend to be 0.65 times their length and 1.1 times their span. Despite this similarity, the ZPG-TBL also demonstrates a tendency towards longer structures with a shorter wall-normal extent. This trend is reversed in the APG-TBL as a tendency develops towards taller structures, consistent with the increased Reynolds stress

**Table 3.** Number and volume of wall-attached $Q_{-attached}$ and near wall $Q_{-nearwall}$ structures for the ZPG-TBL and the APG-TBL relative to the number $N_{Q_-}$ and volume $V_{Q_-}$ of $Q_-$ structures.

| | | $Q_{-attached}$ | $Q_{-nearwall}$ |
|---|---|---|---|
| ZPG-TBL | $N_Q/N_{Q_-}$ | 0.444 | 0.306 |
| | $V_Q/V_{Q_-}$ | 0.59 | 0.02 |
| APG-TBL | $N_Q/N_{Q_-}$ | 0.115 | 0.103 |
| | $V_Q/V_{Q_-}$ | 0.68 | 0.02 |

away from the wall. The mean dimensions associated with the $Q_-$ structures are listed in table 4. On average the spatial extent of these structures are around 5 times shorter and 3 times shorter and narrower, consistent with $Q_-$ structures occupying 5 times less volume than in the ZPG-TBL. The mean structure is also far more isotropic in extent as the streamwise length is reduced.

Variation in the wall-normal extent of $Q_-$ structures with the height of the structure centroid $y_{cent.}$ above the wall can be observed in the joint PDF of $L_y$ and $y_{cent.}$ (see figure 8). The dashed-

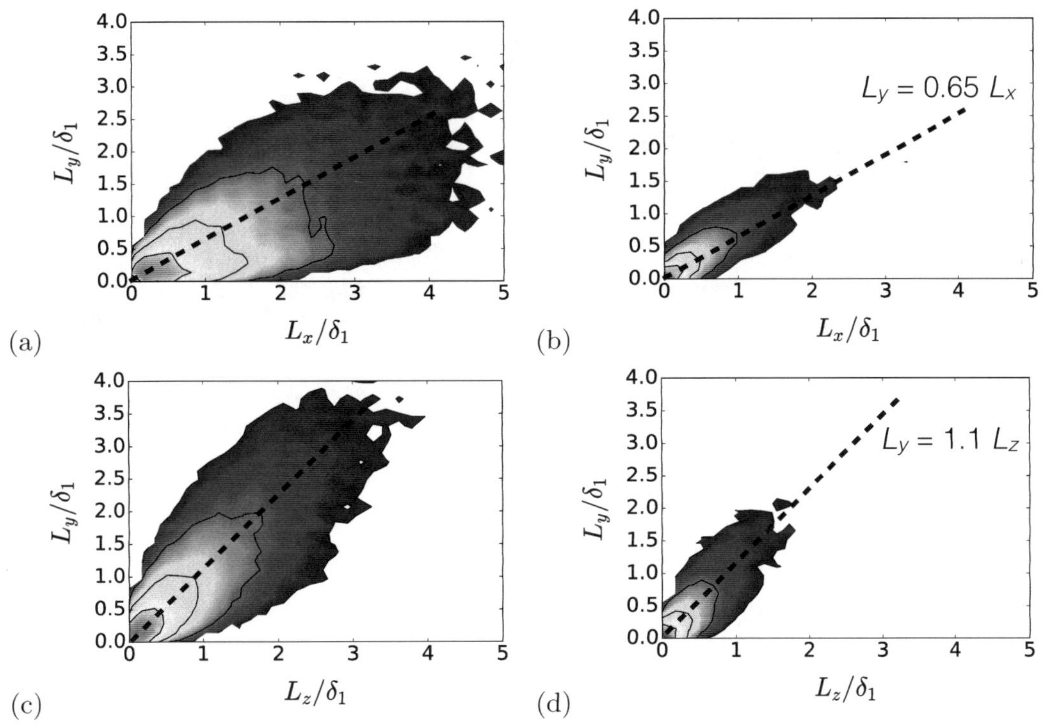


**Figure 7.** Joint PDFs of the streamwise $L_x$ and wall-normal $L_y$ extent of the $Q_-$ structures in the ZPG-TBL (a) and APG-TBL (b), and joint PDFs of the spanwise $L_z$ and wall-normal $L_y$ extent of $Q_-$ structures in the ZPG (c) and APG-TBL (d). Contour levels are shown for 1, 0.1, 0.01, and 0.001.

IOP Conf. Series: Journal of Physics: Conf. Series **1001** (2018) 012001 doi:10.1088/1742-6596/1001/1/012001

**Table 4.** Average wall-normal extent, centroid, length, span and height of $Q_-$ structure in the ZPG-TBL and APG-TBL.

| | ZPG-TBL | APG-TBL |
|---|---|---|
| $\langle y_{min} \rangle$ | $1.02\delta_1$ $(0.24\delta)$ | $0.75\delta_1$ $(0.18\delta)$ |
| $\langle y_{cent.} \rangle$ | $1.20\delta_1$ $(0.28\delta)$ | $0.79\delta_1$ $(0.19\delta)$ |
| $\langle y_{max} \rangle$ | $1.39\delta_1$ $(0.32\delta)$ | $0.83\delta_1$ $(0.20\delta)$ |
| $\langle L_x \rangle$ | $0.66\delta_1$ $(0.11\delta)$ | $0.09\delta_1$ $(0.02\delta)$ |
| $\langle L_y \rangle$ | $0.37\delta_1$ $(0.06\delta)$ | $0.08\delta_1$ $(0.02\delta)$ |
| $\langle L_z \rangle$ | $0.35\delta_1$ $(0.06\delta)$ | $0.08\delta_1$ $(0.02\delta)$ |

line represents the wall-attached structures with the region near the horizontal representing short wall-detached structures. The strong APG-TBL involves a large number of detached $Q_-$ structures as previously discussed, however this also shows that the majority of these structures have a short wall-normal extent. At the location of peak Reynolds stress, $y_{cent.} = \delta_1$, figure 8b indicates that this is near the highest point above the wall where $Q_-$ structures have a reasonable probability of extending down to the wall.

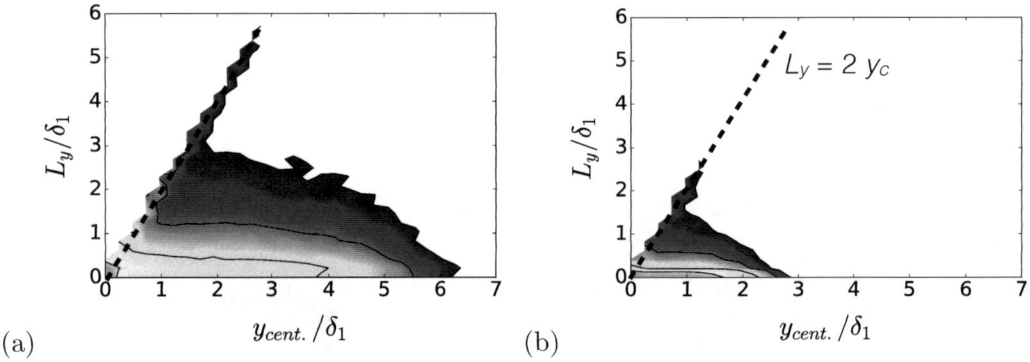


**Figure 8.** Joint PDFs of the structure wall-normal centroid $y_{cent.}$ and the wall-normal extent $L_y$ of the $Q_-$ structures in the ZPG-TBL (a) and APG-TBL (b). Contour levels are shown for 1, 0.1, 0.01, and 0.001.

## 6. Conclusions

Mean Reynolds stress profiles and instantaneous Reynolds stress structures have been investigated in a self-similar APG-TBL at the verge of separation using data from DNS. Comparisons are made with the DNS of a ZPG-TBL developing from the same inlet boundary layer. Structures are detected based on a threshold of the local product of the $uv$ velocity fluctuations relative to the local background turbulence level, and a clustering of adjacent points in the domain that reside in the same $uv$ quadrant. The Reynolds stress in the APG-TBL shows a narrower peak located further from the wall which is responsible for the majority of the production. Unlike the ZPG-TBL the $uv$ component of the Reynolds stress tensor in the APG-TBL is no longer the only component that is responsible for the production of turbulent kinetic energy, although it is still the primary contributor. The correlation between the $u$ and $v$ velocity fluctuations in the APG-TBL is half that of the ZPG-TBL in the near wall region, yet increases to a similar level in the outer region. This indicates less organised near wall ejections and sweeps, and is consistent with an increase in the relative number of $Q_1$ and $Q_3$ events.

ZPG-TBLs contain $Q_2$ ejections and $Q_4$ sweeps that together represent 60% of the detected $\langle uv \rangle$ structures, where as in the APG-TBL, ejections and sweeps only represent just over 50%. In the APG-TBL these structures are found to occupy 5 times less of the boundary layer volume. This difference is associated with a reduction in height and span, and a significantly shorter streamwise extent. The increase in turbulent activity in the outer flow correlates with the increase in the number of wall-detached $Q_-$ structures. The location of the peak Reynolds stress is found to be roughly the furthest distance from the wall that is occupied by a combination of attached and detached structures.

Comparison between the present self-similar APG-TBL and the investigation of the rapidly decelerating APG-TBL of Maciel et al. [9] following the same method of $uv$ structure detection, reveal a significantly lower percentage of wall-attached $Q_-$ and near wall structures in the present self-similar APG-TBL, along with a smaller streamwise extent of the present structures. Given the tendency for the wall-attached structures to be larger in both studies the smaller streamwise extent of the $Q_-$ structures in the self-similar APG-TBL is consistent with the reduced number of wall-attached structures. It is postulated that this difference is due to upstream flow history of the rapidly decelerated APG-TBL, which owing to the sudden onset of the pressure gradient, continues to maintain a number of wall-attached structures from upstream.

**Acknowledgements**

The support of the Australian Research Council (ARC) and the computational resources provided through NCMAS by the Australian National Computational Infrastructure, iVEC and the Pawsey Supercomputing Centre, and PRACE supported by the European Union, for this work is gratefully acknowledged. We would also like to thank VicNode for their technical support in transferring our massive database from Monash University to Madrid. Dr Atkinson was supported by an ARC Discovery Early Career Researcher Award (DECRA) fellowship while undertaking this work. This work was also funded in part by the Coturb program of the European Research Council. We are grateful to P. Schlatter for his careful reviewing of the manuscript.

**References**

[1] Kitsios V, Atkinson C, Sillero J A, Borrell G, Gungor A G, Jiménez J and Soria J 2016 *Int. J. Heat Fluid Fl.* **61A** 129–136
[2] Kline S J, Reynolds W C, Schraub F A and Runstadler P W 1967 *J. Fluid Mech.* **30** 741–773
[3] Wallace J M, Eckelmann H and Brodkey R S 1972 *J. Fluid Mech.* **54** 39–48
[4] Willmarth W W and Lu S S 1972 *J. Fluid Mech.* **55** 65–92
[5] Lozano-Durán A, Flores O and Jiménez J 2012 *J. Fluid Mech.* **694** 100–130
[6] Gungor A G, Maciel Y, Simens M P and Soria J 2014 *J. Phys.: Conf. Ser.* **506** 012007
[7] Kitsios V, Atkinson C, Sillero J, Borrell G, Gungor A, Jiménez J and Soria J 2017 *Whither Turbulence and Big Data in the 21st Century* (Springer) pp 61–75
[8] Bobke A, Vinuesa R, Örlü R and Schlatter P 2017 *J. Fluid Mech.* **820** 667–692
[9] Maciel Y, Simens M and Gungor A 2017 *Flow Turbul. Combust.* **98** 1–20
[10] Townsend A A 1980 *The structure of turbulent shear flow* (Cambridge U. Press)
[11] Mellor G L and Gibson D M 1966 *J. Fluid Mech.* **24** 225–253
[12] George W K and Castillo L 1993 *Near-wall turbulent flows* 901–910
[13] Kitsios V, Sekimoto A, Atkinson C, Sillero J, Borrell G, Gungor A, Jiménez J and Soria J 2017 *J. Fluid Mech.* **829** 392–419
[14] Borrell G, Sillero J A and Jiménez J 2013 *Comput. Fluids* **80** 37–43
[15] Sillero J A 2014 *High Reynolds number turbulent boundary layers* Ph.D. thesis U. Politécnica Madrid
[16] Spalart P and Watmuff J 1993 *J. Fluid Mech.* **249** 337–371
[17] Lighthill M J 1963 *Laminar Boundary Layers* ed Rosenhead L (Oxford U. Press) pp 46–113

IOP Conf. Series: Journal of Physics: Conf. Series **1001** (2018) 012002 doi:10.1088/1742-6596/1001/1/012002

# Flow topology of rare back flow events and critical points in turbulent channels and toroidal pipes

**C. Chin[1], R. Vinuesa[2], R. Örlü[2], J. I. Cardesa[3], A. Noorani[2], P. Schlatter[2] and M. S. Chong[4]**

[1]School of Mechanical Engineering, University of Adelaide, South Australia 5005, Australia
[2]Linné FLOW Centre, KTH Mechanics, SE-100 44 Stockholm, Sweden
[3]School of Aeronautics, Universidad Politécnica de Madrid, Madrid, Spain
[4]Mechanical Engineering Department, University of Melbourne, Victoria 3010, Australia

E-mail: rey.chin@adelaide.edu.au

**Abstract.** A study of the back flow events and critical points in the flow through a toroidal pipe at friction Reynolds number $Re_\tau \approx 650$ is performed and compared with the results in a turbulent channel flow at $Re_\tau \approx 934$. The statistics and topological properties of the back flow events are analysed and discussed. Conditionally-averaged flow fields in the vicinity of the back flow event are obtained, and the results for the torus show a similar streamwise wall-shear stress topology which varies considerably for the spanwise wall-shear stress when compared to the channel flow. The comparison between the toroidal pipe and channel flows also shows fewer back flow events and critical points in the torus. This cannot be solely attributed to differences in Reynolds number, but is a clear effect of the secondary flow present in the toroidal pipe. A possible mechanism is the effect of the secondary flow present in the torus, which convects momentum from the inner to the outer bend through the core of the pipe, and back from the outer to the inner bend through the pipe walls. In the region around the critical points, the skin-friction streamlines and vorticity lines exhibit similar flow characteristics with a node and saddle pair for both flows. These results indicate that back flow events and critical points are genuine features of wall-bounded turbulence, and are not artifacts of specific boundary or inflow conditions in simulations and/or measurement uncertainties in experiments.

## 1. Introduction

A detailed assessment of the near-wall region in turbulent flows is a very complex problem, which involves a number of interesting fundamental questions including its modulation by the flow in the outer region [1, 2, 3, 4]. The transport phenomena [5] present close to the wall, which can be characterised in terms of the wall-shear stress vector field, are relevant to understand a wide range of applications, including cardiovascular flows [6] (and also as discussed by Arzani *et al.* [7] in the context of Lagrangian wall-shear stress structures) and heat transfer (see for instance the studies by Dairay *et al.* [8] and Wu *et al.* [9]). In particular, the presence of regions of instantaneous reverse flow (denoted in the present work as back flow events) in wall-bounded turbulence is a topic of relevance for the understanding of separation mechanisms, both in steady [10] and unsteady [11] aerodynamic applications. A recent review by Carlomagno & Ianiro [12] discussed that flow reversal produced by local pressure gradient fluctuations in the near-wall can initiate secondary vortices. It is reported that this flow reversal is one of two features (the other being vortex generation) that can enhance the heat transfer characteristics in the flow. Moreover, back flow events and the topology of the wall-shear stress have been used to further understand the separation mechanisms from the perspective of dynamical systems [13], control theory [14] and reduced-order modelling [15], among others. Back flow events in turbulent channel flow were characterised in detail by Lenaers *et al.* [16], together with extreme wall-normal velocity fluctuations near the wall. In their direct numerical simulations (DNSs), they found that both the probability of finding a back flow event and the distance from the wall at which reverse flow was observed increased with Reynolds number. These indications show that there is a connection between the modulation of the near-wall region by the large-scale motions in the outer part of the flow and the presence of back flow events. This connection was quantitatively investigated by documenting the back-flow event density beneath large-scale structures residing in the logarithmic layer. Their conclusions were confirmed in a DNS of turbulent channel flow [17] at a slightly higher Reynolds number, where a probability of finding back flow events of around 0.07% was reported at a friction Reynolds number of $Re_\tau = 2,000$. Note that $Re_\tau$ is formed in terms of the channel half-height $h$, the friction velocity $u_\tau = \sqrt{\tau_w/\rho}$ (where $\tau_w$ is the wall-shear stress and $\rho$ is the fluid density) and the kinematic viscosity $\nu$ of the fluid. In both studies a strong connection between the presence of back flow events and energetic large-scale motions in the outer region were reported. These studies confirmed the indications in earlier works [18, 19, 20] regarding the presence of back flow events in wall-bounded turbulence, despite the fact that they had not been identified in certain measurement campaigns [21, 22]. The more recent experimental study by Willert [23], who performed particle image velocimetry (PIV) measurements in zero-pressure-gradient (ZPG) turbulent boundary layers (TBLs) at $Re_\tau = 1,000$ and $2,700$, confirmed the findings by Lenaers *et al.* [16] and Yao *et al.* [17]. Brücker [24] used micro-pillar sensors to measure the topology of the wall-shear stress vector in a ZPG TBL at $Re_\tau \simeq 940$, and identified a probability of finding back flow events of around 0.05%, aligned with the results by Lenaers *et al.* [16] at approximately the same $Re$. Note that micro-pillar sensors, based on correlating the deflection of small flexible pillars on the wall and the shear stress, provide accurate measurements of the $\tau_w$ fluctuations, as well as their spatial correlations (see Refs. [25, 26, 27]). Thus, Brücker [24] was able to determine that back flow events are correlated with strong spanwise gradients of the wall-shear stress, and characterised the topology of critical points (defined as points where both the wall-shear stress and the surface vorticity are zero), which are correlated with large-scale events in the logarithmic layer. This correlation was investigated by Monty *et al.* [28], Chong *et al.* [29] and Cardesa *et al.* [30] using DNS of turbulent channel and pipe flows, and they reported that critical points are formed when large-scale motions transport vorticity towards the wall.

Extreme events close to the wall have been mostly studied in canonical cases, such as turbulent channel flows and ZPG TBLs. The effect of an adverse pressure gradient (APG) was studied by Vinuesa *et al.* [31] using a DNS of the flow around a NACA4412 wing section [10]. In that study it is reported that the APG increases the probability of finding back flow events as the Clauser pressure-gradient parameter $\beta = \delta^*/\tau_w \mathrm{d}P/\mathrm{d}x$ (note that $\delta^*$ is the displacement thickness and $\mathrm{d}P/\mathrm{d}x$ is the streamwise pressure gradient) increases. The APG reduces the velocity gradient at the wall, and produces more energetic large-scale motions in the outer region [32]. Thus, the combination of both effects leads to the increase of back flow events. Their analysis of extreme events also revealed that under very strong APG conditions (*i.e.*, values of $\beta \simeq 35$), the flow is either aligned with the stream or against it, with a reduced spanwise wall-shear component, which is otherwise large at lower values of $\beta$. The characteristics of back flow events in APG TBLs were studied experimentally by Bross and Kähler [33] using time-resolved tomographic PIV measurements, which allowed them to characterise the flow structure around the regions of reverse flow.

Note that periodic boundary conditions were imposed in the spanwise direction (as well as in the streamwise direction in the case of channel flow) in all the simulations discussed above.

Although the existence of back flow events has been established beyond doubt, a perfect flow case to study back flow events would be one in which the imposed periodicity is not only part of the computational setup, but a feature of the flow case itself. Such a geometry and flow case is given by the flow through a toroidal pipe. The relevance of this flow case is related to the fact that it serves as the common asymptotic limit of two distinct classes of important flow cases, *i.e.*, the flow in spatially-developing pipe bends (when entry effects become negligible), and the flow in helical pipes (when the torsion becomes negligible). Since the experimental realisation of the flow through a toroidal pipe is complicated [34], their stability [35] as well as laminar [36] and turbulent [37] base flows (covering a wide range of curvatures) have only recently been explored by means of DNS.

In the present study we analyse the characteristics of back flow events in toroidal pipes using a numerical approach similar to the one reported in Ref. [38], in order to exploit the naturally-occurring periodicity in the flow to address the question of back flow events and critical points. In incompressible wall-bounded flows, the skin-friction and surface-vorticity fields are associated with turbulence transport mechanisms [39, 40]. Besides the physical nature of the periodicity in the torus, flows through curved pipes are characterised by a strong in-plane secondary flow that has a strong impact on the kinematics and dynamics of the flow [41], which makes it an interesting flow case with respect to back flow events and critical points.

## 2. Method

### *2.1. Channel flow simulations*

The turbulent channel flow data used for the present study comes from a direct numerical simulation (DNS). The numerical scheme integrates the fluid flow equations following the method described in Ref. [42]. On the wall-parallel planes, space is discretised using Fourier expansions, while Chebyshev polynomials are used in the wall-normal direction. The data set used here is a smaller version of case L950 in Ref. [43], and the details of the present simulation are summarised in table 1. The mesh resolutions in the streamwise and spanwise directions are denoted by $\Delta_x$ and $\Delta_z$, respectively, and the corresponding lengths of the computational domain are $L_x$ and $L_z$. The streamwise computational domain length is $L_x = 2\pi\delta$. The effects of insufficient computational length are well documented [39, 44] (*e.g.* lack of convergence of higher-order turbulence statistics). However, it is shown that statistics at the near wall with the current domain length $L_x = 2\pi\delta$ remain unaffected [45, 46]. The number of coefficients in the Fourier expansions in those two directions are $N_x = 512$ and $N_z = 511$, whereas $N_y = 385$ denotes the number of Chebyshev polynomials of the wall-normal direction. The superscript '+' is used to denote viscous scaling, *i.e.* in terms of the friction velocity $u_\tau$ or the viscous length $\nu/u_\tau$.

### *2.2. Simulation of the flow through a toroidal pipe*

The DNS of the turbulent flow through a toroidal pipe was performed using the numerical code Nek5000, developed at the Argonne National Laboratory by Fischer *et al.* [47] and based on the spectral-element method (SEM) by Patera [48]. Lagrange interpolants of order $N$ are used to expand the velocity and pressure, respectively, at the Gauss–Lobatto–Legendre (GLL) quadrature points within the spectral elements. The nonlinear terms in the incompressible Navier–Stokes equations are treated explicitly through third-order extrapolation (EXT3), and an implicit third-order backward differentiation (BDF3) scheme is used for the viscous terms. Numerical stability of the SEM was ensured through explicit filtering of 5% of the energy of the highest mode (see the work by Fischer and Mullen [49] for additional details). The simulations were carried out on the Cray XC40 system "Beskow", located at the PDC Center from KTH, running on 4,096 cores.

The toroidal pipe under consideration is shown in figure 1, which is similar to the bent pipes considered by Noorani *et al.* [37]. Figure 1(a) shows the instantaneous streamwise wall-

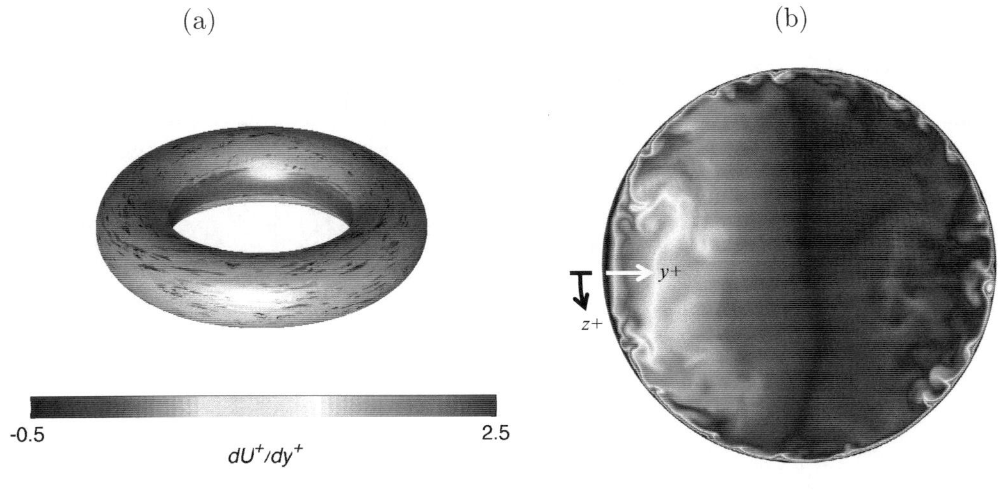


**Figure 1.** (a) Instantaneous inner-scaled streamwise wall-shear stress in the torus. (b) Cross-sectional view of an instantaneous streamwise velocity field, also showing one quarter of the in-plane spectral-element mesh. Note that the individual GLL poins within elements are also shown, and that in this panel dark blue and dark red correspond to $-0.1$ and $1.4$, respectively. The coordinate system is defined as $x$, $y$ and $z$, which are the streamwise (into the page), wall-normal and azimuthal directions, respectively.

shear stress and in figure 1(b) a cross-sectional view of the torus is presented. The coordinate system is defined as $x$, $y$ and $z$, which are the streamwise, wall-normal and azimuthal directions, respectively. This is chosen to be consistent with the channel flow notation. The corresponding mean streamwise, wall-normal and azimuthal velocities are $U$, $V$ and $W$. We consider a bulk Reynolds number $Re_b = 2U_bR_p/\nu$ (based on bulk velocity $U_b$ and pipe diameter $2R_p$) of $19,000$, which according to El Khoury *et al.* [50] yields a corresponding $Re_\tau = 550$ in straight pipes. As will be discussed below, due to the presence of the secondary flow induced by the pipe curvature, the value of $Re_\tau$ depends on the particular azimuthal location, and the averaged value is different from 550. The friction Reynolds number based on the averaged $u_\tau$ is $Re_\tau \approx 650$. Note that the Reynolds numbers are different between the torus and channel as shown in table 1. The geometry is completely defined with the curvature $\kappa = R_p/R_t$ (where $R_t$ denotes the radius of the torus, *i.e.*, the distance from the torus centre to the pipe centerline), which in the present study is set to 0.3. The computational mesh is designed for a straight pipe first [37, 50], following typical guidelines for DNS as discussed for instance in Ref. [39], and then this mesh is bent through analytical morphing in order to obtain the geometry of the toroidal pipe as discussed by Noorani *et al.* [37]. A total of $717,288$ spectral elements with $N = 7$ was used to discretise the domain, which amounts to around 370 million grid points. The in-plane SEM mesh is shown in figure 1(b) and the simulation parameters are summarised in table 1. The simulations are run with constant mass flux, and periodicity is imposed in the streamwise direction. The laminar Poiseuille profile was used as initial condition, and low-amplitude pseudo-random noise was added to trigger transition to turbulence. After around 200 convective time units the turbulence could be considered fully-developed and independent of the initial conditions. Thus, the data under consideration in the present work was obtained after the initial 200 time units of simulation. Note that although Nek5000 solves the Navier–Stokes equations in a Cartesian frame of reference, rotation matrices were employed to express all the tensors in the local toroidal frame of reference as stated previously.

IOP Conf. Series: Journal of Physics: Conf. Series **1001** (2018) 012002 doi:10.1088/1742-6596/1001/1/012002

**Table 1.** Computational parameters for the toroidal pipe and the channel.

| | $Re_\tau$ | $L_x^+$ | $L_z^+$ | $\Delta_x^+$ | $\Delta_z^+$ | $N_x$ | $N_y$ | $N_z$ | $N_y \times N_z$ | $u_\tau \times 10^{-2}$ |
|---|---|---|---|---|---|---|---|---|---|---|
| Channel | 934 | 5870 | 2935 | 11.5 | 5.7 | 512 | 385 | 511 | - | 4.539 |
| Torus | 650 | 13614 | 4084 | [3.62, 11.81] | [1.71, 5.61] | 1936 | - | - | 189696 | 6.84 |

### *2.3. Definition of back flow events*

The classification of back flow events is straightforward for the channel flow, where negative $\mathrm{d}U/\mathrm{d}y$ values correspond to back flow or reverse flow events. This is not the case for the flow through a toroidal pipe as shown in figure 2. Figures 2(a) and (b) show the instantaneous streamwise wall-shear stress for the torus and channel, respectively. It is clear that in the channel flow, the streamwise wall-shear stress profile exhibits an homogeneous pattern in the spanwise direction, whereas in the torus the distribution varies significantly from the inner bend ($z/\delta = 0$ and $2\pi$) to the outer bend ($z/\delta = \pi$) as shown in Ref. [51]. Note that in the context of the present work the variable $\delta$ denotes both the pipe radius $R_p$ in the torus and the channel half-height $h$. The probability density function of the orientation of the wall-shear stress vector is shown, in the form of a wind-rose plot, for the torus and the channel in figures 2(c,e) and (d,f), respectively. The wind rose plots essentially provide information of the magnitude and direction of the wall-shear stress vector. Figures 2(d) shows that the wall-shear stress vector in the channel is oriented predominantly in the streamwise direction, where the probability of the flow being aligned with the streamwise pressure gradient is approximately 12%. On the other hand, in the flow through the toroidal pipe the wall-shear stress vector exhibits three main preferential directions: the probability of the flow being aligned with the streamwise pressure gradient is around 5%, whereas the probability of it forming approximately $\pm 50°$ with respect to it is around 4% on each of the two cases. Therefore, to determine the back flow events in a toroidal flow, the local wall-shear stress angle has to be considered using the streamwise and spanwise wall velocity gradients $\mathrm{d}U/\mathrm{d}y$ and $\mathrm{d}W/\mathrm{d}y$, respectively. Comparing the zoomed-in wind rose plots of the torus (figure 2e) and channel (figure 2f) clearly show that back flow events occur in both flows. The results also show a higher percentage of back flow events in the channel as compared to the torus. The temporal and streamwise average of the local wall-shear stress angle $\psi$ (in degrees) is shown in figure 3(a), using the same angular frame of reference as the one considered in figure 2. The angle $\psi$ at the inner and outer bends is $\psi = 90°$, which implies that in these regions the wall-shear stress vector is aligned with the streamwise direction, and that there is no preferential direction for the spanwise wall-shear stress. The angle $\psi$ is continuously changing from the inner to the outer bend. However, our results clearly show that $\psi$ is symmetrical about either the inner or the outer bend. Since, as observed in figure 2(c), there are three preferential directions of the wall-shear stress, we will split the toroidal wall surface into several different regions to account for the variation in the local flow direction. To determine suitable regions to analyse back flow events, the rate of change of $\psi$ is chosen to discriminate the regions as shown in figure 3(b). The torus wall surface is split into six regions, which are denoted from R1 to R6. Note that R1, R2 and R3 are mirrored regions of R6, R5 and R4 respectively. Coincidently, this is similar to splitting the wall surface into four quadrants ($Q$) along the circumference in the cross section. Each quadrant has an arc length of $L_{arc} = \pi/2$. The four quadrant are: $Q1 =$ R1$(0 < z/R_p < \pi/4)$ + R6 $(7/4\pi < z/R_p < 2\pi)$, $Q2 =$ R2 $(\pi/4 < z/R_p < 3/4\pi)$, $Q3 =$ R5 $(5/4\pi < z/R_p < 7/4\pi)$ and $Q4 =$ R3 $(3/4\pi < z/R_p < \pi)$ + R4 $(\pi < z/R_p < 5/4\pi)$.

IOP Conf. Series: Journal of Physics: Conf. Series **1001** (2018) 012002 doi:10.1088/1742-6596/1001/1/012002

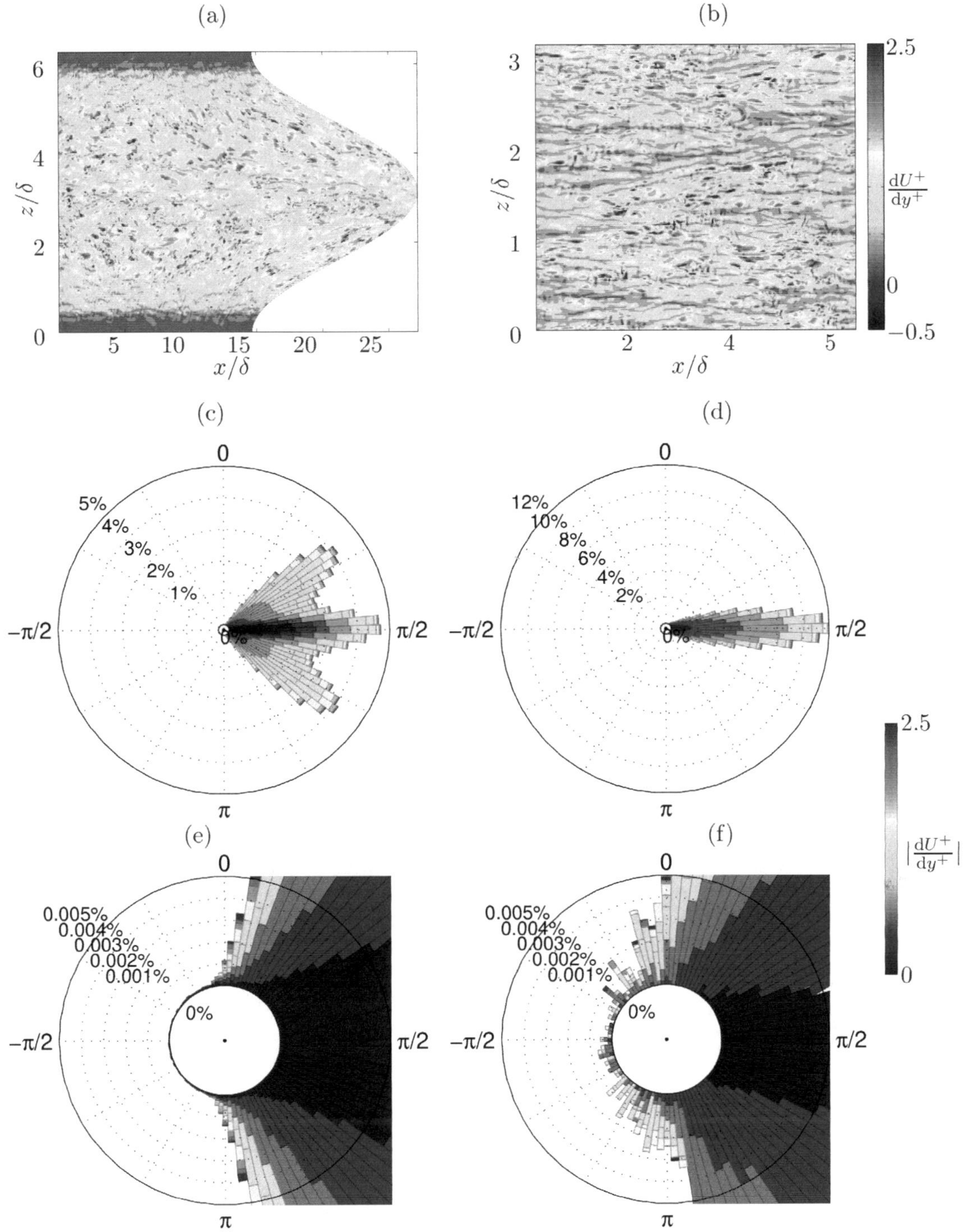


**Figure 2.** Instantaneous streamwise wall-shear stress for (a) torus (unwrapped) and (b) channel. Probability density function of the orientation of the wall-shear stress vector and magnitude for (c) torus with (e) zoomed-in view and for (d) channel with (f) zoomed-in view.

IOP Conf. Series: Journal of Physics: Conf. Series **1001** (2018) 012002 doi:10.1088/1742-6596/1001/1/012002

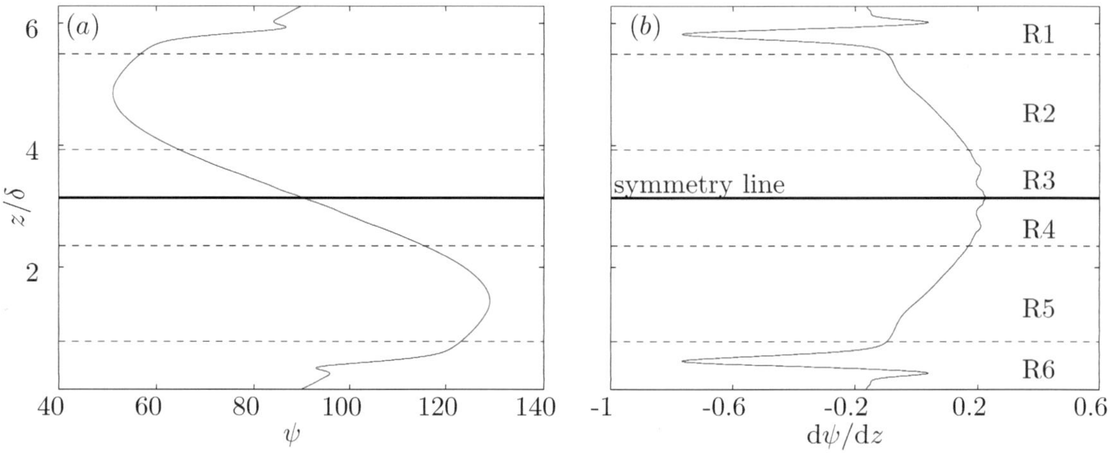


**Figure 3.** (a) Local average wall-shear stress angle $\psi$ (in degrees) for the torus. (b) Derivative of $\psi$ in the azimuthal direction. The solid line denotes the location of the outer bend. The dash lines split the torus into 6 regions, denoted R1 to R6, used for analysis purposes.

*2.4. Definition and characterisation of critical points*

To analyse the no-slip critical points on the wall, one can adopt a frame of reference in the channel flow, where the three orthogonal directions and velocities are labelled as $\mathbf{e}_1$, $\mathbf{e}_2$, $\mathbf{e}_3$ the three unit vectors along the three spatial coordinates. The velocity gradient tensor will be labelled as:

$$A_{ij} = \frac{\partial U_i}{\partial x_j}. \tag{1}$$

It can be shown that only $A_{12}$ and $A_{32}$ do not vanish due to the no-slip, impermeability and incompressibility conditions. Under these conditions, the three similarity invariants of $A_{ij}$ vanish: *i.e.* $P$, $Q$ and $R$ as defined by [52]. For a local flow characterisation, a higher-order truncation is required. We now introduce the no-slip tensor $\mathcal{A}_{ij}$ as:

$$\underline{\mathcal{A}} = \begin{bmatrix} \mathcal{A}_{11} & \mathcal{A}_{12} & \mathcal{A}_{13} \\ \mathcal{A}_{21} & \mathcal{A}_{22} & \mathcal{A}_{23} \\ \mathcal{A}_{31} & \mathcal{A}_{32} & \mathcal{A}_{33} \end{bmatrix} = \begin{bmatrix} \frac{\partial^2 U}{\partial x \partial y} & \frac{1}{2}\frac{\partial^2 U}{\partial y \partial y} & \frac{\partial^2 U}{\partial z \partial y} \\ \frac{\partial^2 V}{\partial x \partial y} & \frac{1}{2}\frac{\partial^2 V}{\partial y \partial y} & \frac{\partial^2 V}{\partial z \partial y} \\ \frac{\partial^2 W}{\partial x \partial y} & \frac{1}{2}\frac{\partial^2 W}{\partial y \partial y} & \frac{\partial^2 W}{\partial z \partial y} \end{bmatrix}. \tag{2}$$

Writing the Taylor expansion of the velocity field at a no-slip wall in matrix form leads to:

$$\begin{bmatrix} \mathring{x} \\ \mathring{y} \\ \mathring{z} \end{bmatrix} = \begin{bmatrix} 0 & A_{12} & 0 \\ 0 & 0 & 0 \\ 0 & A_{32} & 0 \end{bmatrix} + \begin{bmatrix} \mathcal{A}_{11} & \mathcal{A}_{12} & \mathcal{A}_{13} \\ \mathcal{A}_{21} & \mathcal{A}_{22} & \mathcal{A}_{23} \\ \mathcal{A}_{31} & \mathcal{A}_{32} & \mathcal{A}_{33} \end{bmatrix} \begin{bmatrix} x \\ y \\ z \end{bmatrix}, \tag{3}$$

or $\mathring{x}_i = A_{ij} + \mathcal{A}_{ij} x_j$ in more compact form. $\mathring{x}_i = \mathrm{d}x_i/\mathrm{d}\tau \neq \dot{x}_i = \mathrm{d}x_i/\mathrm{d}t$, and that incompressibility at a no-slip wall implies $\mathcal{A}_{21} = \mathcal{A}_{23} = 0$, $\mathcal{A}_{22} = -\left(\mathcal{A}_{11} + \mathcal{A}_{33}\right)/2$.
The two-dimensional skin-friction field at the wall is given as

$$\mathbf{s} = A_{12}\mathbf{e}_1 + A_{32}\mathbf{e}_3. \tag{4}$$

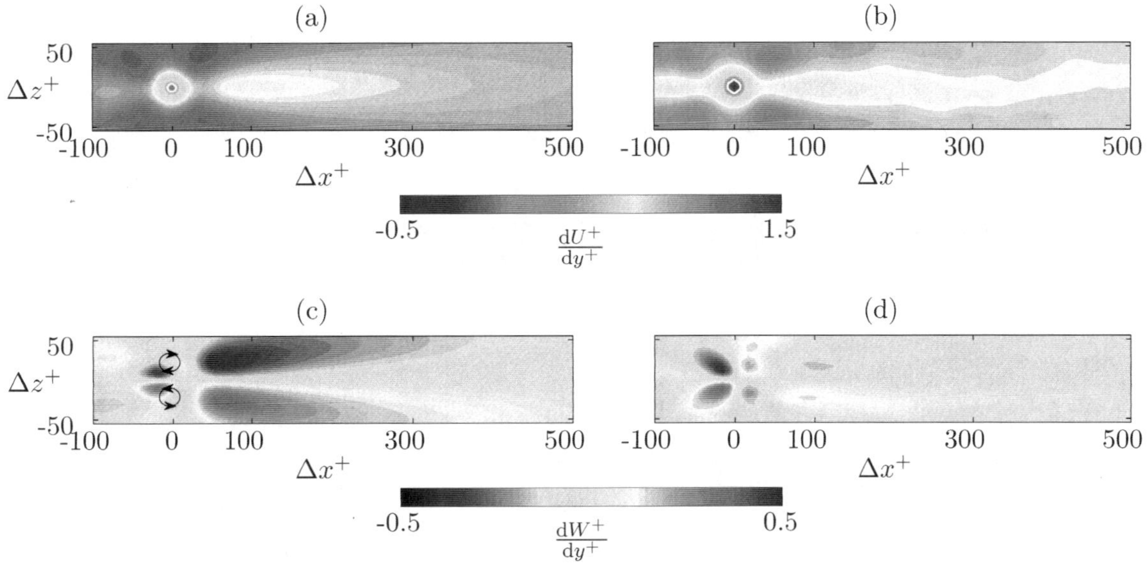


**Figure 4.** Distributions of $\mathrm{d}U^+/\mathrm{d}y^+$ (a & b) and $\mathrm{d}W^+/\mathrm{d}y^+$ (c & d) conditionally averaged to the presence of a back flow event at $\Delta x = \Delta y = 0$. The evaluation was performed over the entire azimuthal direction of the toroidal pipe. Left panels correspond to the torus and right panels to the channel. The white contour in panels (a) and (b) denote zero streamwise wall-shear stress.

The definition of critical points in **s** (see equation 4) and a description of how to locate them in a discrete numerical grid is similar to Ref. [30].

## 3. Results

The conditional averaging of $\mathrm{d}U^+/\mathrm{d}y^+$ and $\mathrm{d}W^+/\mathrm{d}y^+$ in the vicinity of a back flow event is first performed for the entire toroidal pipe without considering the local flow angle. The results are compared with the channel flow and are shown in figure 4. The top row, *i.e.* figures 4(a) and (b), show a comparison of the streamwise wall-shear stress $\mathrm{d}U^+/\mathrm{d}y^+$ in the torus and in the channel. Note that in these figures the centre of the back flow event is located at $\Delta x = \Delta z = 0$. The overall distribution is qualitatively similar in both cases, with low $\mathrm{d}U^+/\mathrm{d}y^+$ values right upstream and right downstream of the back flow event. The back flow events are flanked by regions of strong $\mathrm{d}U^+/\mathrm{d}y^+$ values, and interestingly the torus exhibits consistently larger shear-stress values in the vicinity of the back flow event than the channel. The mean diameter of these back flow events differs between the torus and the channel, with a mean diameter of about 16 viscous units in the torus and around 20 viscous units in the channel. Note that a diameter of around 20 viscous units was also reported by Vinuesa *et al.* [31] in the adverse-pressure-gradient turbulent boundary layer developing on the suction side of a wing section and by Lenaers *et al.* [16] in their channel flow simulations.

Figures 4(c) and (d) show the spanwise wall-shear stress distribution $\mathrm{d}W^+/\mathrm{d}y^+$ for the torus and the channel, respectively. In the case of the torus, the back flow event is flanked by two region of positive (negative) $\mathrm{d}W^+/\mathrm{d}y^+$, immediately followed by small areas of opposite sign in the downstream direction. This pattern suggests the presence of a pair of counter-rotating vortices in the vicinity of the back flow events, as illustrated in figure 4(c), which may be responsible for the formation of the region of reverse streamwise velocity.

In this region around the back flow event ($\Delta x^+ \approx [-50, 20]$, $\Delta z^+ \approx [-25, 25]$), this alternating pattern in $\mathrm{d}W^+/\mathrm{d}y^+$ is also present in the channel, although its topology is slightly different: the

IOP Conf. Series: Journal of Physics: Conf. Series **1001** (2018) 012002 doi:10.1088/1742-6596/1001/1/012002

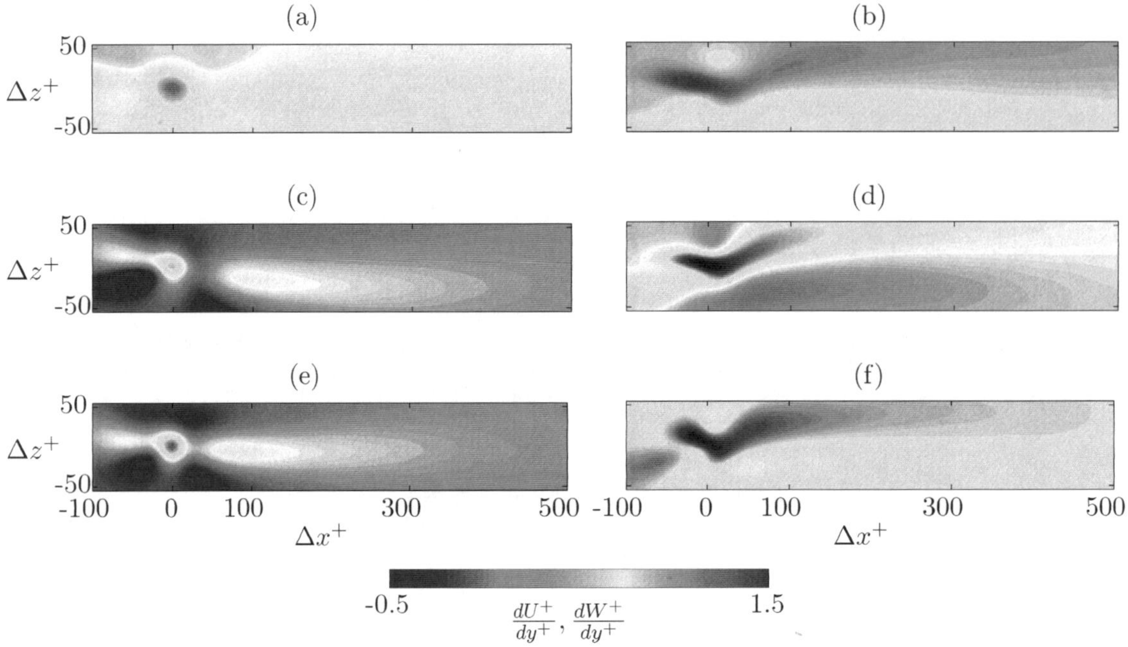


**Figure 5.** (Left panels) streamwise wall-shear stress $dU^+/dy^+$ and (right panels) spanwise wall-shear stress $dW^+/dy^+$. The results are shown for (a,b) region R1, (c,d) region R2 and (e,f) region R3. Regions are as defined in figure 3(b).

regions of positive (negative) spanwise wall-shear stress flanking the back flow event are larger, and are inclined in the streamwise direction, as opposed to what is observed in the torus. It is interesting to note that at further downstream distance ($\Delta x^+ > 50$) of the back flow event in the toroidal pipe, the $dW^+/dy^+$ signature is significantly stronger and longer than the upstream distribution, whereas the opposite is observed in the channel.

Since the results in figure 4 for the torus are a combination of contributions from the different

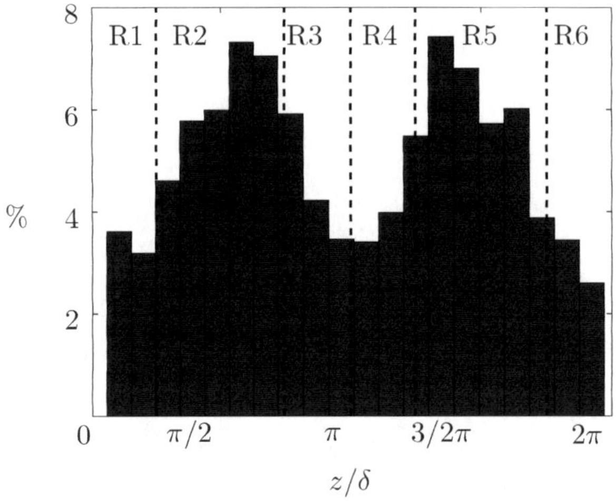


**Figure 6.** Histogram of back flow events in the toroidal pipe along the spanwise direction. R1 to R6 are the different regions as defined in figure 3.

IOP Conf. Series: Journal of Physics: Conf. Series **1001** (2018) 012002 doi:10.1088/1742-6596/1001/1/012002

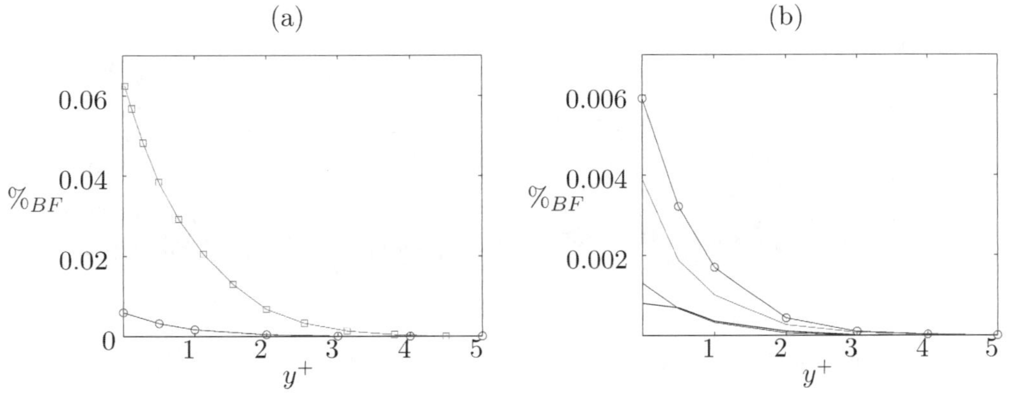


**Figure 7.** (a) Percentage of back flow events $\%_{BF}$ with increasing inner-scaled wall-normal distance for torus (blue line with ○) and channel (red line with □). (b) Back flow events $\%_{BF}$ for torus decomposed into different regions. R1+R6 blue line; R2+R5 red line and R3+R4 black line. The blue line with dots is the same torus data as in (a).

regions defined in figure 3(b) (R1 to R6), it will be natural to investigate the influence of the secondary motion in the flow within the various areas of the torus. Since the regions are symmetrical about the symmetry line (see figure 3), only results from regions R1, R2 and R3 will be presented, which, however, include the contributions from their symmetrical counterparts to increase the sample space. The streamwise wall-shear stress $\mathrm{d}U^+/\mathrm{d}y^+$ from the torus is shown in figure 5(a, c, e) for regions R1, R2 and R3, respectively. This field exhibits significant differences in R1 compared to the field obtained when the contributions from all the regions are combined, as can be observed comparing figures 5(a) and 4(a). As documented by Noorani *et al.* [37], the pipe curvature induces an in-plane secondary flow which convects momentum from the inner to the outer bend through the core of the pipe, and from the outer back to the inner bend through the pipe walls. This secondary flow amounts to around 15% of the bulk velocity for the present curvature, which implies that it has a strong effect on the local flow features. The conditionally-averaged field in R1 reflects the fact that the strong vertical convection produces local laminar-like conditions in the inner bend. This results in a significantly weaker shear-stress field, with a small region of higher $\mathrm{d}U^+/\mathrm{d}y^+$ on one of the sides of the back flow event, induced by the secondary flow returning to the inner bend through the pipe wall. The distribution shown in figure 5(e) for region R3, *i.e.* for the outer bend, resembles the one in figure 4(a), although the latter exhibits smaller shear-stress values due to the fact that it also accounts for the weaker field present in the inner bend. Regarding region R2, the $\mathrm{d}U^+/\mathrm{d}y^+$ field shown in figure 5(c) shares some similarities with the outer bend, although its asymmetry is produced by the strong effect of the secondary flow in that region of the pipe. Figures 5(b, d, f) show the spanwise wall-shear stress $\mathrm{d}W^+/\mathrm{d}y^+$ for regions R1, R2 and R3, respectively. Here the profiles are clearly different from the ones obtained by considering the whole torus, which are shown in figure 4(c). This discrepancy is somewhat expected, since the secondary flow imprints a strong spanwise velocity *footprint* on the flow. The distributions for regions R1, R2 and R3 show a clear spanwise inclination, which is most prominent in figure 5(d), corresponding to region R2. Regions R1 and R3, *i.e.*, the inner and outer bend regions, are therefore slightly less influenced by the spanwise motion imposed by the secondary motion. Figure 6 shows the distribution of back flow events in the various regions (R1 to R6). Here the results further suggest that a strong spanwise secondary motion increases the occurrence of back flow events and in regions R1 and R3, the occurrence of back flow events decreases.

Next, a comparison of the percentage of back flow events ($\%_{BF}$) with increasing wall-normal

IOP Conf. Series: Journal of Physics: Conf. Series **1001** (2018) 012002 doi:10.1088/1742-6596/1001/1/012002

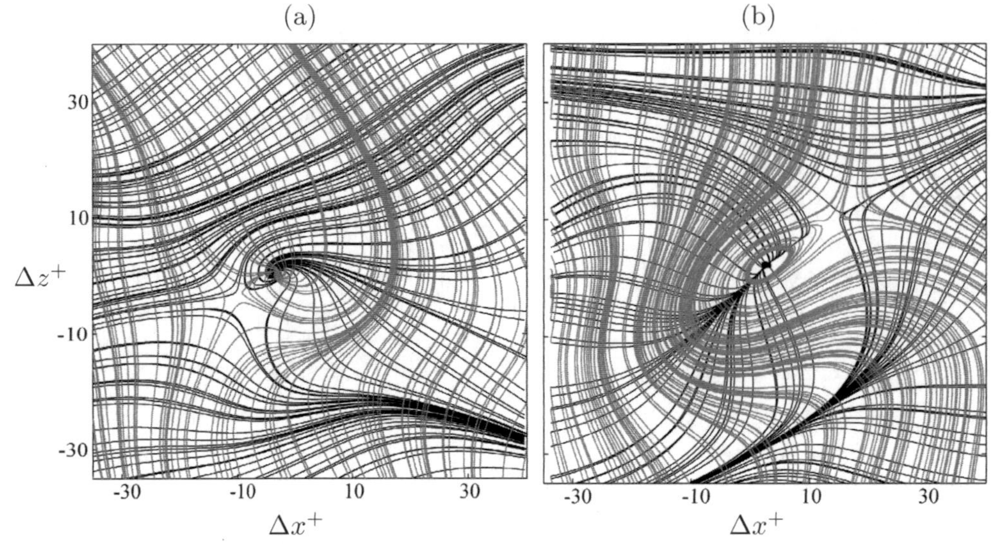


**Figure 8.** (Black) Skin-friction and (red) vorticity streamlines around the vicinity of a critical point for (a) torus and (b) channel. Note that the critical point is located at $\Delta x^+ = \Delta z^+ = 0$ in both cases.

distance $y^+$ is performed for the toroidal pipe and the channel flow, and is shown in figure 7. The percentage of back flow events decreases exponentially with increasing wall-normal distance, an observation in agreement with previously reported results by Lenaers *et al.* [16] and Vinuesa *et al.* [31]. Regarding the results from the torus, figure 7(a) shows that the percentage of back flow events is almost an order of magnitude lower than that of the channel flow. Moreover, the $\%_{BF}$ trend shows a similar exponential decay with $y^+$. A decomposition of the $\%_{BF}$ for the torus into different regions is shown in figure 7(b). For regions R1+R6 and R3+R4, which are least subjected to the influence of the secondary motion, they have very similar $\%_{BF}$ away from the wall. The $\%_{BF}$ for region R2+R5 is consistently larger than in the other regions. The results here suggest that the vortical structures required for the production of back flow events are significantly affected by the secondary flow in the toroidal pipe. Further inspection of figures 4(c) and (d) indicates that the spanwise wall-shear stress pattern responsible for the pair of counter-rotating vortices is weaker in the torus, a fact that is connected with the lower occurrence of reverse flow events. As reported in Ref. [31], APG TBLs exhibit higher occurrence of back flow events than ZPGs at the same Reynolds number. This is due to the fact that the wall-normal convection produced by the APG reduces the near-wall velocity, thus promoting the formation of regions of negative velocity. In the case under study here, the strong secondary flow convects momentum from the inner to the outer bend of the pipe, therefore the flow on the outer bend experiences a strong convection towards the wall, with the opposite effect to that of an APG, thus resulting in a reduction of back flow events. As reported by Lenaers *et al.* [16], back flow events are absent in laminar flows, and are only produced as extreme events in turbulent flows. The very strong convection away from the inner bend produces laminar-like conditions, as can be observed for instance in figure 1(b), therefore justifying the reduction of back flow events in this region. Overall, the present results suggest that the percentage of back flow events drops to around zero for wall-normal locations above $y^+ = 5$.

Figures 8(a) and (b) show the flow field around the vicinity of critical points for the torus and the channel, respectively. It has been shown that critical points exist in pairs and in close proximity to each other [29]. This pair of critical points consists of one node and one saddle. The critical points (nodes) for both flows are located at $\Delta x^+ = \Delta z^+ = 0$. The saddle for the

IOP Conf. Series: Journal of Physics: Conf. Series **1001** (2018) 012002 doi:10.1088/1742-6596/1001/1/012002

**Table 2.** Summary of back flow and critical point statistics in toroidal pipe and channel flow.

| | **Torus** | **Channel** |
|---|---|---|
| Percentage area of back flow (at the wall) | 0.006% | 0.06% |
| Percentage area of back flow for R1+R6 (at the wall) | 0.0008% | - |
| Percentage area of back flow for R2+R5 (at the wall) | 0.0039% | - |
| Percentage area of back flow for R3+R4 (at the wall) | 0.0013% | - |
| Number of back flow events (per $10^6$ viscous unit area) | 0.3 | 1.8 |
| Mean diameter of back flow event (in viscous units) | 16 | 20 |
| Number of critical points (per $10^6$ viscous unit area) | 0.35 | 2 |
| Number of critical points for R1+R6 (per $10^6$ viscous unit area) | 0.144 | - |
| Number of critical points for R2+R5 (per $10^6$ viscous unit area) | 0.136 | - |
| Number of critical points for R3+R4 (per $10^6$ viscous unit area) | 0.07 | - |

torus is at $\Delta x^+, \Delta z^+ \approx (-10, -5)$ and for the channel is at $\Delta x^+, \Delta z^+ \approx (15, 10)$. From figure 8, both flow fields show similar skin-friction streamlines and skin-friction vorticity lines patterns. Even though both flows are very different in nature, the results here suggest that critical point characteristics are similar at the wall and might be an essential component in the dynamics of wall-bounded turbulent flows.

A summary of the statistics for the back flow events and critical points in the toroidal pipe and in the channel is presented in table 2. The toroidal pipe consistently shows a lower occurrence of back flow events and critical points (both in terms of percentage area of the wall and number of events per $10^6$ viscous unit area) when compared to the channel flow. These statistics also show that the number of back flow events is lower than the number of critical points in the two flow cases under study. The further decomposition of the torus statistics ($\%_{BF}$ and number of critical points at the wall) into the three main regions are also shown in table 2. Results clearly indicated higher occurrence of back flow events and critical points where secondary motion is strongest.

## 4. Conclusions

A DNS of the turbulent flow through a toroidal pipe at $Re_\tau \simeq 650$ was performed, using the spectral-element code Nek5000 [47], with the aim of analysing the characteristics of the wall-shear stress vector. Back flow events, *i.e.* regions of reverse flow, were characterised in the torus and compared with the ones obtained in a DNS of turbulent channel flow at $Re_\tau \simeq 934$. Our results show that back flow events are less numerous in the toroidal pipe than in the channel, with probabilities of occurrence of 0.006% and 0.06% in both cases, respectively. The fields of streamwise and spanwise wall-shear stress conditioned to the presence of a back flow event are stronger in the torus, and the diameter of the reverse flow regions is smaller than in the channel. The diameter is around 16 viscous units in the former, while its value is about 20 in the channel, in agreement with the results reported in other channel flow simulations [16] and in APG TBLs [31]. Note that, although the two flows are compared at different Reynolds numbers, the *Re* effect is much weaker than the discrepancy observed in the two cases, as reported in Ref. [16]. These differences are therefore explained by the effect of the secondary flow present in the torus, which convects momentum from the inner to the outer bend through the core of the pipe, and back from the outer to the inner bend through the pipe walls. This secondary flow, which amounts to around 15% of the bulk velocity [37] for the present curvature, inhibits the formation of back flow events in the outer bend through a mechanism opposite to the one reported in APG TBLs

[31]. On the other hand, the significant wall-normal convection from the inner bend produces laminar-like conditions, which also lead to a dramatic reduction in the emergence of reverse-flow regions [16]. It is important to note that although it is common to impose periodicity in the streamwise and spanwise directions in turbulent channel flow simulations, in the torus the flow is naturally streamwise-periodic. Therefore, the present results provide clear evidence that back flow events and critical points are genuine features and not artifacts of specific boundary or inflow conditions in simulations and/or measurement uncertainties in experiments. Future extensions of the present study will utilise the entire flow field of the toroidal pipe to investigate flow structures above these back flow events and critical points. Another interesting possible extension of this study would be to analyse and compare the impact of the secondary flow in turbulent ducts on back flow events and critical points.

## Acknowledgments
This work was supported with supercomputing resources provided by the Phoenix HPC service at the University of Adelaide. Computer time was provided by the Swedish National Infrastructure for Computing (SNIC). This study was funded in part by the Coturb Grant (grant number ERC-2014.AdG-669505) of the European Research Council. RV, RÖ and PS acknowledge the funding provided by the Swedish Research Council (VR) and the Knut and Alice Wallenberg Foundation. MSC and CC acknowledge the financial support of the Australian Research Council. We are grateful to A. Ianiro for his thoughtful review of this manuscript.

## References
[1] Mathis R, Hutchins N and Marusic I 2009 *J. Fluid Mech.* **628** 311–337
[2] Schlatter P and Örlü R 2010 *Phys. Fluids* **22** 051704
[3] Marusic I, Mathis R and Hutchins N 2010 *Science* **329** 193–196
[4] Vinuesa R, Hites M H, Wark C E and Nagib H M 2015 *Phys. Fluids* **27** 105107
[5] Klewicki J, Chin C, Blackburn H M, Ooi A and Marusic I 2012 *Phys. Fluids* **24** 045107
[6] Poon E K, Hayat U, Thondapu V, Ooi A, Asrar Ul Haq M, Moore S, Foin N, Tu S, Chin C, Monty J P, Marusic I and Barlis P 2015 *Coron. Artery Dis.* **26** e43–e54
[7] Arzani A, Gambaruto A M, Chen G and Shadden S C 2016 *J. Fluid Mech.* **790** 158–172
[8] Dairay T, Fortuneé V, Lamballais E and Brizzi L E 2015 *J. Fluid Mech.* **764** 362–394
[9] Wu Z, Laurence D, Iacovides H and Afgan I 2017 *Int. J. Heat Mass Transf.* **110** 193–208
[10] Vinuesa R, Hosseini S M, Hanifi A, Henningson D S and Schlatter P 2017 *Flow Turbul. Combust.* **99** 613–541
[11] Negi P S, Vinuesa R, Schlatter P, Hanifi A and Henningson D S 2017 *In Proc. TSFP-10* 1D–3
[12] Carlomagno G and Ianiro A 2014 *Exp. Therm. Fluid Sci.* **58** 15–35
[13] Surana A, Grunberg O and Haller G 2006 *J. Fluid Mech.* **564** 57–103
[14] Alam M R, Liu W and Hallerb G 2006 *Phys. Fluids* **18** 043601
[15] Brunton S L, Rowley C W and Williams D R 2013 *J. Fluid Mech.* **724** 203–233
[16] Lenaers P, Li Q, Brethouwer G, Schlatter P and Örlü R 2012 *Phys. Fluids* **24** 035110
[17] Yao Y C, Huang W X and Xu C X 2017 *Acta Mech. Sin.* DOI 10.1007/s10409-017-0687-2
[18] Johansson G 1988 *An experimental study of the structure of a flat plate turbulent boundary layer, using laser-Doppler velocimetry* Ph.D. thesis Chalmers U. Technology
[19] Spalart P R and Coleman G N 1997 *Eur. J. Mech. B* **16** 169–189
[20] Hu Z W, Morfey C L and Sandham N D 2006 *AIAA J.* **44** 1541–1549
[21] Eckelmann H 1974 *J. Fluid Mech.* **65** 439–459
[22] Colella K J and Keith W L 2003 *Exp. Fluids* **34** 253–260
[23] Willert C E, Cuvier C, Foucaut J M, Klinner J, Stanislas M, Laval J P, Srinath S, Soria J, Amili O, Atkinson C, Kahler C J, Scharnowski S, Hain R, Schröder A, Geisler R, Agocs J and Röse A 2018 *Exp. Therm. Fluid Sci.* **91** 320–328
[24] Brücker C 2015 *Phys. Fluids* **27** 031705
[25] Brücker C, Bauer D and Chaves H 2007 *Exp. Fluids* **42** 737–749
[26] Große S and Schröder W 2008 *Meas. Sci. Technol.* **19** 015403
[27] Vinuesa R and Örlü R 2017 *Experimental Aerodynamics* ed Discetti S and Ianiro A (CRC Press) chap 12
[28] Monty J, Chin C, Chong M and Marusic I 2012 *In Proc. ERCOFTAC ETMM-9*
[29] Chong M S, Monty J P, Chin C and Marusic I 2012 *J. Turbul.* **13** 1–10

[30] Cardesa J I, Monty J P, Soria J and Chong M S 2014 *J. Phys.: Conf. Ser.* **506** 012009
[31] Vinuesa R, Örlü R and Schlatter P 2017 *J. Turbul.* **18** 170–185
[32] Bobke A, Vinuesa R, Örlü R and Schlatter P 2017 *J. Fluid Mech.* **820** 667–692
[33] Bross M and Kähler C J 2017 *In Proc. TSFP-10* 7A–5
[34] Kühnen J, Holzner M, Hof B and Kuhlmann H 2014 *J. Fluid Mech.* **738** 463–491
[35] Canton J, Schlatter P and Örlü R 2016 *J. Fluid Mech.* **792** 894–909
[36] Canton J, Örlü R and Schlatter P 2017 *Int. J. Heat Fluid Flow* **66** 95–107
[37] Noorani A, El Khoury G K and Schlatter P 2013 *Int. J. Heat Fluid Flow* **41** 16–26
[38] Noorani A and Schlatter P 2016 *Int. J. Heat Fluid Flow* **61** 108–116
[39] Chin C 2011 *Numerical study of internal wall-bounded turbulent flows* Ph.D. thesis University of Melbourne
[40] Chin C, Philip J, Klewicki J, Ooi A and Marusic I 2014 *J. Fluid Mech.* **757** 747–769
[41] Kalpakli Vester A, Örlü R and Alfredsson P H 2016 *Appl. Mech. Rev.* **68** 050802
[42] Kim J, Moin P and Moser R 1987 *J. Fluid Mech.* **177** 133–166
[43] del Álamo J, Jiménez J, Zandonade P and Moser R D 2006 *J. Fluid Mech.* **561** 329–358
[44] Saha S, Chin C, Blackburn H and Ooi A 2011 *Int. J. Heat Fluid Flow* **32** 1083–1097
[45] Chin C, Ooi A S H, Marusic I and Blackburn H M 2010 *Phys. Fluids* **22** 115107
[46] Chin C, Monty J P and Ooi A 2014 *Int. J. Heat Fluid Flow* **45** 33–40
[47] Fischer P F, Lottes J W and Kerkemeier S G 2008 NEK5000: Open source spectral element CFD solver. Available at: `http://nek5000.mcs.anl.gov` Tech. rep. Argonne NL
[48] Patera A T 1984 *J. Comput. Phys.* **54** 468–488
[49] Fischer P and Mullen J 2001 *C. R. Acad. Sci. Paris* **332** 265–270
[50] El Khoury G, Schlatter P, Noorani A, Fischer P, Brethouwer G and Johansson A 2013 *Flow Turbul. Combust.* **91** 475–495
[51] Noorani A, Sardina G, Brandt L and Schlatter P 2016 *J. Fluid Mech.* **793** 248–279
[52] Chong M, Perry A and Cantwell B 1990 *Phys. Fluids* **2** 765–777

IOP Conf. Series: Journal of Physics: Conf. Series **1001** (2018) 012003 doi:10.1088/1742-6596/1001/1/012003

# Intense structures of different momentum fluxes in turbulent channels

**Kosuke Osawa**[1] **and Javier Jiménez**[2]

[1] Department of Mechanical and Aerospace Engineering, Tokyo Institute of Technology, 2-12-1 Ookayama, Meguro-ku, Tokyo, Japan
[2] School of Aeronautics, Universidad Politécnica de Madrid, 28040 Madrid, Spain

E-mail: osawa.k.ad@m.titech.ac.jp

**Abstract.** The effect of different definitions of the momentum flux on the properties of the coherent structures of the logarithmic region of wall-bounded turbulence is investigated by comparing the structures of intense tangential Reynolds stress with those of the alternative flux proposed in [Jimenez (2016) *J. Fluid Mech.* 809:585]. Despite the fairly different statistical properties of the two flux definitions, it is found that their intense structures show many similarities, such as the dominance of 'wall-attached' objects, and geometric self-similarity. However, the new structures are wider, although not taller, than the classical ones, and include both high- and low-momentum regions within the same object. It is concluded that they represent the same phenomenon as the classical group of a sweep, an ejection, and a roller, which should thus be considered as the fundamental coherent structure of the momentum flux. The present results suggest that the properties of these momentum structures are robust with respect to the definition of the fluxes.

## 1. Introduction

Understanding the physical mechanism of momentum transfer in wall turbulence is crucial both theoretically and in applications, because it is directly connected with the mechanical equilibrium of the flow and with the possible control of the wall friction. Traditionally, since the average of the Reynolds shear stress over wall-parallel planes is equal to the total wall-normal transfer of streamwise momentum, its local value ($uv$) has been interpreted as a local momentum flux, and regions of intense Reynolds stress have been identified and studied as coherent structures of the momentum transfer. They are usually referred to as 'quadrant' structures [1, 2, 3], and their spatial and temporal characteristics have come to be considered properties of the momentum transfer itself.

However, it has been repeatedly pointed out that it is the divergence of the Reynolds stresses, and not the stresses themselves that enter the equations of motion [4, 5], and that the definition of local fluxes is ambiguous. Recently, Jiménez [6] revisited this question, and studied in some detail a particular 'optimal' definition of the momentum flux, in the sense of having a minimum integrated norm for a given divergence. There is nothing in this new definition to make it better, or worse, than the classical Reynolds stress. The real interest is that these new fluxes are different from the classical ones, although equally valid, and that any property that is not shared by the two definitions may be considered 'suspect'.

IOP Conf. Series: Journal of Physics: Conf. Series **1001** (2018) 012003 doi:10.1088/1742-6596/1001/1/012003

**Table 1.** Parameters of the channel flow databases. $L_x, L_z, 2h$ are the streamwise, spanwise and wall-normal domain size, respectively. $\Delta x, \Delta z$ are spatial resolutions, in terms of Fourier modes, and $N_y$ is the number of wall-normal grid points. $N_s$ is the number of snapshots used in the analyses.

| $Re_\tau$ | $L_x/h$ | $L_z/h$ | $\Delta x^+$ | $\Delta z^+$ | $N_y$ | $N_s$ | Symbol | Ref. |
|---|---|---|---|---|---|---|---|---|
| 934 | $8\pi$ | $3\pi$ | 11 | 5.7 | 385 | 20 | none | [8] |
| 2003 | $8\pi$ | $3\pi$ | 12 | 6.1 | 633 | 9 | $\circ$ | [9] |

Somewhat disturbingly, [6] found that the instantaneous fields and many of the statistical properties of the new fluxes differ considerably from those of the Reynolds stress, raising some doubts about whether the previously mentioned studies of coherent structures and, indeed, a lot of the effort dedicated to modelling locally the Reynolds stress tensor, may only apply to a particular flux definition, and be therefore largely irrelevant.

An obvious alternative would be to abandon the use of fluxes and stresses, and to fall back on their divergence. This approach was discussed in some detail in [6], and has been attempted before, mostly from the point of view of modelling [7]. It changes the character of the analysis, because fluxes are large-scale quantities, while their divergence is associated with smaller scales. The question is similar to whether to focus on the pressure or on the pressure gradient. The gradient appears in the equations of motion and determines the acceleration of the fluid, but the pressure is more directly related to the velocities. Think of Bernouilli's equation.

To assess the 'robustness' of the older stress analyses, we repeat in this study the identification and classification of intense flux structures using the new optimal definition, and compare it with the coherent structures based on the classical Reynolds stress.

The organisation of this paper is as follows. The simulations used for the analysis, and the identification criteria for the structures, are introduced in §2. The geometry of the identified structures is discussed in §3, and their relative position with respect to high- and low-momentum regions and to other flux structures is presented in §4 and §5. Finally §6 examines the conditional velocity fields relative to the new structures, and §7 concludes.

## 2. Numerical experiments and structure identification

We use data from direct numerical simulations of a turbulent channel flow at two Reynolds numbers, periodic in the two wall-parallel directions and of half-height $h$. The streamwise, wall-normal and spanwise directions are $x, y, z$, respectively, and $u, v, w$ are the corresponding fluctuating velocity components. The ensemble-averaged mean streamwise velocity is $U(y)$. Variables with a '+' superscript are normalized in wall units, defined from the kinematic viscosity, $\nu$, and from the friction velocity, $u_\tau$, which is in turn defined from the shear stress at the wall as $u_\tau^2 = \nu \partial_y U|_{y=0}$. The friction Reynolds number is $Re_\tau = u_\tau h/\nu$. Overlines, $\overline{()}$, denote $y$-dependent ensemble averages, and primes stand for root-mean-square fluctuation intensities.

The simulations use Fourier spectral discretisation in $(x, z)$, dealiased by the 2/3 rule. The wall-normal discretisation is Chebychev spectral for the lower-Reynolds-number case [8], and compact finite differences for the higher one [9]. Time stepping is third-order semi-implicit Runge–Kutta, keeping the mass flux constant. Some numerical parameters are collected in table 1, and further details can be found in the original publications.

We focus on the structures of the classical Reynolds stress, $uv$, and of the corresponding optimal flux, $\phi_{12}$. Details of the computation of the $\phi_{ij}$ tensor can be found in [6], and will not be repeated here. Basically, it satisfies

$$\partial_j \phi_{ij} = \partial_j (u_i u_j), \tag{1}$$

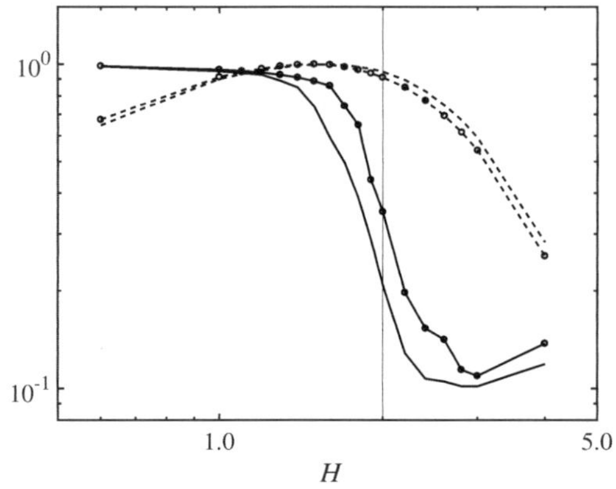


**Figure 1.** Percolation diagram for the identification of *Ops*. The solid lines are the ratio of the volume of the largest object to the volume of all identified objects; the dashed ones are the ratio of the number of identified objects to the maximum number of objects. The vertical line is the chosen threshold, $H = 2$. Symbols denote the two Reynolds numbers, as in table 1.

while minimising $\int \phi_{ij}\phi_{ij}\, \mathrm{d}^3\boldsymbol{x}$, where the integral extends over the whole channel, and indices range from 1 to 3 (or equivalently, from $x$ to $z$). Repeated indices imply summation. In general, because the difference between $\phi_{ij}$ and the classical Reynolds stresses is divergence-free, they generate the same integrated flux across any closed surface. In particular, $\phi_{12}$ plays the same role as $uv$, transferring streamwise fluctuating momentum along $y$, and the instantaneous integral of the two quantities over wall-parallel planes is identical, $\overline{(\phi_{12})}^+_{xz} = \overline{(uv)}^+_{xz}$.

Coherent structures of these quantities are identified by the same method used in [2] for the quadrant structures of $uv$. Namely, the intense regions of a generic quantity $q$ are extracted by thresholding,

$$|q| > H f(y), \tag{2}$$

with respect to some $y$-dependent flow statistics $f(y)$. Each set of mutually contiguous points satisfying (2) is defined as a structure. The constant $H$ is chosen using the percolation analysis developed in [10]. The threshold used for $uv$ is the same as in [2],

$$|uv| > 1.75\, u'v'. \tag{3}$$

After some experimentation, the threshold for $\phi_{12}$ was chosen to be

$$|\phi_{12}| > 2.00\, \phi'_{12}, \tag{4}$$

which is seen in figure 1 to be in the midst of the percolation transition. It was checked that the results presented below are qualitatively independent of the threshold as long as it is within the transition range $1.4 < H < 2.4$. It was also checked that using the different thresholding quantity, $f(y) = (\overline{\phi_{12}^2})^{1/2} = (\phi'^2_{12} + \overline{\phi_{12}}^2)^{1/2}$, had a small effect on the properties of the structures. It is important to recall that, although quantities such as $u$ denote zero-mean fluctuations, neither $\phi_{12}$ nor $uv$ have zero mean. In what follows, the coherent structures identified for $uv$ and $\phi_{12}$ are denoted as *Qs* and *Ops*, respectively.

## 3. The geometry of individual structures

Table 2 shows the number and volume fraction occupied by the objects identified by the procedure explained above, oriented in each half of the channel with respect to their closest

**Table 2.** Number fraction of identified objects with respect to their total number, and volume fraction with respect to the flow domain. Superscripts show the sign of the mean momentum flux within each structure.

| $Re_\tau$ | $N^-_{Qs}$ | $N^+_{Qs}$ | $V^-_{Qs}$ | $V^+_{Qs}$ | $N^-_{Ops}$ | $N^+_{Ops}$ | $V^-_{Ops}$ | $V^+_{Ops}$ |
|---|---|---|---|---|---|---|---|---|
| 934 | 0.60 | 0.40 | 0.080 | 0.001 | 0.64 | 0.36 | 0.073 | 0.006 |
| 2003 | 0.61 | 0.39 | 0.073 | 0.002 | 0.67 | 0.33 | 0.068 | 0.008 |

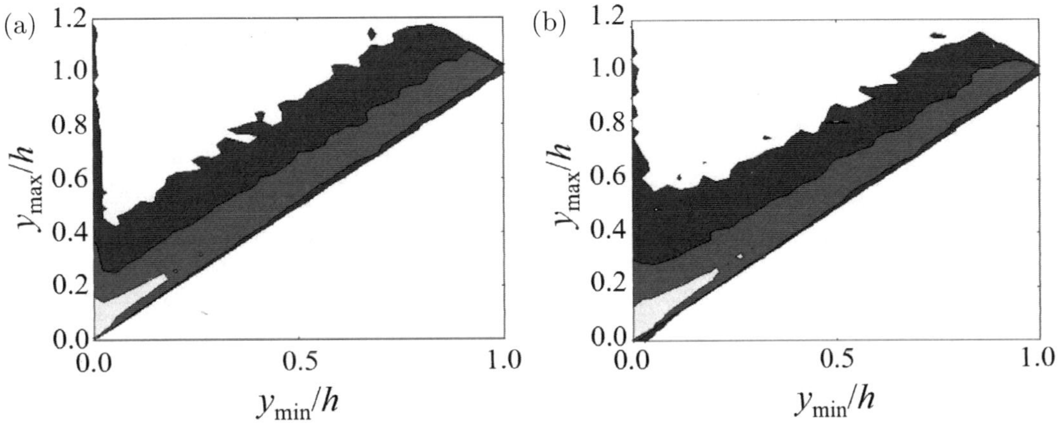


**Figure 2.** Joint PDF of the distance from the closest wall to the bottom ($y_{min}$) and top ($y_{max}$) of the structures. (a) $Qs$, (b) $Ops$. Contours include 40%, 90% and 98.8% of data. $Re_\tau = 934$.

wall. With this notation, the average momentum flux is negative, and objects carrying a net negative momentum flux will be called co-gradient. Individual objects can carry either a net positive or negative flux, but table 2 shows that co-gradient ones are more common, both for $uv$ and for $\phi_{12}$. They also tend to be larger than counter-gradient ones, as seen by the larger relative difference between the volume contributions of the two kinds, compared to the number frequencies. In the case of $Qs$, for which the sign of the velocity fluctuations is fixed by their definition, co-gradient objects are classified as 'sweeps' or 'ejections' [1] depending on whether their streamwise velocity fluctuation is positive of negative, respectively. In the case of $Ops$, that distinction is not clear, and the relation of $Ops$ with the two types of $Qs$ will be one of the questions to be discussed below.

Each object is circumscribed within a parallelepipedic box aligned to the Cartesian axes, which is used to define both its dimensions, $\Delta_x, \Delta_y$ and $\Delta_z$, and its position. Figure 2 shows the joint probability density function (PDF) of the minimum and maximum distance of the objects from the nearest wall, $y_{min}$ and $y_{max}$. Lozano-Durán et al. [2] showed that the coherent structures of $uv$ can classified into two families based on their minimum distance from the wall. The 'attached' structures in the left-most side of the figure 2(a) have $y^+_{min} < 20$, but some of them are very large, growing to be as tall as the channel half-height. The 'detached' structures in the diagonal band of figure 2(a) never approach the wall, and their height $\Delta_y = y_{max} - y_{min}$ depends little on the wall distance. The same classification into attached and detached families applies to vorticity and velocity structures [11, 12] with minor variations in the limiting $y_{min}$ that separates the two families. In the logarithmic layer of channels, around 60% of the Reynolds stress is carried by attached $Qs$, while detached objects carry a much smaller fraction of the total stress. In fact, it was shown in [2, 13] that detached objects are random fluctuations which are

IOP Conf. Series: Journal of Physics: Conf. Series **1001** (2018) 012003 doi:10.1088/1742-6596/1001/1/012003

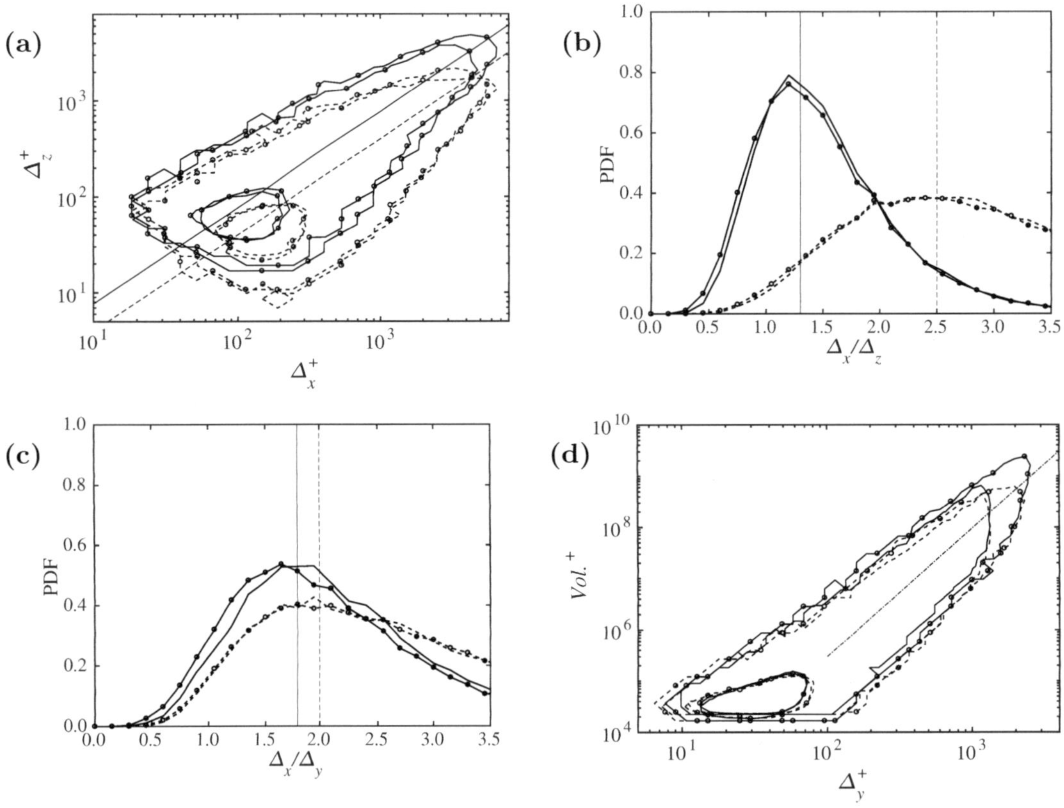


**Figure 3.** (a) Joint PDF of the wall-parallel dimensions of attached structures $(\Delta_x, \Delta_z)$. Contours include 50% and 99.8% of the data. The two diagonals are $\Delta_x = 1.3\Delta_z$ and $\Delta_x = 2.5\Delta_z$. (b) PDF of the wall-parallel aspect ratio $\Delta_x/\Delta_z$ of tall attached objects $(y^+_{min} > 100)$. The two vertical lines mark the peak value of the PDF, obtained by parabolic fitting, $\Delta_x/\Delta_z = 1.3$ and $\Delta_x/\Delta_z = 2.5$. (c) As in (b) for the wall-normal aspect ratio $\Delta_x/\Delta_y$ of tall attached objects. The vertical lines are $\Delta_x/\Delta_y = 1.8$ and $\Delta_x/\Delta_y = 2$. (d) Joint PDF of the volume and height of attached structures. Contours include 50% and 99.8% of data. The diagonal line is $Vol. \propto \Delta_y^{2.5}$. In all figures, solid and dashed lines correspond to *Ops* and *Qs*, respectively, and symbols denote the two Reynolds numbers, as in table 1.

too small to couple with the mean shear, and that they are present in all shear flows. They are isotropically oriented, and do not collectively support a mean momentum flux. Attached objects are also common features of shear flows, and they can even be identified in flows without walls [13]. They are large enough to couple with the mean shear, and are responsible for most of the mean momentum flux. This coupling also means that large co-gradient objects extract energy from the mean flow, which allows them to grow even larger. In essence, 'attached' objects are defined by being large, rather than by being attached to the wall, but the geometry of the flow over walls is such that they cannot grow to be large enough to couple with the shear without also hitting the wall [12].

As can be seen in figure 2(b), *Ops* also separate into attached and detached families. In the logarithmic layer, attached *Ops* carry 30% of the mean momentum flux, which is close to the overall contribution of all the *Ops*. Co-gradient detached *Ops* carry approximately 7% of the flux, which is is partially compensated by about –2% from the counter-gradient ones. As in the case of *Qs*, attached *Ops* are dominant, and we mostly focus on co-gradient attached objects in the following.

Specially relevant are attached objects that are large enough to reach above the buffer layer, $\Delta_y^+ > 100$, and which are therefore largely inviscid. Most of the discussion in the previous paragraphs applies to these 'tall' objects. They were also the main focus of the discussion in [11, 2, 13], and we will mostly be interested in them in what follows.

It should be mentioned at this point that the smaller contribution of the *Ops* to the total flux, when compared to the *Qs*, is mainly, but not wholly, due to the choice of the identification threshold $H$. The total contribution of *Qs* to the momentum flux varies from 95% to 30% as the threshold changes across the percolation range from $H = 1$ to $H = 3$, while that of the *Ops* only varies from 65% to 20% as $H$ changes from 1.4 to 2.4. Overall, $\phi_{12}$ is less intermittent than $uv$, and a given fraction of the momentum flux requires a larger volume in the former than in the latter [6].

The dimensions of attached *Qs* and *Ops* are compared in figure 3(a), which shows the joint PDF of the logarithms of the wall-parallel sizes of the attached objects, as determined from their bounding box. The PDF follows a linear trend in both cases, but with a different slope for *Qs* and for *Ops*. The PDFs of the aspect ratio $\Delta_x/\Delta_z$, shown in figure 3(b) for tall attached objects, indicate that the *Ops* are almost twice wider than the *Qs*. A similar linear relation applies to the wall-parallel and wall-normal dimensions of the objects, confirming that both *Qs* and *Ops* form self-similar families in the logarithmic layer, but figure 3(c) shows that the wall-normal aspect ratio of *Qs* and *Ops* is very similar. *Ops* are wider than *Qs* but not taller.

Figure 3(d) shows the joint PDF of the volume and height of attached *Qs* and *Ops*. For the tall objects defined above, the relation between $\Delta_y$ and the volume approximately follows a power law, similar in both structures. The exponent changes slightly with the Reynolds number, and with the range of wall-distances considered, but it is always close to 2.5, with a slight tendency to higher exponents in *Ops* than in *Qs*. If interpreted as a fractal dimension, it implies fairly full shell- or flake-like objects [2].

In all these results, it is important to note the excellent collapse of the two Reynolds numbers used in the study.

## 4. Relation to low- and high-momentum regions

In the rest of the paper, we will only consider tall attached objects ($\Delta_y^+ > 100$) that do not extend beyond the logarithmic layer ($\Delta_y/h < 0.2$). This is the same restriction applied to *Qs* in [2], and will facilitate the comparison.

The first point to consider is the relation of *Qs* and *Ops* to the high- and low-momentum regions (streaks) of the flow. In the case of *Qs*, it follows from their definition that sweeps ($u > 0$) reside in high-speed regions, and ejections ($u < 0$) in low-speed ones. It was found in [2] that sweeps and ejections tend to be paired side by side, with each pair straddling the border between a high- and a low-speed streak. In the case of *Ops*, figure 4 is a snapshot of $u$ at $y^+ = 100$, including contours of $\phi_{12}$ at the reference threshold used to define structures in (4). In contrast to *Qs*, the figure shows that many *Ops* intersect both a high- and a low-momentum region. This can be quantified by the volume fraction occupied by low-momentum points within each object,

$$F = \frac{\int_{\Omega_{u<0}} \mathrm{d}V}{\int_{\Omega_{u\geq 0}} \mathrm{d}V + \int_{\Omega_{u<0}} \mathrm{d}V}, \tag{5}$$

where $\Omega$ is the interior of the object and subscripts represent conditioning by the sign of $u$. Objects with $F = 0$ are completely within a high-momentum region, and those with $F = 1$ are within a low-momentum one.

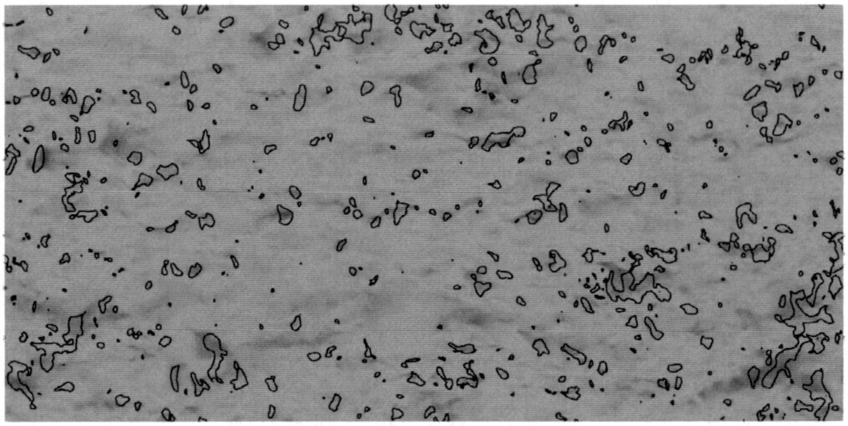

**Figure 4.** (Colour online) The colour map is the instantaneous streamwise fluctuating velocity, and the line contours is the identification threshold for *Ops* at $y/h = 0.10$. Flow is from left to right. $Re_\tau = 934$.

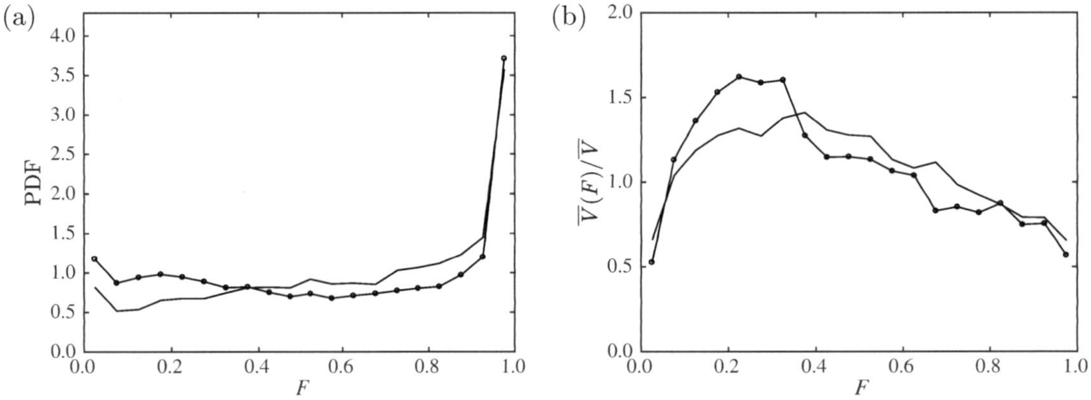


**Figure 5.** (a) PDF of the low-momentum fraction, $F$, as defined in (5), for tall attached *Ops* in the logarithmic layer. (b) Average volume of *Ops* as a function of $F$, where the overline denotes averaging over all tall attached objects in the logarithmic layer. Symbols denote the two Reynolds numbers, as in table 1.

For the reasons mentioned above, most *Qs* are in one of these two limits, but the PDF in figure 5(a) shows that the distribution of the logarithmic-layer *Ops* is more uniform. Except for an excess of purely low-momentum *Ops*, their PDF is approximately uniform. Moreover, Figure 5(b) shows that 'pure' *Ops* tend to be smaller than those with mixed momentum regions, so that their contribution to the overall object volume is less than implied by their PDF. In fact, the number fraction of the pure *Ops* ($F < 5\%$ or $F > 95\%$) is 26% of the total, but their contribution to the volume is only 14%. Moreover, when all the tall attached objects are considered, including those taller than $\Delta_y/h = 0.2$, the volume fraction of the pure *Ops* is only 3%. The result is that, as opposed to *Qs*, the *Ops* cannot be classified into sweeps and ejections.

The reason for the excess of low-momentum *Ops* is unclear, but it is known that low-momentum regions in wall-bounded flows contain more coherent vortices than high-momentum

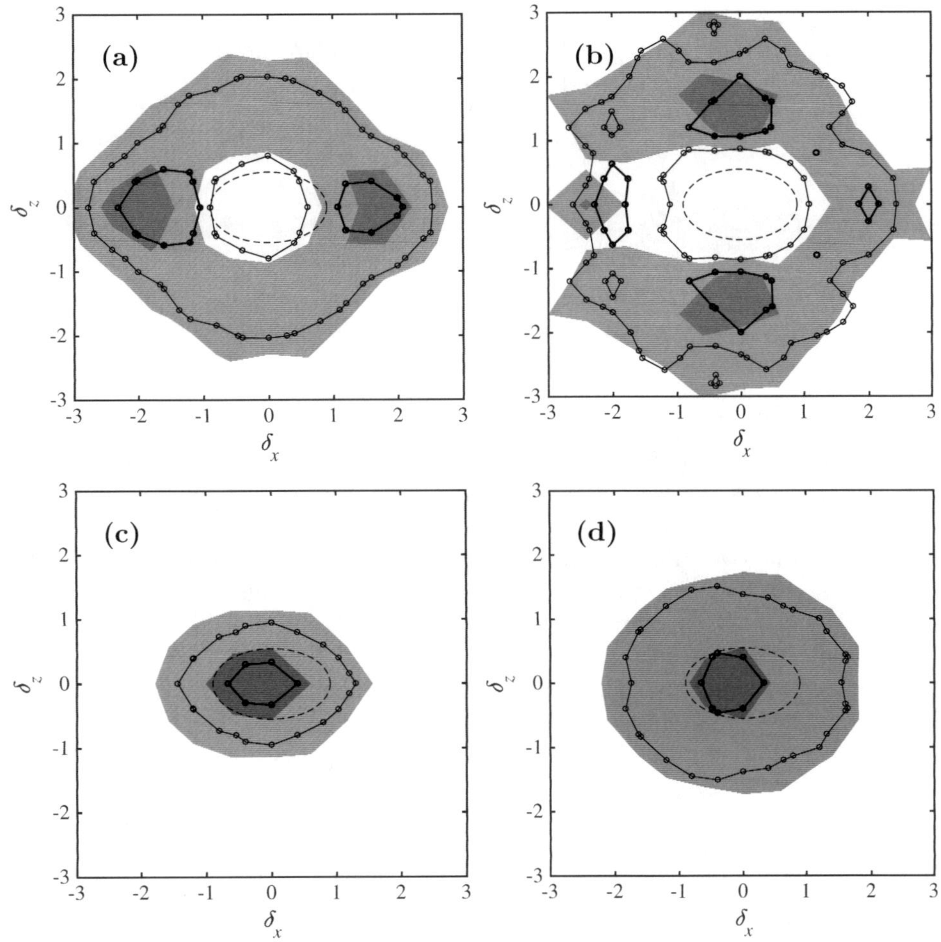


**Figure 6.** Two-dimensional pdf of the relative position of: (a) Closest $Op$ relative to another $Op$. (b) Closest low-momentum $Op$ ($F \geq 0.5$) relative to a high-momentum $Op$ ($F < 0.5$). (c) Closest $Q$ relative to an $Op$. (d) Closest sweep-ejection $Q$-pair relative to an $Op$. Bold line contours and darker filled areas contain 10% of the data. Thinner and lighter ones contain 50% of the data. Shaded contours are $Re_\tau = 934$ and lines are $Re_\tau = 2003$. The dashed ellipses correspond to the mean size of the reference object.

ones [14], and that ejections, which are low-$u$ objects, are more common than sweeps above the buffer layer [2]. There are several possible explanations for this inhomogeneity, going from artefacts due to using a uniform threshold across a non-uniform flow plane, to actual physics. The most likely reason, which is supported by informal visual inspection of individual flow fields, is that low-velocity regions are associated with positive wall-normal velocities, which bring active turbulence from the high-shear region near the wall, while high-velocity ones receive relatively quiescent flow from the low-shear region in the centre of the channel.

IOP Conf. Series: Journal of Physics: Conf. Series **1001** (2018) 012003 doi:10.1088/1742-6596/1001/1/012003

## 5. Geometric relations between objects

The relationships among different objects is discussed next by means of the PDF of their relative position, which is defined for object $j$ relative to object $i$ as

$$\boldsymbol{\delta}^{(ij)} = \frac{\boldsymbol{x}_c^{(j)} - \boldsymbol{x}_c^{(i)}}{\overline{\Delta_y}^{(ij)}}, \tag{6}$$

with components $\delta_k^{(ij)}, k = x \ldots z$. Positions are referred to the centre, $\boldsymbol{x}_c^{(i)}$, of the circumscribing box of object $i$, and lengths are normalised with the mean wall-normal height of the two objects, $\overline{\Delta_y}^{(ij)} = (\Delta_y^{(i)} + \Delta_y^{(j)})/2$. As in [2], only objects with similar size,

$$1/2 < \Delta_y^{(j)}/\Delta_y^{(i)} < 2, \tag{7}$$

are considered to be related, but we follow in this work the practice in [13] of compiling the PDFs using only the closest object to the reference one. The statistical spanwise symmetry of the flow is used to improve the convergence of the PDFs.

Figure 6(a) shows the two-dimensional PDF of the relative position of the closest *Op* relative to another *Op*. Related *Ops* tend to be located streamwise from one another, separated by the mean streamwise length of the reference *Op*, and 85% of the *Ops* find another *Ops* satisfying condition (7) at $Re_\tau = 2003$.

This arrangement is similar to the relative position of *Qs* of the same kind (e.g. sweep-sweep), but *Qs* of different kinds, such as a sweep and its closest ejection tend to form pairs aligned spanwise from one another. In the logarithmic layer, 80% of the *Qs* form such pairs, or similar composite structures [2]. We have seen that *Ops* cannot be classified into sweeps and ejections, but the relative position of predominantly low-momentum *Ops* ($F > 0.5$) with respect to high-momentum ones ($F \leq 0.5$) is shown in figure 6(b). There is also in this case a tendency to spanwise organisation, but the convergence of the PDF is poor because only 41% of the low-momentum *Ops* have a high-momentum one satisfying (7).

The relative position between *Ops* and *Qs* is shown in figure 6(c). Most *Ops* (98%) have at least a $Q$ satisfying (7), and 50% of these *Qs* are closer than the mean diameter of the reference *Op*. Essentially *Ops* and *Qs* are collocated.

Finally, the distance from *Ops* to the closest sweep-ejection pair, as defined in [2], is examined in figure 6(d). In this case, the centre of the pair is defined as

$$\boldsymbol{x}_c^{(p)} = \frac{\boldsymbol{x}_c^{(Q_2)} + \boldsymbol{x}_c^{(Q_4)}}{2}, \tag{8}$$

and its size is defined as the mean of the wall-normal size of the two components of the pair. Most (87%) *Ops* have a closest $Q$-pair satisfying (7) and figure 6(d) shows that most $Q$-pairs are found within the mean length and width of the reference *Op*. As with individual *Qs*, *Ops* and $Q$-pairs overlap.

This result is consistent with section 3, which showed that attached *Ops* are twice wider than *Qs*, but not taller, and with section 4, where we found that most *Ops* intersect both high- and low-momentum regions. Together with the results in figure 6, the logical conclusion is that tall attached *Ops* correspond to the sweep-ejection pairs discussed in [2]. This will be further confirmed by the conditional fields in the next section.

## 6. Conditionally averaged fields

In this section, we consider the conditionally averaged flow field around *Ops*. As was the case with the comparable conditional fields around $Q$-pairs in [2, 13], this implies defining both a

IOP Conf. Series: Journal of Physics: Conf. Series **1001** (2018) 012003 doi:10.1088/1742-6596/1001/1/012003

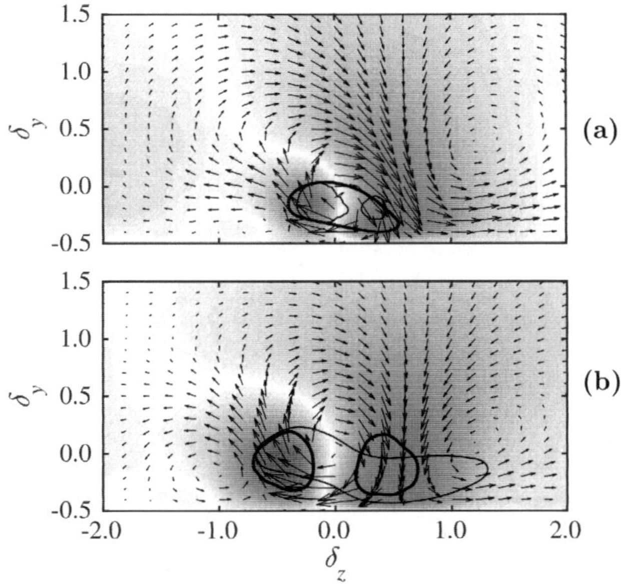


**Figure 7.** (Colour online) (a) Cross section at $\delta_x = 0$ of the conditionally averaged field around tall attached *Ops* in the logarithmic region; looking in the direction of the flow. The colour background is $\widetilde{u}$, increasing from blue to red; arrows are the velocity in the cross-plane; the heavier contour is $\widetilde{\phi_{12}}^+ = -2.1$, and the lighter double one is $\widetilde{uv}^+ = -2.1$. (b) As in (a), for the conditionally averaged field around tall attached sweep-ejection pairs in the logarithmic region. Symbols are as in (a), but the heavier double contour is $\widetilde{uv}^+ = -1.3$, and the lighter one is $\widetilde{\phi_{12}}^+ = -1.3$. $Re_\tau = 934$.

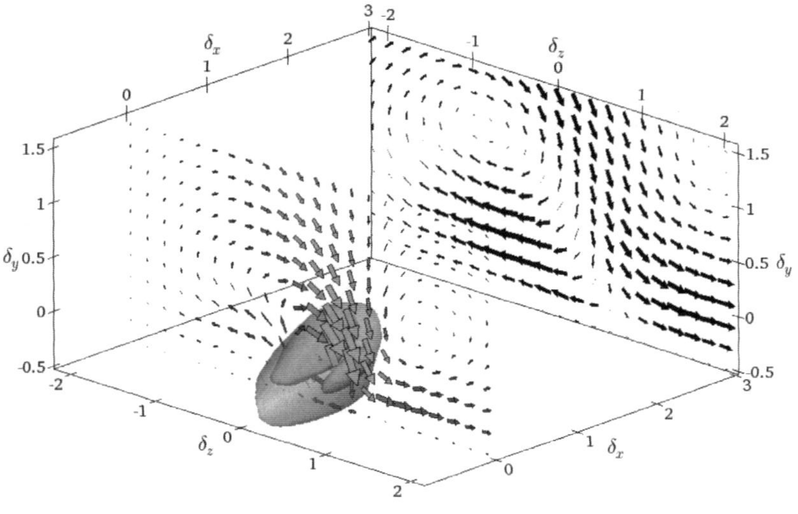


**Figure 8.** (Colour online) Three-dimensional plot of the conditionally averaged field around tall attached *Ops* in the logarithmic layer. The translucent isosurface is $\widetilde{\phi_{12}}^+ = -2.1$, coloured by $\widetilde{u}$; the yellow solid isosurface is $\widetilde{uv}^+ = -2.1$; arrows are the conditional cross-flow velocity: grey ones at $\delta_x = 0$, and black ones at $\delta_x = 3$. Flow is from lower-left to upper-right. $Re_\tau = 934$.

common centre and a common scale for the individual objects, so that the conditional average is a scale-less function of a dimensionless vector coordinate $\boldsymbol{\delta}$. Thus, the conditional field for a set of individual fields $q^{(i)}(\boldsymbol{x})$, $i = 1 \dots N$, is

$$\widetilde{q}(\boldsymbol{\delta}) = N^{-1} \sum_i q^{(i)}(\Delta_y^{(i)} \boldsymbol{\delta} + \boldsymbol{x}_c^{(i)}), \tag{9}$$

where $\boldsymbol{x}_c^{(i)}$ is the centre of the box circumscribing the $i$-th object, so that

$$\boldsymbol{\delta} = (\boldsymbol{x} - \boldsymbol{x}_c^{(i)})/\Delta_y^{(i)}. \tag{10}$$

Individual fields are centred, rescaled with the height, $\Delta_y^{(i)}$, of its object, and averaged. In computing the conditional fields around $Q$-pairs in [2, 13] it was found convenient to take advantage of the statistical spanwise symmetry of the flow to reorient each pair so that the sweep is always on the same orientation with respect to the flow direction. We have seen that sweeps and ejections are not relevant for *Ops*, but individual *Ops* are reoriented so that the highest $u$ is always on the right-hand side, looking in the direction of the flow. To avoid the indeterminancy of this reorientation for *Ops* with $F \approx 0$ or $F \approx 1$, which we saw in figure 5 to be high in number but low in volume, only objects with $0.2 < F < 0.8$ are considered for the conditional fields.

Figure 7(a) shows a section through the $\delta_x = 0$ cross-flow plane of the conditional field around tall attached *Ops* in the logarithmic layer. The colour indicate $\widetilde{u}$ and the arrows are the cross-flow velocity. The average position of the *Op* is given by the heavy contour of $\widetilde{\phi_{12}}$, which intersects both the high- and the low-$u$ region. It has a single peak near the centre of the conditional box, whose position is approximately independent of the choice of the $\delta_x$ slice and of the contour magnitude. The position of the associated *Qs* is marked by the lighter contours of $\widetilde{uv}$, which has two peaks corresponding to the average location of the sweep and of the ejection. The cross-flow velocities form a circular roller around the box centre, reminiscent of the conditional rollers associated with sweep-ejection pairs in [2, 13].

For completeness, figure 7(b) shows the conditionally averaged velocity field around a sweep-ejection pair, confirming the similarity of the flow around *Ops* and $Q$-pairs. It includes the mean position of the associated *Op*, in the form of a line contour of $\widetilde{\phi_{12}}$, and it is clear from these two figures that *Ops* and sweep-ejection pairs are collocated on average.

Figure 8 shows the three-dimensional counterpart to the conditional field in figure 7(a). It is clear that there is a strong sweep-ejection pair at the same location as the conditional *Op*. As seen from the arrows at the two cross planes, the circular motion in fig. 7(a) is part of an inclined quasi-streamwise roller, whose orientation can be evaluated by tracking the maximum of the averaged fluctuation vorticity[1], $|\widetilde{\omega}|$. A least-square fit to these local maxima in $0 \le |\delta_x| \le 1$ results in a roller inclined at 24° to the wall, and tilted 7.9° towards the low momentum side. These angles are approximately twice larger than those of other streamwise eddies in the literature [15, 16], but comparable, for example, to the inclination of the roller of large detached pairs in [13]. They are also defined by a different procedure than those in most previous reports, including the present restriction to the logarithmic layer, and it was noted in [16] that the inclination of eddies depends on how they are defined, and on the length scales being considered.

## 7. Conclusions

We have presented in this paper a detailed comparison between the intense structures (*Ops*) of the alternative momentum flux $\phi_{12}$ defined in [6], and the classical quadrant structures (*Qs*)

[1] Note that $\widetilde{\omega}$ is different from the vorticity of the conditional velocity $\nabla \times \widetilde{u}$, because of the variable rescaling of the spatial coordinates

based on the tangential Reynolds stress $uv$. Since both fluxes are equivalent in the mean, the emphasis has been on identifying similarities and differences between their intense structures, with a view to ascertaining which of their properties are associated with a particular flow definition, and which ones are more general, and therefore presumably characteristic of the momentum flux as such.

The two types of structures share many characteristics. For example, objects large enough to attach to the wall are responsible for most of the momentum transfer in both cases, and those mostly located within the logarithmic layer form self-similar families characterised by fixed aspect ratios, rather than by a fixed size.

However, the tall attached *Ops* are twice wider than the *Qs*, although not taller, and they intersect both high- and low-momentum regions, while *Qs*, because of their definition, tend to be associated with one or with the other. A consequence is that it is difficult to classify *Ops* into the equivalent of the sweeps and ejections found for *Qs*.

In fact, the PDFs of the relative positions of *Ops* with respect to other *Ops*, as well as with respect to *Qs* and to sweep-ejection pairs, strongly support the idea that most *Ops* are manifestations of the same phenomenon as the sweep-ejection pairs. Both are associated with an inclined conditional roller in the border between a high- and a low-speed streak.

Our results suggest that the coherent structures of the momentum flux and the associated streamwise rollers are robust with respect to the definition of the fluxes, and are truly representative of flow features in the logarithmic layer. They also suggest that sweeps and ejections should not be studied as individual objects, but as members of a composite structure containing both. It should finally be mentioned that all these results are found to be independent of the Reynolds number for the two cases considered in this work.

**Acknowledgments**
This work was supported by the Coturb project of the European Research Council, and performed in part during the 2017 Turbulence Summer Workshop at the UPM. The stay of Mr. Osawa at Madrid was partially funded by the Erasmus Mundus EASED programme (Grant 2012-5538/004-001) coordinated by CentraleSupélec. We are grateful to Dr. Lozano–Durán for his careful refereeing of an initial version of this manuscript.

## References

[1] Wallace J M, Eckelmann H and Brodkey R S 1972 *J. Fluid Mech.* **54** 39–48
[2] Lozano-Durán A, Flores O and Jiménez J 2012 *J. Fluid Mech.* **694** 100–130
[3] Lozano-Durán A and Jiménez J 2014 *J. Fluid Mech.* **759** 432–471
[4] Tsinober A 1990 *Phys. Fluids A* **2** 484–486
[5] Hamman C W, Klewicki J C and Kirby R M 2008 *J. Fluid Mech.* **610** 261–284
[6] Jiménez J 2016 *J. Fluid Mech.* **809** 585–600
[7] Wu J, Zhou Y, Lu X and Fan M 1999 *Phys. Fluid* **11** 627–635
[8] Del Alamo J C, Jiménez J, Zandonade P and Moser R D 2004 *J. Fluid Mech.* **500** 135–144
[9] Hoyas S and Jiménez J 2006 *Phys. Fluids* **18** 011702
[10] Moisy F and Jiménez J 2004 *J. Fluid Mech.* **513** 111–133
[11] del Álamo J C, Jiménez J, Zandonade P and Moser R D 2006 *J. Fluid Mech.* **561** 329–358
[12] Jiménez J 2018 *J. Fluid Mech.* **842** P1–100
[13] Dong S, Lozano-Durán A, Sekimoto A and Jiménez J 2017 *J. Fluid Mech.* **816** 167–208
[14] Tanahashi M, Kang S J, Miyamoto T, Shiokawa S and Miyauchi T 2004 *Int. J. Heat Fluid Flow* **25** 331–340
[15] Jeong J, Hussain F, Schoppa W and Kim J 1997 *J. Fluid Mech.* **332** 185–214
[16] Sillero J A, Jiménez J and Moser R D 2014 *Phys. Fluids* **26** 105109

IOP Conf. Series: Journal of Physics: Conf. Series **1001** (2018) 012004 doi:10.1088/1742-6596/1001/1/012004

# Extremely high wall-shear stress events in a turbulent boundary layer

**Chong Pan[1], Yongseok Kwon[2]**

[1] Fluid Mechanics Key Laboratory of Education Ministry, Beijing University of Aeronautics and Astronautics, 100191 Beijing, PR China

[2] School of Aeronautics, Universidad Politécnica de Madrid, 28040 Madrid, Spain

E-mail: panchong@buaa.edu.cn

**Abstract.** The present work studies the fluctuating characteristics of the streamwise wall-shear stress in a DNS of a turbulent boundary layer at $Re_\tau$=1500 from a structural view. The two-dimensional field of the fluctuating friction velocity $u'_\tau(x,z)$ is decomposed into the large- and small-scale components via a recently proposed scale separation algorithm, Quasi-bivariate Variational Mode Decomposition (QB-VMD). Both components are found to be dominated by streak-like structures, which can be regarded as the wall signature of the inner-layer streaks and the outer-layer LSMs, respectively. Extreme positive/negative wall-shear stress fluctuation events are detected in the large-scale component. The former's occurrence frequency is nearly one order of magnitude higher than the latter; therefore, they contribute a significant portion of the long tail of the wall-shear stress distribution. Both two-point correlations and conditional averages show that these extreme positive wall-shear stress events are embedded in the large-scale positive $u'_\tau$ streaks. They seem to be formed by near-wall 'splatting' process, which are related to strong finger-like sweeping (Q4) events originated from the outer-layer positive LSMs.

## 1. Introduction

Wall-shear stress is of obvious importance in wall-bounded turbulence. It sets the boundary condition for the sustenance and dissipation of near-wall turbulence [1–3], thus determines the viscous scales that well characterizes the scaling of the inner-layer cycle. On the practical side, the skin friction reduction over the surface of transportation vehicles is the persistent pursuit of engineers [4].

To date, various works have built up the empirical or semi-theoretical formulation between the mean wall-shear stress, $\overline{\tau}_w$, and Reynolds number ($Re$) in smooth-wall-bounded turbulence [5–8]. Nevertheless, less progress on the statistics and dynamics of wall-shear stress fluctuations has been made in the past. The law of the wall [9] assumes a $Re$-independency of the statistics of wall-shear stress. This is consistent with early experiments [10–13], which reported compatible values of high-order moments (variance, skewness and flatness) of the streamwise component of the fluctuating wall-shear stress $\tau'_x$ that were less sensitive to $Re$ if taking the measurement uncertainty into consideration. Traditionally, the root-mean-squared value was regarded to be $\tau_{x,rms} \approx 0.4\overline{\tau}_w$ [13], but recent experiments and DNS [14–19] have shown a slow variation of $\tau_{x,rms}$ with respect to $Re$. Based on assessing different DNS datasets of a canonical turbulent boundary layer with zero pressure gradient, Schlatter and Örlu [8] proposed an empirical scaling of $\tau_{x,rms} \sim 0.018 \ln Re_\tau$, and Diaz-Daniel et al. [20] recently extended it to the spanwise component $\tau_{z,rms}$.

The $\ln Re_\tau$ variation of $\tau_{x,rms}$ and $\tau_{z,rms}$ is consistent with the so-called high $Re$ effect on the near-wall turbulence, which is majorly characterized as the increasing influence of the outer-layer large-scale and very-large-scale motions (LSMs and VLSMs) on the near-wall dynamics in high $Re$ scenario [21–29]. Considering the traditional viewpoint that the generation of wall-shear stress is directly linked to the near-wall processes [4, 30, 31], the existing drag reduction techniques aimed at controlling the inner-layer cycles might be challenged by this high $Re$ effect [32, 33]. In fact, the near-wall motions were found to have less importance in skin-friction generation with the increase of $Re$ [34, 35]. This gives new impetus to further investigate the dynamical correlation between shear stress fluctuations on the wall and the turbulent structures with various sizes in the flow.

In this aspect, Marusic and Heuer [36] for the first time measured the fluctuating wall-shear stress in an atmospheric surface layer at $Re_\tau \sim O(10^5)$, and calculated the cross-correlation between the fluctuating wall-shear stress and the streamwise velocity in the logarithmic region. Abe et al. [16] identified the footprint of VLSMs in both the physical and the spectral space of two-dimensional (2D) wall-shear stress field in a DNS channel flow up to $Re_\tau = 640$. Marusic et al. [37] and Mathis et al. [38] developed a predictive model for the wall-shear stress distribution based on the concept of the amplitude modulation of near-wall small-scale streamwise velocity by the large-scale outer layer structures. However, de Giovanetti et al. [39] recently showed that in a channel flow up to $Re_\tau \approx 4000$, the largest amount of skin friction is generated by the self-similar energy-containing motions in the form of Townsends attached eddies, while the contribution of LSMs and VLSMs are of limited importance. This is inconsistent with the idea that the near-wall dynamics are increasingly affected by the large-scale outer-layer structures at higher $Re$, and deserves to be further studied.

Various researchers [14, 17, 19, 20, 31] have reported that $\tau_x'$ generally follows a log-normal distribution. The positively skewed long tail of the p.d.f. of $\tau_x'$ indicates the existence of extreme positive $\tau_x'$ events, which seems to increase with $Re$ even if normalized by the r.m.s. value $\tau_{x,rms}$. Diaz-Daniel et. al. [20] argued that this might suggest that the wall-shear stress is a highly intermittent signal. They then proposed a theoretical model to relate the intermittency, which is represented by low-pass-filtered $\tau_w'$, to the velocity fluctuations away from the wall within the framework of attached eddy hypothesis.

To our understanding, the outer-layer large-scale structures are more likely to be responsible for the intermittent behaviour of the wall-shear stress, since the large-scale amplitude modulation coefficient was observed to continuously increase when approaching to the wall [25, 29, 40]. This conjecture inspires the present work to study the extreme high wall-shear stress events from a structural view. For such a purpose, the instantaneous $\tau_x'$ field in a DNS dataset of a spatially developing turbulent boundary layer at $Re_\tau = 1500$ was examined. Unlike previous spectral-based scale separation analysis, a recently proposed method of Quasi-bivariate Variational Mode Decomposition (QB-VMD) is used to decompose the instantaneous wall-shear stress field into the large- and small-scale components. It is shown that those frequently occurred extreme wall events, which contain high positive magnitudes of $\tau_x'$, are mainly carried by the former. Two-point correlation and conditional average analysis further suggest that a 'splatting' scenario of strong Q4 events (sweeping motion of high speed fluids towards the wall, defined by $u > 0$ and $v < 0$, see [41] for more details), which are found to be tightly related to the LSMs with positive momentum fluctuation, might be responsible for the generation of these extreme wall events.

## 2. DNS dataset

The DNS dataset studied here was obtained by Sillero et al. [42, 43] in Universidad Politécnica de Madrid (UPM), who simulated a spatially developing turbulent boundary layer over a smooth wall with a periodic boundary condition in the spanwise direction and non-periodic inflow/outflow condition in the streamwise direction. Fractional-step method was used to solve the incompressible Navier-Stokes equations expressed in primitive variables, with spectral expansion in the spanwise direction and compact finite difference in the other two directions [44]. This simulation covered a range of $Re_\tau = 980 \sim 2000$. To our knowledge, it is one of DNS of turbulent boundary layer that achieved the highest $Re$.

For the present study, 12 snapshots of volumetric velocity fields of the UPM DNS dataset (http://torroja.dmt.upm.es/ftp/blayers/) were analysed. Instantaneous fields of the streamwise wall-shear stress $\tau_{w,x}(x,z)$ were estimated by the slope of the linear trend line fitted to the wall-normal profile of the instantaneous streamwise velocity $U^+(y^+)$ within the viscous sublayer (below $y^+ < 5$) at each $(x,z)$ station. Note that the spanwise component $\tau_{w,z}$ is not concerned in the present work. The yielded mean value $\overline{\tau}_{w,x}$ in the range of $Re_\tau = 1000 \sim 1900$ was in good agreement with the empirical formulation of Nagib et al. [45] with relative difference smaller than 1%, while the variation of the r.m.s. value $\tau_{x,rms}$ as a function of $Re_\tau$ showed the same trend with the formulation of $\tau_{x,rms} = 0.298 + 0.018 \ln Re_\tau$ suggested by Schlatter and Örlu [8], but the magnitude was about 4% lower overall.

The whole $\tau_{w,x}(x,z)$ field was then sliced to a streamwise field of view (FOV) of $8\delta$ ($\delta$ is the local boundary layer thickness, defined by $99\%U_\infty$ criteria) centering at the streamwise station where $Re_\tau = 1500$. This streamwise FOV is both large enough to accommodate instantaneous large-scale structures and small enough so that the variation of $Re_\tau$ is negligible (less than 5%). The spanwise FOV is about $10\delta$. $\tau_{w,x}$ was further converted to the instantaneous friction velocity $u_\tau$ via $u_\tau = \sqrt{\tau_{w,x}/\rho}$.

Figure 1 gives one snapshot of the fluctuating $u'_\tau(x,z)$ for illustration, in which a recurring pattern of streak-like structures of positive and negative $u'_\tau$ is clearly visualized. The inner-scaled dimensions of these $u'_\tau$ structures are about $O(10^3)$ in length and $O(10^2)$ in width, as indicated by the spectral peak in the pre-multiplied spectra $k_x k_z \Phi_{u_\tau u_\tau}$ shown in figure 3(a). This is consistent with those of the buffer-layer streaks in the flow, again demonstrating the distinct correlation between the inner-layer cycle and the wall-shear stress. Moreover, figure 1 shows a clear tendency of the organization (or concatenation) of multiple successive (or neighbouring) $u'_\tau$ streaks into a rather large group. Such a tendency indicates the footprint of outer-layer large-scale structures on the wall, which plays a role of synchronizing the inner-layer structures to form a scenario of the co-existence of small- and large-scale wall-shear stress fluctuations.

## 3. Scale-based decomposition

Another observation from figure 1 is that the small-scale negative $u'_\tau$ structures seem to be more apt to organize into longer structures with rather strong coherence, while the small-scale positive ones tend to be more like isolated patches with less probability to organize into longer (and wider) structures. The zoom-in plot of figure 2(a) gives a clear illustration. To further credit this observation, the small- and large-scale components need to be separated from the $u'_\tau(x,z)$ field. A recently proposed QB-VMD algorithm [46] is used for this purpose. This algorithm is a two-dimensional update of one-dimensional (1D) VMD [47]. The essence of VMD is to obtain a set of 1D Intrinsic Mode Functions (IMFs) via an optimization procedure to achieve the most compact bandwidth. Comparing to the traditional Empirical Mode Decomposition (EMD), VMD has a rigorous mathematical foundation, and is free from the so-called 'mode mixing' problem. QB-VMD inherits this value by treating the 2D snapshot as a collection of slices along the primary direction with distinct scale separation. Additionally, the continuity of the decomposed length scale along the non-primary direction is kept by an additional carrier frequency re-balancing strategy. These features make it suitable for decomposing non-stationary non-equilibrium 2D dataset with one primary direction. More details about this algorithm can

IOP Conf. Series: Journal of Physics: Conf. Series **1001** (2018) 012004 doi:10.1088/1742-6596/1001/1/012004

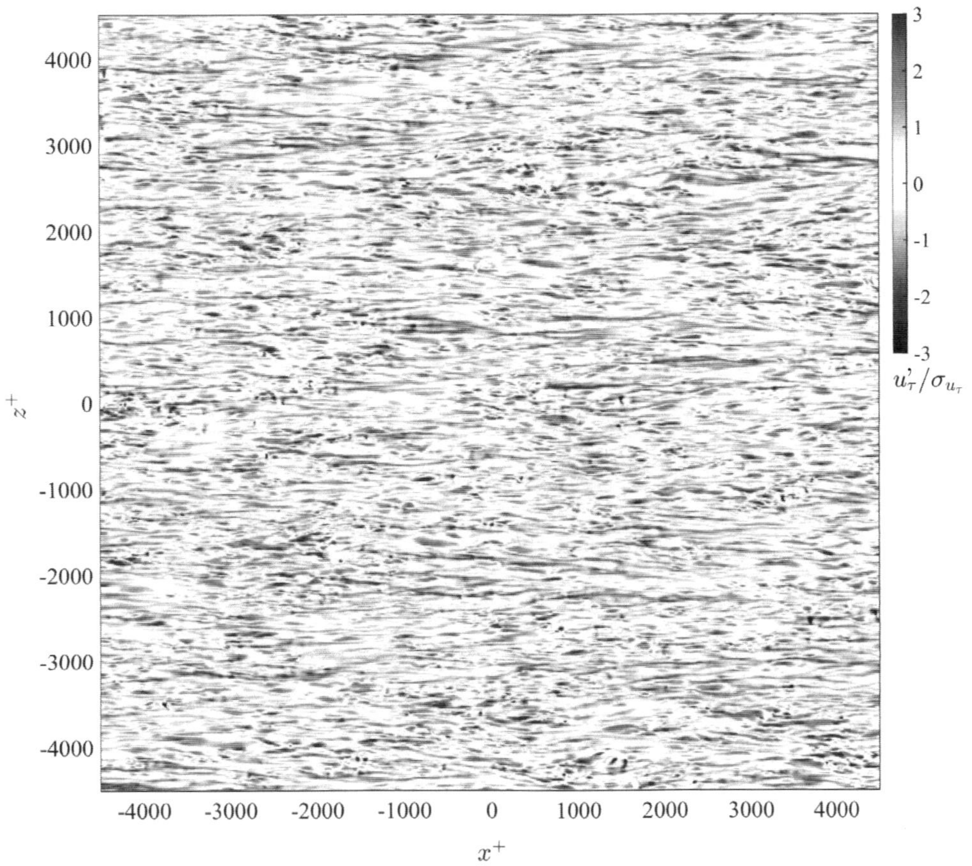


**Figure 1.** Snapshot of the fluctuating friction velocity $u'_\tau(x, z)$. $u'_\tau$ is normalized by $u_{\tau,rms}$. The origin of the $x$ axis is shifted to the station where $Re_\tau$=1500, the same in the following.

be referred to Wang et al. [46].

For the present $u'_\tau(x, z)$ field, QB-VMD took the spanwise direction as the primary decomposition direction. The justification is that either the small- or large-scale streaky structures in the $u'_\tau(x, z)$ fields are quasi-periodically spaced along the spanwise direction. Each snapshot of $u'_\tau(x, z)$ was decomposed into five individual IMFs and one residual mode, among which the first-rank IMF (with the largest scale) itself was regarded as the large-scale component $u'_{\tau,L}$, while the sum of the rest was considered as the small-scale component $u'_{\tau,S}$. It has been checked that the yielded $u'_{\tau,L}$ and $u'_{\tau,S}$ are rather robust to the number of the decomposed IMFs $K$, once only the first IMF is attributed to the large-scale component and $K \geq 4$.

The decomposed $u'_{\tau,L}$ and $u'_{\tau,S}$ fields are illustrated in figure 2(b, c) as a zoom-in plot. Both $u'_{\tau,L}(x, z)$ and $u'_{\tau,S}(x, z)$ are dominated by streaky structures with different spanwise length sales. By comparing to the original $u'_\tau$ field in figure 2(a), the $u'_{\tau,L}$ component is believed to capture the large-scale characteristics, which can be interpreted as the synchronization or amalgamation of finer streaky structures. Moreover, figure 2(b) suggests that the positive large-scale streaks in $u'_{\tau,L}$ are generally shorter than the negative ones, consistent with the observation in the full-order $u'_\tau$ fields in both figure 1 and figure 2(a).

The skewness of the large- and small-scale $u'_\tau$ components are $S_{u'_{\tau,L}}$=0.203 and $S_{u'_{\tau,S}}$=0.189, respectively. Comparing to the magnitude of the full-order component, i.e. $S_{u'_\tau}$=0.487, the

IOP Conf. Series: Journal of Physics: Conf. Series **1001** (2018) 012004 doi:10.1088/1742-6596/1001/1/012004

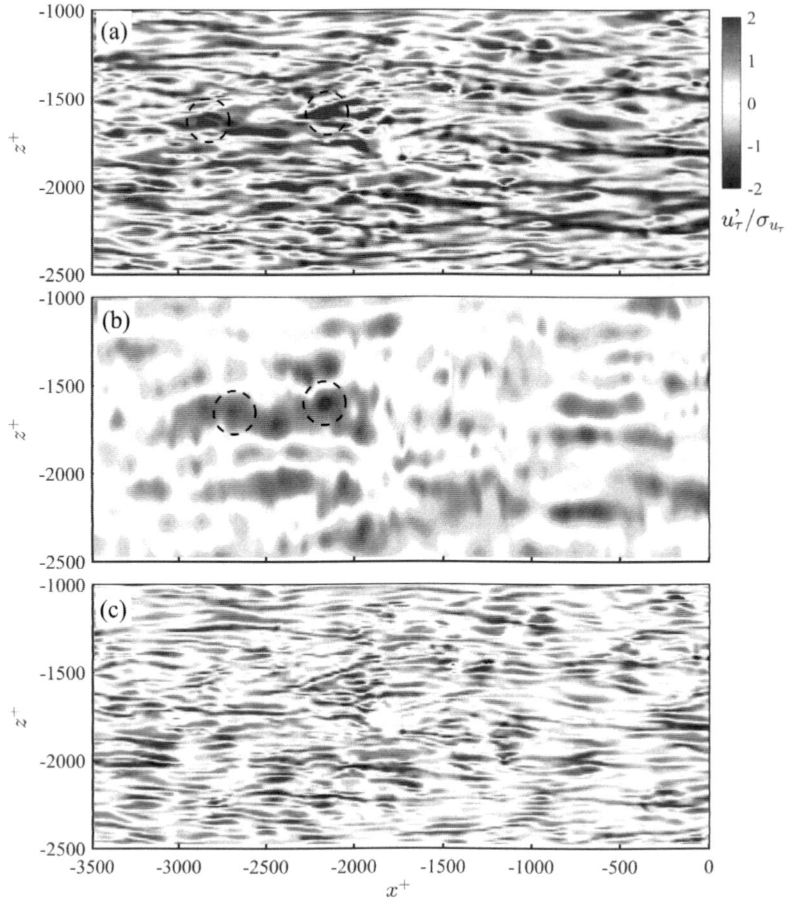


**Figure 2.** Comparison of the original $u'_\tau$ field in (a) to the large-scale $u'_{\tau,L}$ field in (b) and the small-scale $u'_{\tau,S}$ field in (c).

skewness of the scale-separated components get remarkably decreased. Nevertheless, $S_{u'_{\tau,L}}$ is still positive, indicating the positively skewed nature of the large-scale wall-shear stress. This is somehow consistent with the qualitative observation of the existence of the asymmetry of the length scale between the positive and negative $u'_{\tau,L}$ streaks. Note that such a weak asymmetry was also observed in the large-scale $u'$ component in the buffer layer by Agostini [33], who used bi-dimensional-EMD (B-EMD) for scale decomposition.

The pre-multiplied spectra of $u'_{\tau,S}$ and $u'_{\tau,L}$, as shown in figure 3(b), are separated from each other at about $\lambda_z^+ = 200$. The former retains the energetic spectra patch of the original $u'_\tau$ field centering at $\lambda_x^+ \sim O(10^3)$ and $\lambda_z^+ \sim O(10^2)$, while the latter is primarily confined in the large $(\lambda_x, \lambda_z)$ domain, with a distinct spectral peak at $\lambda_x^+ \sim 1600$ and $\lambda_z^+ \sim 360$ in inner-scaling or $\lambda_x/\delta \sim 1.1$ and $\lambda_z/\delta \sim 0.24$ in outer scaling. Note that this spectral peak is not present in the original spectra. Moreover, a self-similar scaling of $\lambda_x \approx 4.5\lambda_z$ is found to well characterize the ridge of the energetic $k_x k_z \Phi_{u_{\tau,L}u_{\tau,L}}$ patch in the large-scale component, and the energy-containing length scale extends to $\lambda_x \sim 2\delta$ and $\lambda_z \sim 0.45\delta$. Such a self-similar scaling seems to indicate the wall-signature of attached eddies whose longitudinal/lateral length scales linearly grow in wall-normal direction.

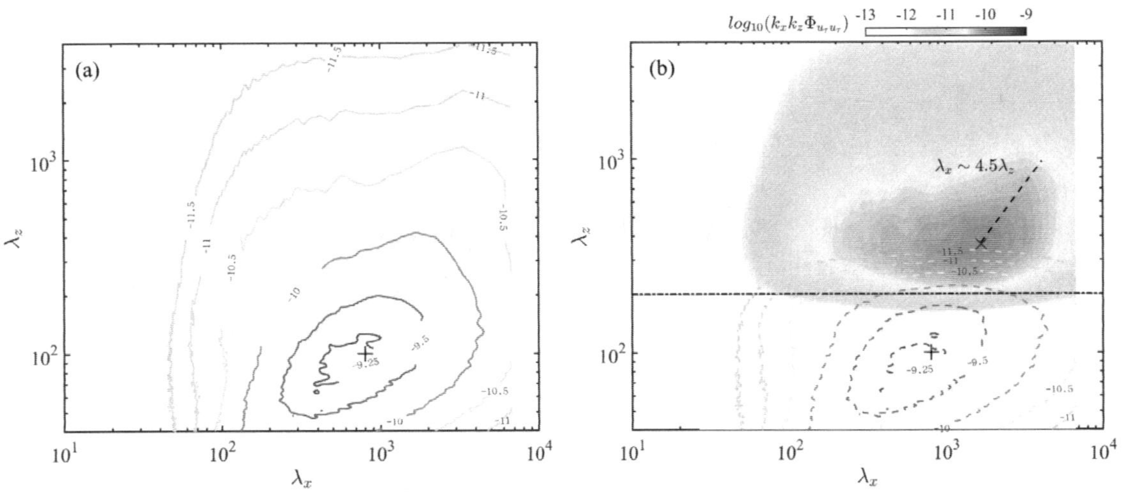

**Figure 3.** Pre-multiplied spectra $k_x k_z \Phi_{u_\tau u_\tau}$ of (a) the original $u'_\tau$ fields and (b) the QB-VMD separated large-scale $u'_{\tau,L}$ fields (indicated by pseudo-color plot) and small-scale $u'_{\tau,S}$ fields (contour plot). In each plot, $k_x k_z \Phi_{u_\tau u_\tau}$ is normalized by the square of the r.m.s of $u'_\tau$, $u'_{\tau,L}$ or $u'_{\tau,S}$, respectively.

In the QB-VMD operation, the length scales along the primary (spanwise) direction is necessarily decomposed by definition. This is the reason that the spectra of $u'_{\tau,S}$ and $u'_{\tau,L}$ are well separated into distinct ranges of the spanwise length scales. The slight overlap of the spectra at the separation boundary of $\lambda_z^+ = 200$ might be attributed to the non-stationary nature of spanwise scales in individual snapshots. This stresses the difficulty in drawing a pre-determined sharp cut-off to differentiate small and large scales.

On the other hand, the scales along the non-decomposed (streamwise) direction need not be decomposed and the spectra of $u'_{\tau,S}$ still contain non-negligible energy in the large $\lambda_x$ domain. Inspection of the instantaneous $u'_{\tau,S}$ field (figure 2c) and the non-decomposed $u'_\tau$ field (figure 2a) suggests that the large $\lambda_x$ contents in $k_x k_z \Phi_{u_{\tau,S} u_{\tau,S}}$ may not indicate the failure of the scale separation, but imply the actual existence of longer streaks with the spanwise length scales being $\lambda_z^+ \sim O(10^2)$ (despite the spatial growth of the boundary layer). Moreover, it has been tested that both pseudo-bi-dimensional-EMD (PB-EMD) [48] and B-EMD [49] lead to similar scale-separated spectra (not presented here for simplicity). Such a consistency indicates that the fluctuations with the large-scale $\lambda_x$ in $u'_{\tau,S}$ fields are not enforced by the QB-VMD algorithm. Previous studies [25, 40] on the amplitude modulation usually applied a cut-off threshold to single-point time-series signal of $u'(t)$ via Taylors frozen hypothesis. It is argued that the length scale along the homogeneous spanwise direction might be a more ideal candidate criteria for the scale separation. However, on considering that the present studied $Re_\tau$ does not guarantee a full inner- and outer-scale separation, this conjecture needs to be further studied at higher $Re$.

## 4. Extreme wall-shear stress events

In order to examine the average shape of extreme shear stress fluctuation events, two-point correlation maps of the instantaneous friction velocity fluctuations are utilized. However, the conventional correlation maps of $u'_\tau$ (figure 4a) are not adequate for such purpose, since their shape is dominated by the more frequently occurring small-scale streaks. Instead, the correlation maps computed from $u'_{\tau,L}$ fields are used because, as illustrated by the circled regions in figures

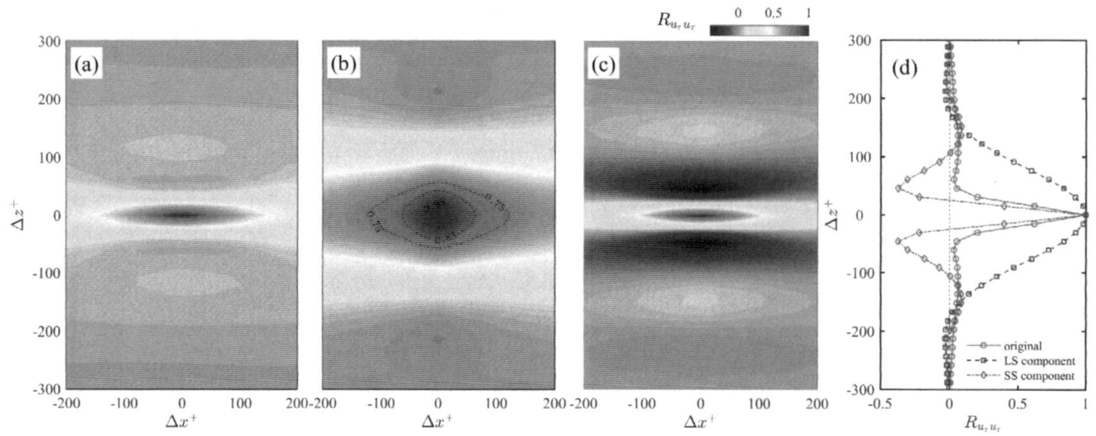


**Figure 4.** Two-point correlation map of (a) the original $u'_\tau$ fields, (b) the large-scale $u'_{\tau,L}$ fields and (c) the small-scale $u'_{\tau,S}$ fields. In (d), three correlation maps are compared at $\Delta x^+ = 0$, LS is short for large scale and SS for small scale.

2(a,b), the extreme shear stress fluctuation events tend to be large-scale. Note that the extremely positive shear stress fluctuation events seem to appear more frequently than their negative counterparts.

The two-point correlation map of the $u'_{\tau,L}$ field, as shown in figure 4(b), presents a round-shaped region with high magnitude of correlation coefficient ($R_{u_{\tau,L}u_{\tau,L}} > 0.85$, see figure 4d for illustration) around the central auto-correlation peak. This round-shaped region is embedded into the streamwise-elongated positive correlation strip, whose spanwise width is about 360 wall units. Considering that two-point correlation reveals the average geometry of coherent fluctuations, such a pattern implies that the spot-like extreme wall-shear stress fluctuation events are a common feature in instantaneous $u'_{\tau,L}$ fields. The absence of a similar high correlation region in the small-scale $R_{u_{\tau,S}u_{\tau,S}}$ map (figure 4c) implies that these events might be primarily associated with the large-scale flow structures away from the wall. Again, it has been checked that the switch of the decomposition method to either PB-EMD or B-EMD will not qualitatively change the signature of extreme $u_{\tau,L}$ events in the correlation map.

Based on the above observation, extreme wall-shear stress events are then detected from the instantaneous $u'_{\tau,L}$ fields. The idea is to detect isolated regions of $u'_{\tau,L}$ whose magnitudes are larger than a threshold. Meanwhile, the isolated regions should have shapes close to circle and are not too small in area. As a first attempt, the magnitude threshold is set as $|u'_{\tau,L}|/u_{\tau,L,rms} \geq H = 4$ (so that the extreme positive and negative $u'_{\tau,L}$ events are both detected), the round shape threshold is $AR \leq 4$ and the area threshold is $r^+ \geq 40$, with $AR$ and $r^+$ being the aspect ratio (the ratio between the major and minor axis) and the equivalent radius of the isolated candidates of extreme $u'_{\tau,L}$ events, respectively.

In twelve analyzed DNS snapshots, 1625 extreme positive $u'_{\tau,L}$ events and 270 extreme negative $u'_{\tau,L}$ events are detected. Figure 5 compares the conditionally averaged fields of $\tilde{u}'_\tau$, $\tilde{u}'_{\tau,L}$ and $\tilde{u}'_{\tau,S}$ around the detected extreme positive $u'_{\tau,L}$ event, while figure 6 shows the conditionally averaged fields around the negative one. In the large-scale component, the extreme $\tilde{u}'_{\tau,L}$ event (in figure 5b and figure 6b) resembles the two-point correlation map shown in figure 4(b), i.e. an elliptical region of high magnitude of positive/negative $\tilde{u}'_{\tau,L}$ slightly elongated along the streamwise direction, is embedded into a large-scale $\tilde{u}'_{\tau,L}$ streak with the same fluctuation sign. In contrast, no distinct structures in the conditionally averaged small-scale $\tilde{u}'_{\tau,S}$ fields (in figure

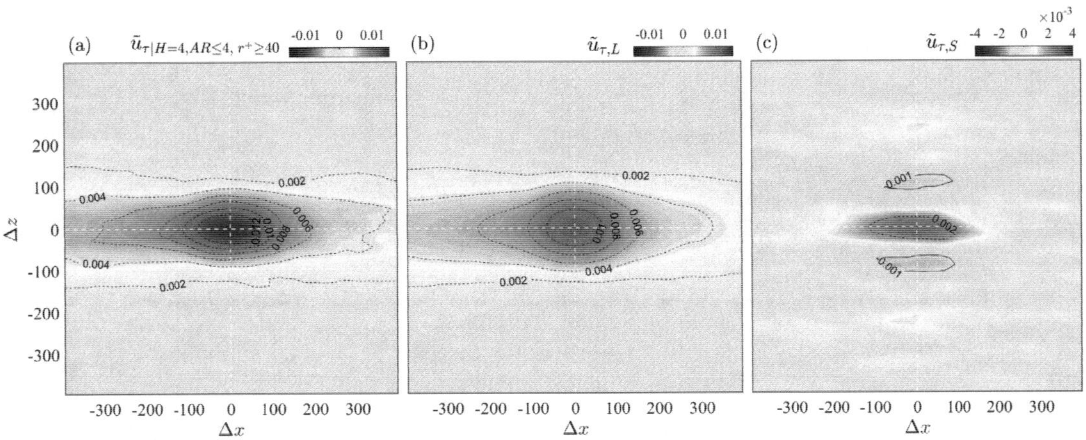

**Figure 5.** Conditionally averaged fluctuating friction velocity field associated with an extreme positive $u'_{\tau,L}$ event detected at $\Delta x$=0, $\Delta z$=0. (a) the original $\tilde{u}'_\tau$, (b) the large-scale $\tilde{u}'_{\tau,L}$, (c) the small-scale $\tilde{u}'_{\tau,S}$.

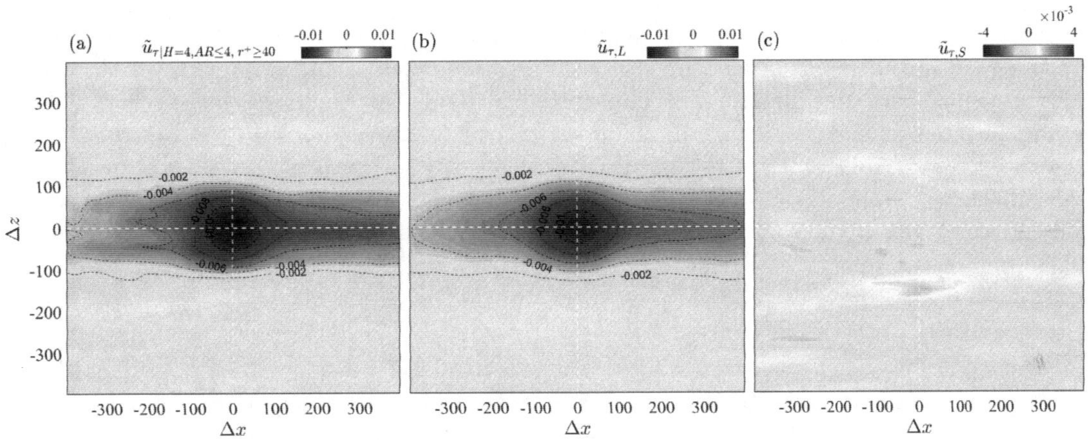

**Figure 6.** Conditionally averaged fluctuating friction velocity field associated with an extreme negative $u'_{\tau,L}$ event detected at $\Delta x$=0, $\Delta z$=0. (a) the original $\tilde{u}'_\tau$, (b) the large-scale $\tilde{u}'_{\tau,L}$, (c) the small-scale $\tilde{u}'_{\tau,S}$.

5c and figure 6c) are observed. This implies that the extreme events are mainly associated with the large-scale component, and it is themselves that contribute to the round-shaped pattern in the full-order $\tilde{u}'_\tau$ field (shown in figure 5a and figure 6a).

One more finding revealed by figures 5(b) and 6(b) is that on the downstream of the large-scale extreme positive wall-shear stress fluctuation events, the positive $\tilde{u}'_{\tau,L}$ streak gets thinner in width and weaker in strength. As for the extreme negative event, a reverse trend is observed: the negative $\tilde{u}'_{\tau,L}$ streak gets a mild growth downstream of the extreme negative event.

Interestingly, Lee and Sung [50] reported asymmetrically biased behavior of VLSMs with positive and negative streamwise fluctuating velocity along the streamwise direction by calculating the conditional two-point correlation of the positive and negative $u$ fluctuations at $y/\delta = 0.18$ in a DNS of a turbulent boundary layer: the positive VLSMs are biased towards

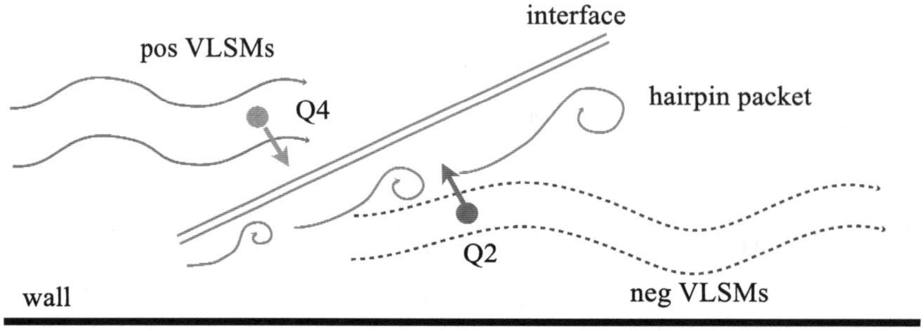


**Figure 7.** A schematic sketch to show the spatial relationship between the hairpin packet and the positive and negative VLSMs.

the upstream, while the negative ones are biased towards the downstream (see figure 28 in [50] for example). Such a biased pattern is similar to that of the conditionally averaged $\tilde{u}'_{\tau,L}$ shown in figure 5(b) and figure 6(b). Lee and Sung [50] attributed it to the preferential spatial alignment of the vortex packet structures and the Q4/Q2 events, where Q2 events are ejections of low speed fluids away from the wall ($u < 0$ and $v > 0$). In their view, the strong Q4 events belong to positive VLSMs (with positive $u$) in the upstream of the inclined interface of the vortex packets, and the Q2 events are linked to negative VLSMs (with negative $u$) in the downstream of the packets' interface. Such a spatial alignment is illustrated in figure 7. Since the large-scale wall-shear stress is an integrated signature of all the flow structures above the wall, the positive/negative spot-like extreme wall events might be directly related with these strong Q4/Q2 events. This will be discussed later. Here, it is stressed that if the vortex packet model offers a reasonable explanation for the asymmetric geometry of the conditionally averaged positive/negative $\tilde{u}'_{\tau,L}$ streaks, one of the inferences is that one decaying positive $\tilde{u}'_{\tau,L}$ streak associated with a spot-like extreme positive event should appear in the upstream of one negative $\tilde{u}'_{\tau,L}$ streak. Such a pattern is clearly observed in figure 2(b).

## 5. Flow structures associated with extreme wall-shear stress event

The conditionally averaged 3D velocity field around extreme negative/positive $u'_{\tau,L}$ events is calculated and shown in figures 8 and 9, respectively. Figure 8 indicates that the extreme negative $u'_{\tau,L}$ event is linked to a large-scale low-speed streak, which is originally attached to the wall but gets lifted up downstream of the extreme wall event.

In contrast, the conditionally averaged flow field (shown in figure 9) associated with the extreme positive $u'_{\tau,L}$ events is characterized by a finger-shaped large-scale negative $\tilde{v}$ region extending from the outer layer to the wall. Quadrant analysis (not shown here) evidences that this $\tilde{v}$ structure corresponds to a large-scale strong Q4 event that transfers outer-layer high-momentum fluids downward (as can be seen in figure 10), which forms a round-shaped thin layer of positive $\tilde{u}$ close to the wall. Note that this strong Q4 event is consistent with the so-called 'splatting' process suggested by Agostini et al. [29, 33], it also seems to describe the same structure as the near-wall 'pocket' named by Falco [51] or the 'dark spot' by Taimon et al. [52] and Lian [53].

If the large-scale $u'_{\tau,L}$ streaks are indeed the wall signature of outer-layer LSMs or VLSMs, the strong Q4 event will introduce a strong top-down interaction between the outer- and inner-region. The connection of the strong Q4 event with positive outer-layer LSMs can be somehow inferred from the volumetric $\tilde{u}$ plot in figure 10, in which one large-scale streak with high momentum is revealed in the wall-parallel plane of $y^+ = 180$. It moves downstream at a rather high speed, and

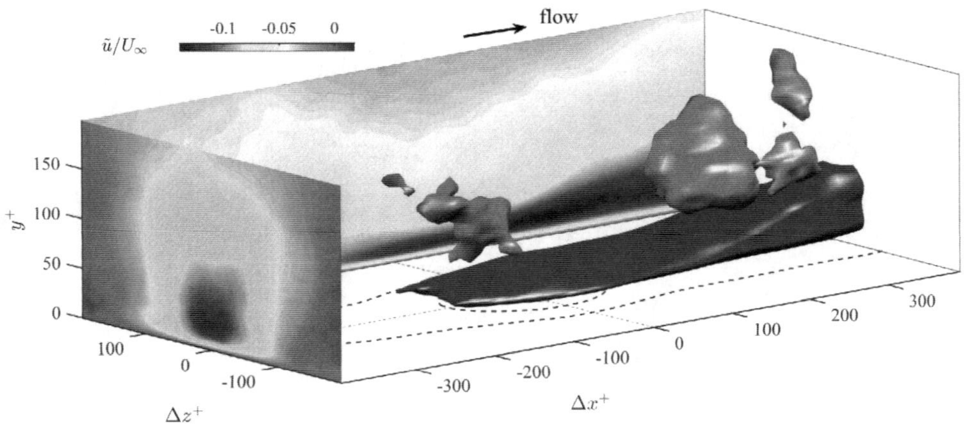


**Figure 8.** Conditionally averaged full-order fluctuating velocity field around an extreme negative $u'_{\tau,L}$ event detected at $\Delta x$=0, $\Delta z$=0. Blue iso-surface represents $65\%\tilde{u}^{[-]}_{max}$ and red iso-surface represents $65\%\tilde{v}^{[+]}_{max}$, with $\tilde{u}^{[-]}_{max}$ being the maximum negative value of $\tilde{u}$ in the presented volume and $\tilde{v}^{[+]}_{max}$ being the maximum positive $\tilde{v}$. Contour on the wall is the conditionally averaged $\tilde{u}'_\tau$ as shown in figure 6(a). Pseudo-color maps on two sliced planes show the planar distribution of $\tilde{u}$ at either $\Delta z^+ = 0$ or $\Delta x^+ = 350$.

leaves a footprint in behind that reaches into the inner-layer. Meanwhile, the strong Q4 event is seen to be directly 'shot' from this positive LSM. Note that in the conditionally averaged field, the strong Q4 event (in figure 9) appears to lift up from the wall and extend downstream, but the real dynamic sequence is that the positive LSM first 'shoots' the strong Q4 event, and then leaves the site at a faster convection speed; while the 'bullet' of the Q4 event decelerates when descending towards the wall, and finally forms a splatting in the near-wall region and causes the extreme positive shear stress fluctuation events on the wall.

To our understanding, the origin of this strong Q4 event is still not clear. One possible explanation is the instability that is associated with the decelerating front of the positive streak-like LSMs, which creates a negative local velocity gradient in the streamwise direction. In this aspect, the existing knowledges on the streak instability in transitional flow [54–56] can provide useful guidance, but this is beyond the topic of the present work.

## 6. Summary

To sum up, the fluctuating characteristics of the streamwise wall-shear stress in a DNS turbulent boundary layer is studied at $Re_\tau$=1500. With the help of QB-VMD, the 2D fields of the fluctuating friction velocity $u'_\tau(x, z)$ are decomposed into the large- and small-scale components, $u'_{\tau,L}$ and $u'_{\tau,L}$, based on the length scale in the spanwise direction, both of which are found to be dominated by streak-like structures elongated along the streamwise direction. Spectral analysis indicates that these two components are actually the wall signature of the inner-layer streaks and the outer-layer LSMs, respectively. A self-similar behaviour of the energetic longitudinal/lateral length scales in the large-scale component is observed, implying the footprints of self-similar attached eddies.

Extreme positive wall-shear stress fluctuation events are detected in the large-scale $u'_{\tau,L}$ component. Most of them present circular (or elliptical) shape if the contamination of small-scale fluctuation is removed. Both non-conditional two-point correlations and conditional averages show that these extreme positive events are embedded in the positive large-scale $u'_{\tau,L}$ streaks.

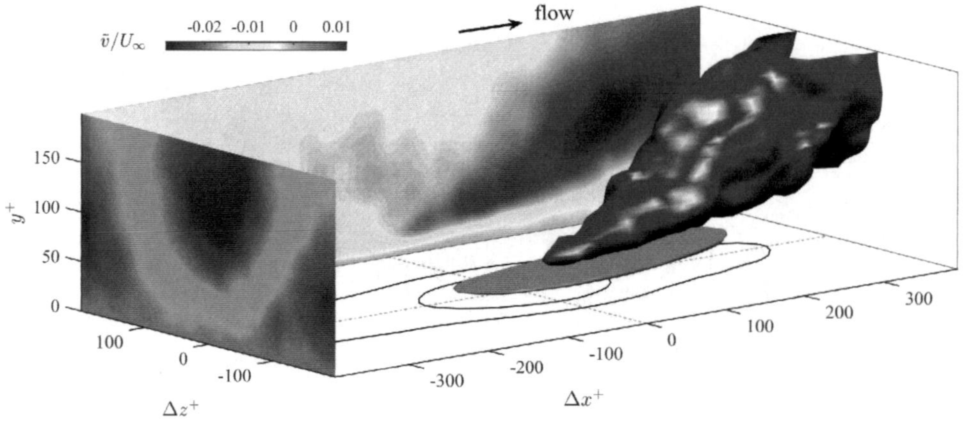


**Figure 9.** Conditionally averaged full-order fluctuating velocity field around an extreme positive $u'_{\tau,L}$ event detected at $\Delta x$=0, $\Delta z$=0. Blue iso-surface represents $65\%\tilde{v}^{[-]}_{max}$ and red iso-surface represents $65\%\tilde{u}^{[+]}_{max}$, with $\tilde{v}^{[-]}_{max}$ being the maximum negative value of $\tilde{v}$ in the presented volume and $\tilde{u}^{[+]}_{max}$ being the maximum positive $\tilde{u}$. Contour on the wall is the conditionally averaged $\tilde{u}'_\tau$ as shown in figure 5(a). Pseudo-color maps on two sliced planes show the planar distribution of $\tilde{v}$ at either $\Delta z^+ = 0$ or $\Delta x^+ = 350$.

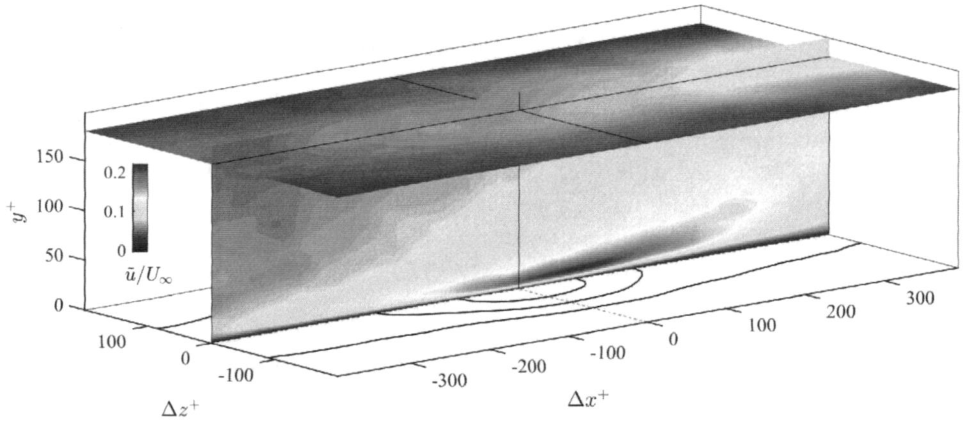


**Figure 10.** Conditionally averaged full-order fluctuating velocity field around an extreme negative $u'_{\tau,L}$ event detected at $\Delta x$=0, $\Delta z$=0. Contour on the wall is the conditionally averaged $u_\tau$ as shown in figure 5(a). Pseudo-color maps on two sliced planes show the planar distribution of $\tilde{u}$ at either $\Delta z^+ = 0$ or $\Delta y^+ = 180$.

The conditionally averaged flow field above the extreme positive wall-shear stress fluctuation events is characterized as a strong finger-like Q4 event, which seems to be 'shot' from outer-layer positive LSM, forming a 'splatting' of high-speed fluids in the near-wall region. Since these extreme positive wall events contribute a significant portion of the long tail of the wall-shear stress distribution, they need to be well treated in drag reduction control strategy. The origin of the strong Q4 events, their *Re*-dependency and the related spanwise wall-shear stress fluctuation features deserve to be further studied.

On the other hand, extreme negative wall-shear stress fluctuation events embedded in negative

$u_{\tau,L}$ streaks are also observed. Despite of their spot-like shapes, they are unlikely to be related with the 'splatting' process. Since their occurrence frequency is nearly one-order smaller than that of the extreme positive events, these extreme negative wall events seem to be less important in terms of skin friction control.

## Acknowledgements

This work was supported by the National Natural Science Foundation of China (11372001, 11672020 and 11490552). The first author would like to thank Prof. Javier Jiménez for his kind invitation to the third Madrid Turbulence Workshop, which was supported by the Coturb project of the European Research Council. We thank Dr. R. Ostilla-Mónico for his careful review of the manuscript.

## References

[1] Jiménez J and Pinelli A 1999 *J. Fluid Mech.* **389** 335–359
[2] Panton R L 2001 *Prog. Aerosp. Sci.* **37** 341–383
[3] Schoppa W and Hussain F 2002 *J. Fluid Mech.* **453** 57–108
[4] Orlandi P and Jiménez J 1994 *Phys. Fluids* **6** 634–641
[5] Fernholz H H and Finleyt P J 1996 *Prog. Aerosp. Sci.* **32** 245–311
[6] Monkewitz P A, Chauhan K A and Nagib H M 2007 *Phys. Fluids* **19** 115101
[7] Chauhan K A, Monkewitz P A and Nagib H M 2009 *Fluid Dyn. Res.* **41** 021404
[8] Schlatter P and Örlu R 2010 *J. Fluid Mech.* **659** 116–126
[9] Smits A J, McKeon B J and Marusic I 2011 *Ann. Rev. Fluid Mech.* **43** 353–375
[10] Eckelmann H 1974 *J. Fluid Mech.* **65** 439–459
[11] Chambers F W, Murphy H D and Mceligot D M 1983 *J. Fluid Mech.* **127** 403–428
[12] Madavan N K, Deutsch S and Merkle C L 1985 *J. Fluid Mech.* **156** 237–256
[13] Alfredsson P H, Johansson A V, Haritonidis J H and Eckelmann H 1988 *Phys. Fluids* **31** 1026–1033
[14] Miyagi N, Kimura M, Shoji H, Saima A, Ho C M, Tung S and Tai Y C 2000 *Int. J. Heat Fluid Fl.* **21** 576–581
[15] Colella K J and Keith W L 2003 *Exp. Fluids* **34** 253–260
[16] Abe H, Kawamura H and Choi H 2004 *J. Fluid Eng.* **126** 835–843
[17] Hu Z W, Morfey C L and Sandham N D 2006 *AIAA J.* **44** 1541–1549
[18] Grosse S and Schroder W 2009 *J. Turbul.* **10** 1–12
[19] Örlu R and Schlatter P 2011 *Phys. Fluids* **23** 021704
[20] Diaz-Daniel C, Laizet S and Vassilicos J C 2017 *Phys. Fluids* **29** 055102
[21] Kim K C and Adrian R J 1999 *Phys. Fluids* **11** 417–422
[22] del Alamo J C and Jiménez J 2003 *Phys. Fluids* **15** L41–L44
[23] Guala M, Hommema S E and Adrian R J 2006 *J. Fluid Mech.* **554** 521–542
[24] Hutchins N and Marusic I 2007 *J. Fluid Mech.* **579** 1–28
[25] Mathis R, Hutchins N and Marusic I 2009 *J. Fluid Mech.* **628** 311–337
[26] Hellstrom L H O, Sinha A and Smits A J 2011 *Phys. Fluids* **23** 011703
[27] Talluru K M, Baidya R, Hutchins N and Marusic I 2014 *J. Fluid Mech.* **746** R1
[28] Duvvuri S and McKeon B J 2015 *J. Fluid Mech.* **767** R4
[29] Agostini L, Leschziner M and Gaitonde D 2016 *Phys. Fluids* **28** 015110
[30] Kravchenko A G, Choi H and Moin P 1993 *Phys. Fluids A* **5** 3307–3309
[31] Sheng J, Malkiel E and Katz J 2008 *Exp. Fluids* **45** 1023–1035
[32] Gatti D and Quadrio M 2013 *Phys. Fluids* **25** 125109
[33] Agostini L, Touber E and Leschziner M A 2014 *J. Fluid Mech.* **743** 606–635
[34] Hwang Y Y 2013 *J. Fluid Mech.* **723** 264–288
[35] Deck S, Renard N, Laraufie R and Weiss P E 2014 *J. Fluid Mech.* **743** 202–248
[36] Marusic I and Heuer W D C 2007 *Phys. Rev. Lett.* **99** 114504
[37] Marusic I, Mathis R and Hutchins N 2011 *J. Phys.: Conf. Ser.* **318** 012003
[38] Mathis R, Marusic I, Chernyshenko S I and Hutchins N 2013 *J. Fluid Mech.* **715** 163–180
[39] de Giovanetti M, Hwang Y and Choi H 2016 *J. Fluid Mech.* **808** 511–538
[40] Schlatter P and Örlu R 2010 *Phys. Fluids* **22** 051704
[41] Wallace J M 2016 *Ann. Rev. Fluid Mech.* **48** 131–158
[42] Sillero J A, Jiménez J and Moser R D 2013 *Phys. Fluids* **25** 105102
[43] Sillero J A, Jiménez J and Moser R D 2014 *Phys. Fluids* **26** 105109
[44] Borrell G, Sillero J A and Jiménez J 2013 *Comput. Fluids* **80** 37–43

[45] Nagib H M, Chauhan K A and Monkewitz P A 2007 *Philos. T. Roy. Soc. A* **365** 755–770
[46] Wang W K, Pan C and Wang J J 2018 *Exp. Fluids* **59** 1
[47] Dragomiretskiy K and Zosso D 2014 *IEEE T. Signal Proces.* **62** 531–544
[48] Huang N E 2001 Computer implemented empirical mode decomposition method, apparatus, and article of manufacture for two-dimensional signals, US Patent 6311130
[49] Al-Baddai S, Al-Subari K, Tome A M, Sole-Casals J and Lang E W 2016 *Inform. Sciences* **348** 305–321
[50] Lee J H and Sung H J 2011 *J. Fluid Mech.* **673** 80–120
[51] Falco R E 1983 *21st Aerospace Sciences Meeting* p 377
[52] Taimon A M, Kunen J M G and Oom G 1986 *J. Fluid Mech.* **163** 459–478
[53] Lian Q X 1999 *Acta Mech. Sinica* **15** 193–200
[54] Brandt L and de Lange H C 2008 *Phys. Fluids* **20** 024107
[55] Ricco P, Luo J S and Wu X S 2011 *J. Fluid Mech.* **677** 1–38
[56] Deng S C, Pan C and Wang J J 2014 *Acta Mech. Sinica* **30** 636–645

IOP Conf. Series: Journal of Physics: Conf. Series **1001** (2018) 012005 doi:10.1088/1742-6596/1001/1/012005

# Coherent structure dynamics and identification during the multistage transitions of polymeric turbulent channel flow

**Lu Zhu and Li Xi**

Dept. Chemical Eng., McMaster U., 1280 Main St. W., Hamilton, ON L8S 4L7, Canada

E-mail: `zhul13@mcmaster.ca`

**Abstract.** Drag reduction induced by polymer additives in wall-bounded turbulence has been studied for decades. A small dosage of polymer additives can drastically reduce the energy dissipation in turbulent flows and alter the flow structures at the same time. As the polymer-induced fluid elasticity increases, drag reduction goes through several stages of transition with drastically different flow statistics. While much attention in the area of polymer-turbulence interactions has been focused on the onset and the asymptotic stage of maximum drag reduction, the transition between the two intermediate stages – low-extent drag reduction (LDR) and high-extent drag reduction (HDR) – likely reflects a qualitative change in the underlying vortex dynamics according to our recent study [1]. In particular, we proposed that polymers start to suppress the lift-up and bursting of vortices at HDR, leading to the localization of turbulent structures. To test our hypothesis, a statistically robust conditional sampling algorithm, based on Jenong and Hussain [2]'s work, was adopted in this study. The comparison of conditional eddies between the Newtonian and the highly elastic turbulence shows that (i) the lifting "strength" of vortices is suppressed by polymers as reflected by the decreasing lifting angle of the conditional eddy and (ii) the curvature of vortices is also eliminated as the orientation of the head of the conditional eddy changes. In summary, the results of conditional sampling support our hypothesis of polymer-turbulence interactions during the LDR-HDR transition.

## 1. Introduction

It is widely known that adding a small amount of polymers into Newtonian turbulence will significantly modify the flow statistics and structures. As a result, the friction factor of the flow is considerably reduced by polymers [3, 4]. In certain flow setups, such as the flow in a straight channel, polymer-induced drag reduction can reach up to 80%. This phenomenology is highly valuable in the development of new flow control strategies for enhancing the transportation efficiency of fluids. One example of the application of polymer additive drag reduction is the Trans-Alaska Pipeline system which saves the pump power by injecting polymers in to the pipe [5]. However, although the polymer additives drag reduction has been intensively studied in the past 60 years, the complex mechanisms behind them are still not fully understood.

In polymeric turbulence, several flow stages occur sequentially as the elasticity increases: the onset of drag reduction (ODR), low-extent drag reduction (LDR), high-extent drag reduction (HDR) and maximum drag reduction (MDR). Before ODR, the effect of polymers on the flows is indiscernible and the mean flow is statistically indistinguishable from those of Newtonian turbulence. Further raising the elasticity leads to the enhancement of drag reduction which eventually converges to an universal upper bound, i.e. MDR. Between the ODR and MDR, recent experimental and numerical studies suggest the existent of two intermediate stages: LDR and HDR [4, 6, 7]. Out recent study indicated that these two stages are distinguished by the regions where flow statistics are affected by polymers: at the LDR stage polymer effects are constrained to the buffer layer while they extend to the whole wall layer at the HDR stage [1].

Same as many other drag reduction strategies, polymer additives drag reduction is the result of the disruption of the turbulence generation cycle. A widely accepted explanation of the polymer-turbulence interaction is that polymers can damp the intensity of the near-wall streamwise vortices in turbulent flows [3]. Dubief et.al. [8] found that the polymer forces applied to the near vortices tend to damp the ejection and sweeping process. Meanwhile, Ptasinski et.al. [9] suggested that polymers could absorb and redistribute the turbulence kinetic energy since the polymer work in the shear stress balance is negative. The mechanism of polymer damping vortices is in good agreement with observations from DNS simulations and experiments [3] and is sufficient to explain the occurrence of drag reduction at ODR. However, this mechanism is not sufficient to explain the drastic changes of flow statistics and coherent structures at the LDR-HDR transition.

The qualitatively different behaviors between these two intermediate stages suggest the existent of another mode of polymer-turbulence interaction that starts during the LDR-HDR transition. In addition to the indiscriminate suppression of vortex intensity, several recent studies observed additional effects of polymers on the coherent structures. Yarin et.al. [10] studied the thin vortex filaments and found that the generation of horseshoe and hairpin vortexes in the near-wall region is prevented by polymers in the high elasticity regime. Moreover, Biancofiore et.al. [11] observes that the critical amplitude of perturbation to trigger a sustained turbulence increases with elasticity at the high elasticity regime, but is constant at the low elasticity regime. This phenomenon is related to the considerable suppression of the lifting strength of near-wall streaks in high elasticity regime. In addition, our recent study [12] on the laminar-turbulence transition of polymeric flows found that polymers can stabilize the primary streak-vortex structures and suppress the bursting of vortexes. These studies all relate the polymer-turbulence interaction at the HDR stage to the lifting process of coherent structures and its following bursting event.

In our recent study [1], a systematic study on the statistical and dynamic changes of turbulence during the LDR-HDR transition was done and a new mechanism which links the LDR-HDR transition to the modification of coherent structures was proposed. In our hypothesis, the lifting and breakdown of vortices and streaks are suppressed by polymers after the LDR-HDR transition. As a result, the bursting process is weakened which prevents the transportation of energy from the buffer layer to the log-law layer and leads to the decreasing of turbulent intensity in the log-law layer. This mechanism is consistent with all known observations of the HDR stage, e.g. turbulence localization and the changing flow statistics in the log-law layer. However, direct evidences are still needed. The conditional sampling approach allows us to understand the statistical properties of vortices and offer an accessible way to study the polymer-turbulence interaction.

The conditional sampling method was initially used in experimental studies of turbulence to obtain quantitative information of a turbulent flow; the readers are referred to Antonia's comprehensive review [13] for more details. In essence, the conditional sampling method is used to obtain the best estimation of certain targeted events. In early experimental studies, the quadrant (based on Q2 and Q4 events) [14] and the Variable Interval Time Average (VITA; based on the large variances of the streamwise velocity) [15] schemes gained considerable attention. But these methods are originally not appropriate to reveal the spatial details of the flow field [16]. Later, the development of numerical simulation allows researchers to have fully 3D representation of turbulence and then spatial conditional sampling techniques were proposed. In most of these spatial conditional sampling techniques, three steps are involved: (i) applying a predetermined event identification criterion to the flow field of turbulence, (ii) extract individual coherent structures from the target field, and (iii) determining the reference points for alignment and implementing zone average.

For step (i), the velocity-gradient-tensor-based algorithms are one of the most widely adopted local criteria (i.e. applied to individual points) in the area of vortex identification [17]. The simplest criterion in this category is vorticity. However, the vorticity is not able to distinguish the difference between the pure shear and the real swirling motion [2, 17]. Then, Hunt et.al.[18] introduced the $Q$ quantity

$$Q \equiv 1/2(\|\mathbf{\Omega}\| - \|\mathbf{S}\|), \tag{1}$$

to describe the swirling motion, i.e. the so-called $Q$-criterion. The strain-rate tensor $\mathbf{S}$ and the vorticity tensor $\mathbf{\Omega}$ in equation (1) are the symmetric and antisymmetric parts of the velocity-gradient tensor $\nabla \boldsymbol{v}$, respectively. In the current study, The $Q$-criterion is adopted to identify the vortex structures in polymeric turbulence.

Step (ii) is of particular importance to obtain representative structures in the flows. For this step, a common approach in the literature takes advantage of the spatial/temporal separation of structures. For instance, applications based on the VITA and quadrant schemes [19, 20] employ the spatial continuity to recognize individual coherent structures. For those algorithms, a careful selection of a cutoff threshold is needed to decrease the potential influence of structure percolation. On the other hand, Hussain et.al. [21] chose the local extrema of vorticity as the sampling events. In their approach, the local extrema are picked without predefining the cutoff threshold which could effectively avoid the percolation issue. Interestingly, Jeong et.al. [2] extended Hussain et.al. [21]'s 2D local extreme conditional sampling to the 3D spatial fields by computing the local extrema of the $\lambda_2$ eigenvalue at each streamwise plane. This method is designed to capture the centre line of streamwise vortices and is robust to the cutoff threshold.

As for Step (iii), the geometry center is normally adopted as the reference point for alignment [2, 19, 21, 22]. However, a smearing problem may raise due to the shape variation of vortex structures. To deal with the smearing issue, a filter is usually set up to discard structures that differ greatly from the target events [16, 19]. Furthermore, Jeong et.al. [2] shifted the reference point according to cross-correlation between realizations.

In polymeric turbulence, the near-wall vortices are elongated and weakened by polymers [3], meanwhile, the long streamwise vortices dominate the near-wall flow field [23]. Therefore, we adopt the Jeong et.al. [2]'s method which is highly efficient in capturing the streamwise vortices. In this study, the original Jeong et al.'s method is adapted to the polymeric case and conditional vortex structures therein are sampled. Comparison is then made with the Newtonian turbulence. The results offer new insight into the coherent structure modification by polymers during the LDR-HDR transition.

## 2. Methodology

### *2.1. Direct numerical simulation*

In this study, Direct Numerical Simulation (DNS) is adopted to simulate the polymeric turbulence in a plane Poiseuille geometry with a fixed pressure drop. The flow geometry is shown in figure 1, where $x$, $y$ and $z$ denote the streamwise, wall-normal and spanwise directions, respectively. The periodic boundary condition is applied to $x$- and $z$-directions while the no-slip boundary condition is used in the $y$-direction. The governing equations are

$$\frac{\partial \boldsymbol{v}}{\partial t} + \boldsymbol{v} \cdot \nabla \boldsymbol{v} = -\nabla p + \frac{1}{\mathrm{Re}} \nabla^2 \boldsymbol{v} + \frac{2(1-\beta)}{\mathrm{ReWi}} \left(\nabla \cdot \boldsymbol{\tau}_p\right), \tag{2}$$

$$\nabla \cdot \boldsymbol{v} = 0, \tag{3}$$

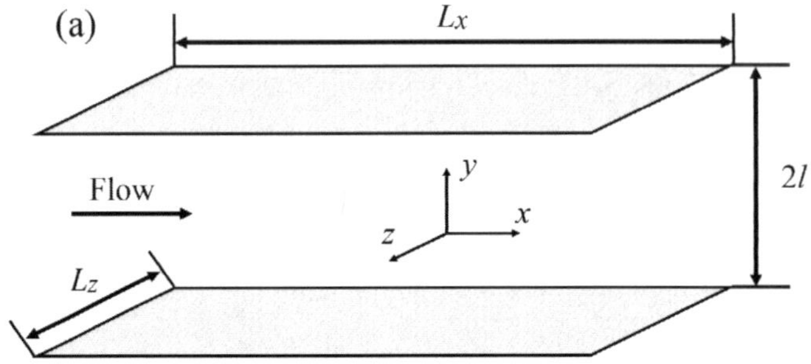


**Figure 1.** The conceptual plot of the flow geometry

$$\frac{\boldsymbol{\alpha}}{1-\frac{tr(\boldsymbol{\alpha})}{b}}+\frac{\mathrm{Wi}}{2}\left(\frac{\partial\boldsymbol{\alpha}}{\partial t}+\boldsymbol{v}\cdot\nabla\boldsymbol{\alpha}-\boldsymbol{\alpha}\cdot\nabla\boldsymbol{v}-(\boldsymbol{\alpha}\cdot\nabla\boldsymbol{v})^{T}\right)=\frac{b\boldsymbol{\delta}}{b+2}, \tag{4}$$

$$\boldsymbol{\tau}_p=\frac{b+5}{b}\left(\frac{\boldsymbol{\alpha}}{1-\frac{tr(\boldsymbol{\alpha})}{b}}-\left(1-\frac{2}{b+2}\right)\boldsymbol{\delta}\right). \tag{5}$$

In the above equations, the Reynolds number is defined as $\mathrm{Re}\equiv\rho U_{\mathrm{CL}}l/\eta$ where $\rho$, $U_{\mathrm{CL}}$, $l$, and $\eta$ are the fluid density, laminar center-line velocity, half-channel height, and fluid viscosity, respectively; under this definition, the friction Reynolds number $\mathrm{Re}_\tau=\sqrt{2\mathrm{Re}}$. The Weissenberg Wi number, defined as $\mathrm{Wi}\equiv 2\lambda U/l$ ($\lambda$ is the polymer relaxation time), measures the level of elasticity. $\beta\equiv\eta_s/\eta$ is the ratio between the solvent and solution viscosities. The FENE-P constitutive equations (4)-(5) [24] is adopted to determine the polymer stress tensor $\boldsymbol{\tau}_p$ in equation (2), where $\boldsymbol{\alpha}$ is the polymer conformation tensor and the maximum extensibility parameter $b$ constrains the square length of the polymer chains $\mathrm{tr}(\alpha)$.

A series of Wi under $\mathrm{Re}=3711$ (i.e., $\mathrm{Re}_\tau=84.85$) is investigated. The rheological parameters $\beta$ and $b$ are fixed to 0.97 and 5000, respectively. The streamwise and spanwise periods are $L_x^+\times L_z^+=4000\times 800$, where the superscript "+" indicates quantities in turbulent inner scales. A Fourier-Chebyshev-Fourier pseudo-spectral scheme is applied for spatial discretization. Meanwhile, a third-order semi-implicit backward-differentiation-Adams-Bashforth scheme[25] is used to integrate equations in time. In addition, an artificial diffusion term $1/(\mathrm{ScRe})\nabla^2\boldsymbol{\alpha}$ with the Schmidt number $\mathrm{Sc}=0.5$ is introduced in equation (4) to achieve better numerical stability.

### *2.2. Conditional sampling*

In this study, vortexes are identified by adopting the $Q$-criterion (equation (1)) [17]. $Q=0$ indicates a pure shear flow while a large negative and positive $Q$ correspond to regions dominated by extensional and rotational flows, respectively. The vortex structures are chosen by satisfying $Q>0.7Q_{rms}$ (where $Q_{rms}$ is the root-mean-square of $Q$).

The conditional sampling algorithm based on Jeong et.al.[16]'s method involves several steps: (1) detect vortex region which satisfies $Q>0.7Q_{rms}$ in the 3D instantaneous flow field; (2) calculate the local maxima of the accepted regions at each y-z plane – the local maxima are regarded as the x-centrepoints of streamwise vortices; (3) group the local maxima to individual vortices by adopting a cone detection method, connecting these points form the centrelines of vortices; (4) categorize vortices into two categories according to the senses of rotation (clockwise/anticlockwise), the sense is determined by the sign of streamwise vorticity at the vortex axis; (5) realizations are discarded if they do not satisfy: (a) the vortex streamwise length $l_x^+\geq 50$, and (b) the average height of the vortex axis $h_y^+\leq 50$; (6) the x-centrepoint

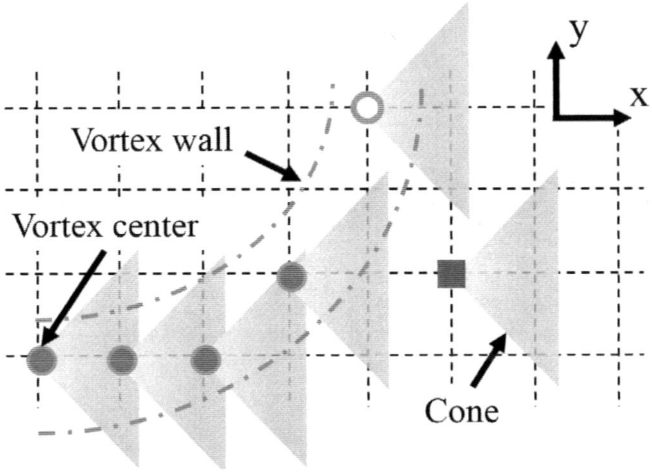


**Figure 2.** Schematic diagram of the cone algorithm method

located at $y^+ = 50$ is chosen as the reference point for alignment and realizations with the same sense of rotation are aligned and averaged to obtain the conditional eddy.

In order to extract individual vortices, the cone detective method is involved. In this method, two x-centrepoints are considered to belong to the same vortex when they satisfy: (a) the two points locate at two adjacent y-z planes; (b) the downstream point is the closest local maximum on its plane to the upstream point; and (c) the distance between the downstream point and the projection of the upstream point is smaller than a threshold, i.e., the downstream point locates within a confining cone extending from the upstream point. A schematic plot of the cone detective method is shown in figure 2. Dots in this figure are the x-centrepoints while the dot dash lines indicate the wall of a vortex. Centrepoints are grouped by the detective method if they fall into the cone (the triangles in figure 2). However, note that this method is inefficient to detect vortices with a high deviation of vortex line from the streamwise direction, e.g. vortex with a strong lifting tendency as shown in the figure.

In Jeong et.al.'s original method [16], the geometry center of vortex axis is chosen to be the reference point for realization alignment. However, since the lengths of streamwise vortexes are different, fixing the reference point to the geometry center will lead to the misalignment of heads and tails of vortices, as shown in figure 3a, and cause the smearing problem. This is worse for high Wi cases where the vortices have a broader range of size and shapes. Instead, we move the reference point for realization alignment to the x-centrepoint at $y^+ = 50$. The new reference point ensures the precise alignment of the heads of vortices (figure 3b). Benefiting from this modification, some important dynamics of vortices, such as the lifting process, can be accurately captured. In addition, the choice of the reference point relaxes the strict constraint of vortex length in step (5) of the conditional sampling algorithm, as the variation of vortex length now has less effects on the conditional eddies (especially in the head region) compared with the original method.

## 3. Results and discussion

The presence of multiple stages in the polymeric turbulence suggests the existence of multiple types of polymer-turbulence interactions at different stages. In this study, we focus on the change of interaction involved in the LDR-HDR transition. As has been discussed in the literature[3, 4], the statistical quantities of turbulence in LDR and HDR stages exhibit qualitatively different

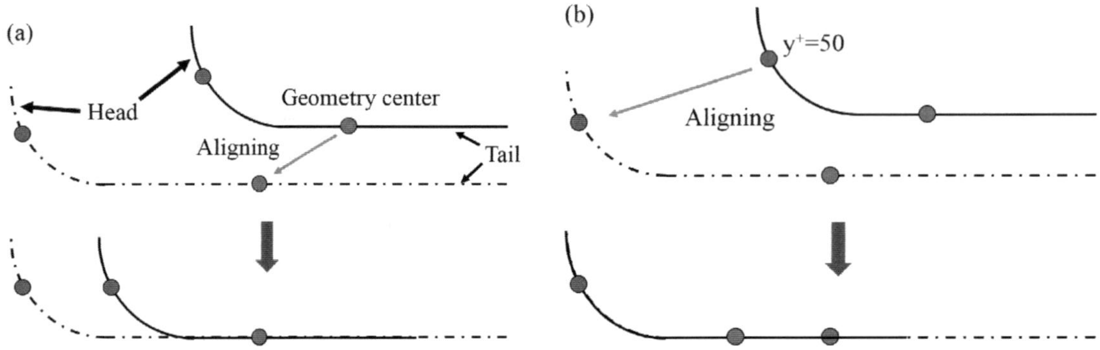


**Figure 3.** Selection of the reference point at (a) the geometry center and (b) the head

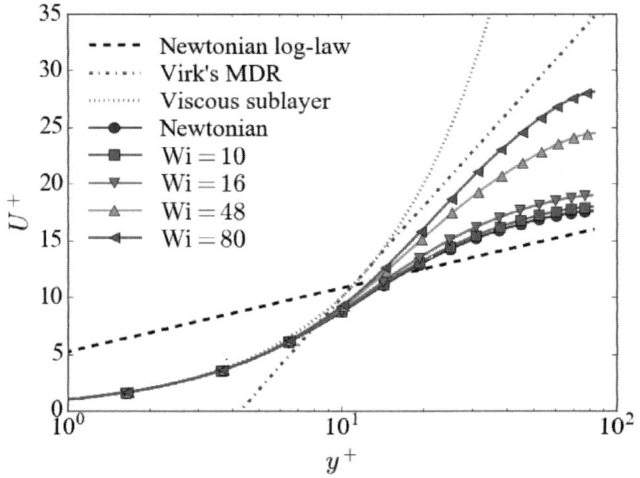


**Figure 4.** Mean velocity profiles at $\mathrm{Re}_\tau = 86.15$

behaviors. Figure 4 shows the mean velocity profiles of Newtonian, LDR (Wi = 16), HDR (Wi = 48) and MDR (Wi = 80) at $\mathrm{Re}_\tau = 86.15$. Starting at the ODR (Wi $\leq$ 10), the velocity profiles lift up as Wi increases and reach an upper bound at the MDR stage. The identification between LDR and HDR depends on the influenced zones of polymers. For the LDR case (Wi = 16), the effect of polymers is constrained in the buffer layer ($y^+ \approx 5 \sim 30$): the velocity profile rises in buffer layer but stays parallel to the Newtonian profile outside the buffer layer region ($y^+ > 30$). By contrast, the slopes of the mean velocity profiles at HDR and MDR stages differ from the Newtonian profile in the whole channel.

The systematic analysis of the statistical and dynamic changes during the LDR-HDR transition is the focus of a separate study of ours [1]. Therein, we hypothesized that at the LDR-HDR transition polymers start to suppress the bursting of turbulence by preventing vortices from lifting up. The typical vortex structures extracted from the instantaneous flow field before and after the LDR-HDR transition are presented in figure 5. Before the LDR-HDR transition, vortexes tend to aggregate into vortex packets. A hairpin vortex is observed at the downstream end of this vortex packet. As the head of the hairpin vortex lifts up, the vortex eventually breaks

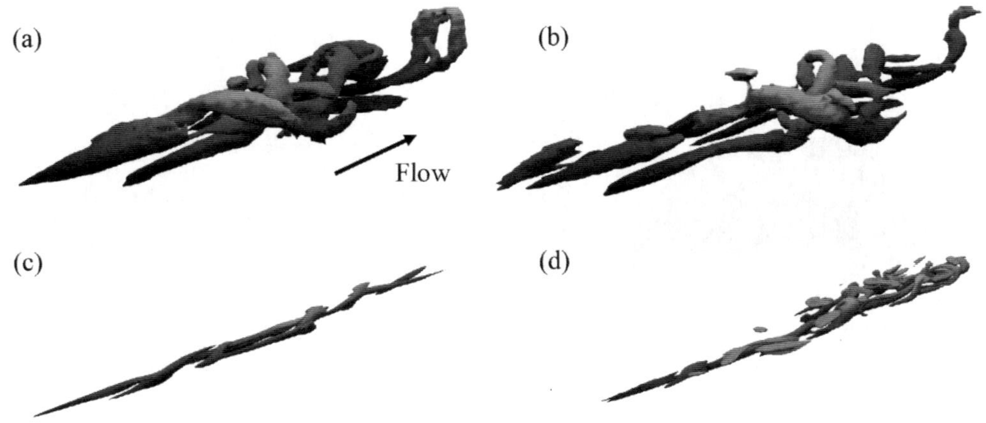


**Figure 5.** Typical instantaneous vortex structures in cases before and after the LDR-HDR transition: (a) and (b) – before, time interval $\Delta t = 8$; (c) and (d) – after, time interval $\Delta t = 32$

down after a period of time. In addition to the hairpin, other type of vortices (e.g. the pseudo-streamwise vortices) also have a high tendency of lifting. In fact, the lift-up and bursting of vortices play an important role in the instability-based vortex regeneration cycle [26, 27]. However, in the high elasticity turbulence (HDR and MDR), the lift-up strength of vortexes is weakened and the hairpin vortices are eliminated by polymers, as shown in figure 5c. Instead, longer and smoother streamwise vortices with weaker lifting strength dominate the near-wall coherent structures. Note that these streamwise vortices usually organize to a ultra-long (the streamwise length $l_x^+ \geq 400$) streamwise vortex string in which the head of a upstream vortex overlaps the tail of an adjacent downstream vortex. The vortex string is similar to those vortices generated by the parent-offspring vortex regeneration cycle[28]. Eventually, as the strength of vortices exceeds a certain threshold, the vortex string suddenly bursts and forms a group of vortices in the adjacent region (figure 5d).

The modification of the vortex organization pattern is due to the change of vortex regeneration mechanism. In our hypothesis, since the instability-based mechanism is interrupted by polymers in HDR and MDR stage due to the prevention of lifting and bursting of vortices, the other mechanism – the parent-offspring vortex mechanism, becomes exposed.

To understand the statistical characteristics of vortices at different stages, the aforementioned conditional sampling algorithm, improved based on ref. [16], is implemented. Figure 6 shows the probability density functions (PDF) of streamwise lengths of vortices at different $y^+$. The PDF distribution of the polymeric flow (Wi = 96) covers a wider range from 0 to over 800 wall units while the Newtonian turbulence only expands to 500 wall units. Also, the peak of the PDF distribution adheres to the wall ($y^+ = 30$) in the Newtonian case while it approaches the center of the channel ($y^+ = 50$) in the polymeric case. To alleviate the potential smearing problem (figure 3), a minimum streamwise length cutoff of vortexes is set to filter vortices with a short streamwise length. In Jeong et al's conditional sampling of the Newtonian flow, the cutoff ($l_x^+ = 200$) is larger than the average length of vortices. As a consequence, the relatively large cutoff biases the sample by eliminating many qualified structures, which limits the representativeness of the conditional eddies. Despite the long cutoff, the issue of uneven vortex length still apparently caused smearing in their conditional eddy. The new criterion for selecting the reference point (the x-centrepoint located at $y^+ = 50$) in our improved algorithm significantly alleviates the smearing issue with a much smaller cutoff ($l_x^+ = 50$). As a consequence, more realizations are included in constructing the conditional eddy which improves the representativeness of the conditional eddy.

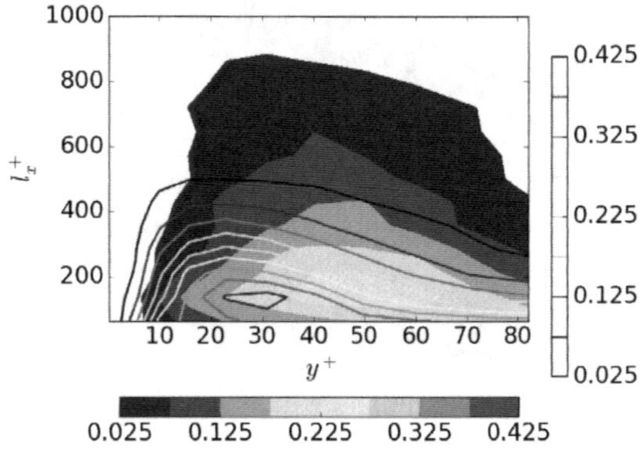


**Figure 6.** Probability density functions of vortex streamwise lengths: the color flood is for the Wi=96 case and lines are for the Newtonian case.

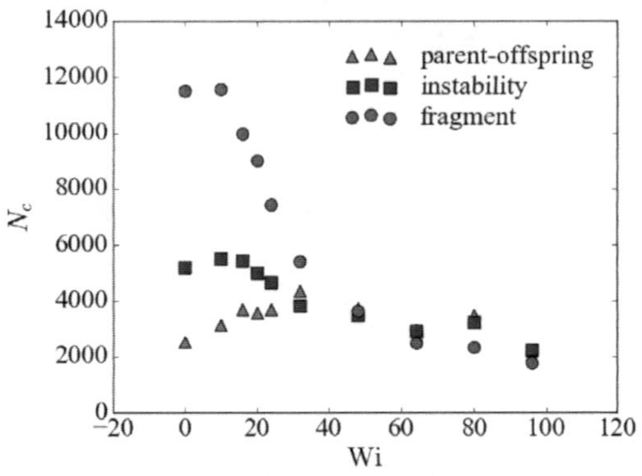


**Figure 7.** The average number of centre points in each vortex class increasing Wi.

Based on their streamwise length, vortices can be grouped into two categories: major vortices ($l_x \geq l_{\text{cutoff}}$) and fragments ($l_x \leq l_{\text{cutoff}}$). On the other hand (as reviewed above), the vortex regeneration mechanisms in wall-bounded flow can be categorized into two types [29]: the parent-offspring and instability-based mechanisms. Noting that the parent-offspring mechanism depends on the direct contact between the parent and the offspring vortices while the steak instability does not, major vortices may thus be categorized according to their generation mechanisms based on their spatial proximity to other vortices. Pseudo-streamwise vortices overlapping with another vortex with the opposite sense of rotation are categorised to the parent-offspring vortex class and other vortices are categorised to the instability vortex class. By employing this classification, we are able to investigate the changing vortex regeneration mechanism at different stages. The average number of vortex centrepoints $N_c$ of these two classes for different Wi is plotted in figure 7. Since the threshold to determine fragments is arbitrary, a number of thresholds are tested and the trend of the profiles is robust to the threshold. In figure 7, the cutoff threshold is 150.

The critical Wi of the PO-LDR, LDR-HDR and HDR-MDR transitions in figure 7 are 10, 24 and 80, respectively. The number of centrepoints in the instability class starts to decrease at Wi = 20 which is close to the LDR-HDR transition, suggesting a strong relation between the LDR-HDR transition and the suppression of the instability mechanism. The parent-offspring vortices, however, does not decrease until well after Wi = 32. In other words, the parent-offspring mechanism is more persistent to the strong polymer effect at HDR. On the other hand, fragments continuously decrease after the onset of drag reduction as polymers suppress the overall intensity of vortices. These observations are in good agreement with our proposed polymer-turbulence interaction during the LDR-HDR transition. However, readers should note that figure 7 is an imprecise estimation of vortex number at different classes. For example, the poor capability of current vortex tracking method in tracking curved vortexes will overestimate the number of vortices in the fragment class. To overcome this issue, the detective algorithm needs to be improved to accommodate higher vortex curvatures, which will be the focus of our future work.

The conditional eddies are obtained in cases before (Newtonian) and after (Wi = 96) the LDR-HDR transition and are presented in figure 8. As a reminder, vortex realizations are

IOP Conf. Series: Journal of Physics: Conf. Series **1001** (2018) 012005 doi:10.1088/1742-6596/1001/1/012005

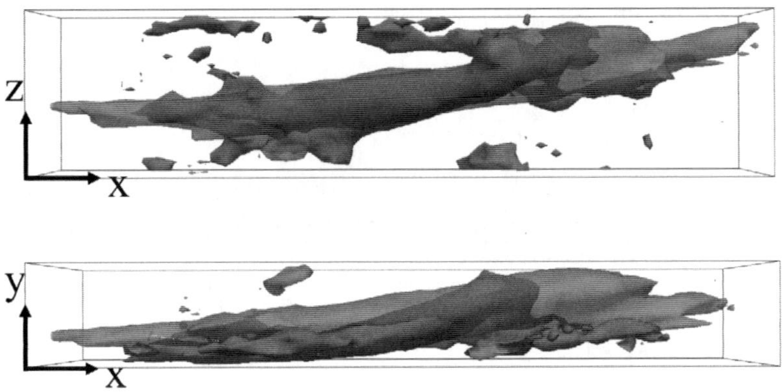


**Figure 8.** (a) The top and (b) side views of the conditional eddies of the Newtonian (red) and Wi = 96 (blue) turbulence.

separated into two classes according to the sign of streamwise vorticity before zone averaging. In figure 8, only the counter-clockwise conditional eddy are shown since the clockwise conditional eddy is its mirror image.

Figure 8(a) shows the top view of the conditional eddies. In general, the length of the conditional eddy at Wi = 96 is longer than that of the Newtonian case, which is consistent with observations in instantaneous flow field images. Comparing the two cases, the tilting angles are similar, but the head of the Newtonian eddy bends further to the positive z direction, which is attributed to the prevalence of highly curved vortexes, e.g. hairpins. In figure 8(b), the lifting angle of the conditional eddy before the transition (Newtonian) is larger than that after the transition (Wi = 96), which agrees with our previous discussion of polymers suppressing the lifting process of vortices. Also, in both cases, an additional iso-surface with opposite streamwise vorticity appears under the head of the main body. Its existence indicates a high probability of another vortex with the opposite sense of rotation showing up under the head of the upstream vortex. Note that in the Newtonian case, the additional structure appears irregular; only in the high-Wi case does the additional eddy become more coherent. This change is possibly a reflection of the increasing importance of the parent-offspring mechanism.

## 4. Conclusions

Despite recent efforts in the literature, a complete picture of the polymer-turbulence interaction is still missing. Especially, the interaction responsible for the LDR-HDR transition remains a puzzle. A hypothesis was put forward in our recent study on the LDR-HDR transition[1], in which polymers suppress vortex regeneration through streak instability by preventing the lift-up and bursting of vortices. In this study, a conditional sampling algorithm improved from that of ref. [16] is employed to compare the vortex dynamics before and after the LDR-HDR transition.

Observations of the conditional eddies obtained support our hypothesis. The lifting angle of the Newtonian eddy is considerably larger than that of the Wi = 96 case. Also, the different shapes of the vortex head suggest that highly curved vortices (e.g., hairpins) existing in the Newtonian and LDR cases are suppressed by polymers at the HDR and MDR stages.

The near-wall vortices are also divided into three classes: one is fragments and the other two are major vortices generated by the streak-instability and parent-offspring mechanisms. The number/size of vortices generated by the streak-instability mechanism drops during the LDR-HDR transition ($Wi = 20$) while those in the parent-offspring class remain frequent until well beyond $Wi = 32$. This result also agrees well with our hypothesis of the modification of vortex regeneration mechanism.

## Acknowledgments

Financial support from the Natural Sciences and Engineering Research Council of Canada (NSERC) through its Discovery Grants Program (No. RGPIN-2014-04903) is acknowledged. The computation is made possible by the facilities of the Shared Hierarchical Academic Research Computing Network (SHARCNET: `www.sharcnet.ca`) and through the computing resources allocation awarded by Compute/Calcul Canada. The authors thank Javier Jiménez and Miguel P. Encinar for fruitful discussions. This work was also funded in part by the Coturb program of the European Research Council.

## References

[1] L. Zhu, H. Schrobsdorff, T. M. Schneider and L. Xi 2017 submitted.
[2] Jeong J and Hussain F 1995 *J. Fluid Mech.* **285** 69–94
[3] White C M and Mungal M G 2008 *Annu. Rev. Fluid Mech.* **40** 235–256.
[4] Xi L and Graham M D 2010 *J. Fluid Mech.*, **647** 421–452.
[5] Burger E D, Chorn L G and Perkins T K 1980 *J. Rheol.*, **24** 603–626.
[6] Li W. and Graham M. D. 2007 *Phys. Fluids* **19** 083101.
[7] K. D. Housiadas, A. N. Beris and R. A. Handler, 2005 *Phys. Fluids* **17** 035106.
[8] Dubief Y, Terrapon V E , White C M, Shaqfeh E S, Moin P and Lele S K 2005 *Flow Turbul. Combust.* **74** 311–329.
[9] Ptasinski P K, Boersma B J, Nieuwstadt F T M, Hulsen M A, Van den Brule B H A A and Hunt J C R 2003 *J. Fluid Mech.* **490** 251–291.
[10] Yarin A L 1997 *J. Non-Newtonian Fluid* **69** 137–153.
[11] Biancofiore L Brandt L and Zaki T A 2017 *Phys. Rev. Flu.* **2** 043304.
[12] Bai X. and Xi L. 2017 submitted.
[13] Antonia R A 1981 *Ann. Rev. Fluid Mech.* **13** 131–156.
[14] Willmarth W W and Lu S S 1972 *J. Fluid Mech.* **55** 65–92.
[15] Blackwelder R F and Kaplan R E 1976 *J.Fluid Mech.* **76** 89–112.
[16] Jeong J, Hussain F, Schoppa W and Kim J 1997 *J. Fluid Mech.* **332** 185–214.
[17] Kol V 2007 *Int. J. Heat Fluid Fl.* **28** 638–652.
[18] Hunt J C R, Wray A A and Moin P 1988 *Proc. Center Turbul. Res. Summer School* 193–208 (Stanford U.).
[19] Lozano-Durn A, Flores O and Jiménez J 2012 *J. Fluid Mech.* **694** 100-130.
[20] Chu D C and Karniadakis G E 1993 *J. Fluid Mech.* **250** 1–42.
[21] Hussain A K M F and Hayakawa M 1987 *J. Fluid Mech.* **180** 193–229.
[22] Kim J and Moin P 1986 *J. Fluid Mech.* **162** 339–363.
[23] Graham M D 2004 *Rheology reviews* **2** 143-170.
[24] Bird R B, Armstrong R C and Hassager O 1987 *Dynamics of polymeric liquids. Vol. 2: Kinetic Theory* (New York: John Wiley).
[25] Peyret R 2013 *Spectral methods for incompressible viscous flow* (New York: Springer).
[26] Robinson S K 1991 *Ann. Rev. Fluid Mech.* **23** 601–639.
[27] Hamilton J M , Kim J and Waleffe F 1995 *J. Fluid Mech.* **287** 317–348.
[28] Brooke J W and Hanratty T J 1993 *Phys. Fluids A* **5** 1011–1022.
[29] Schoppa W and Hussain F 2002 *J. Fluid Mech.* **453** 57–108.

IOP Conf. Series: Journal of Physics: Conf. Series **1001** (2018) 012006 doi:10.1088/1742-6596/1001/1/012006

# Linear estimation of coherent structures in wall-bounded turbulence at $Re_\tau = 2000$

**S Oehler[1], A Garcia–Gutiérrez[2] and S Illingworth[1,]**

[1] Dept. Mechanical Engineering, The University of Melbourne, Victoria 3010 Australia
[2] School of Aeronautics, Universidad Politécnica de Madrid, 28040 Madrid, Spain

E-mail: oehlers@student.unimelb.edu.au, sillingworth@unimelb.edu.au

**Abstract.**
The estimation problem for a fully-developed turbulent channel flow at $Re_\tau = 2000$ is considered. Specifically, a Kalman filter is designed using a Navier–Stokes-based linear model. The estimator uses time-resolved velocity measurements at a single wall-normal location (provided by DNS) to estimate the time-resolved velocity field at other wall-normal locations. The estimator is able to reproduce the largest scales with reasonable accuracy for a range of wavenumber pairs, measurement locations and estimation locations. Importantly, the linear model is also able to predict with reasonable accuracy the performance that will be achieved by the estimator when applied to the DNS. A more practical estimation scheme using the shear stress at the wall as measurement is also considered. The estimator is still able to estimate the largest scales with reasonable accuracy, although the estimator's performance is reduced.

## 1. Introduction

Linear estimation has been used to successfully detect perturbations and coherent structures in wall bounded shear flows at laminar [1–3], transitional [4–6], and turbulent [7, 8] Reynolds numbers. Estimation is useful for obtaining information about quantities that cannot be measured directly, and can provide feedback in a control setup. The design of a linear estimator is a well-understood topic in control theory, and therefore the implementation is relatively straightforward. However, a model of the flow is required. A suitable model for this purpose is the linearized Navier-Stokes equations.

Previous studies have analysed the linearized Navier-Stokes equations to understand the growth and decay of perturbations. The majority of studies have focused on *laminar* shear flows by considering the temporal development of small perturbations about the laminar velocity profile; and the response to external forcing [9–12]. Of particular significance was the realization that, even for stable flows, perturbations can experience significant transient growth before ultimately decaying [13,14]. Linear modelling for *fully-developed* turbulent shear flows, although able to claim a relatively long history [15, 16], has received less attention. In recent years, though, there has been a renewed interest in linear models for turbulent shear flows, motivated by increasing evidence that linear mechanisms play an important role here as well.

Existing research on linear mechanisms in fully-developed turbulent shear flows can be divided into two broad camps: those that, following earlier work [17], include an eddy viscosity in the linear operator [18–21]; and those that do not [22–25].

This study investigates model-based estimation of channel flow at $Re_\tau = 2000$ using a linear, Navier-Stokes-based model (introduced in §2). The estimator uses time-resolved velocity measurements (or shear-stress measurements) at a single wall-normal location to estimate the time-resolved velocity field at other wall-normal locations. Like many previous studies [17–21] the model includes an eddy viscosity in the linear operator; but in contrast to them it does not discard the remaining nonlinear terms, instead absorbing them into an unknown (but non-zero) forcing term following [22].

The work is composed of three parts. The first part of the study serves to consolidate earlier results [8] which were obtained for channel flow at $\mathrm{Re}_\tau = 1000$. The present study builds on that work in three ways. First, applying the techniques to a different dataset at a different Reynolds number provides an important consolidation and confirmation of those earlier results. Second, with access to significantly more time-resolved data at $\mathrm{Re}_\tau = 2000$ we obtain significantly better convergence in the results. This will be of particular importance when comparing the results obtained in DNS with those predicted by the linear model. Third, with access to all wall-normal locations we are able to compare velocity fields not only in single wall-normal planes (as done in [8]), but also across all wall-normal heights. This allows us to qualitatively assess the structures that are estimated by the linear model when it knows only the time-resolved velocity field at a single wall-normal height. In the second part of the study we look at making the measurement scheme more practical. Measuring the velocity in the interior of a flow is difficult to achieve experimentally. For this reason wall-mounted shear stress measurements have been used in a number of previous studies on flow estimation and flow control [1–7, 26–28]. Therefore in the second part of the study we replace the (less practical) velocity measurements used in the first part with (more practical) wall shear stress measurements. Having looked at estimator performance at individual wall-normal heights in the first two parts, the third and final part looks at estimator performance over a range of estimation and measurement planes (for velocity measurements only). Specifically, we quantify the estimator performance for all pairs of measurement and estimation locations over the entire channel half-height. Although less practical (since the sensor could be anywhere), this is interesting because it shows us what is possible with a single sensor.

## 2. Methods

### 2.1. Linear model

We consider a fully developed turbulent channel flow at $\mathrm{Re}_\tau = 2000$. Here $\mathrm{Re}_\tau$ is the friction Reynolds number defined as $u_\tau h/\nu$, where $\nu$ is the kinematic viscosity, $h = 1$ the channel half height, $u_\tau = \sqrt{\tau_w/\rho}$ the friction velocity, $\tau_w$ the wall shear stress, and $\rho$ the density. Thus, the spatial variables are normalised by $h$, velocities by $u_\tau$, time by $h/u_\tau$ and the pressure $p$ by $\rho u_\tau^2$. The streamwise, spanwise, and wall-normal spatial coordinates of the flow are represented by $[x, y, z]$ and velocities by $\boldsymbol{u} = [u, v, w]$.

The channel flow dynamics are represented by a linear model for perturbations ($\boldsymbol{u}$) about a full-developed mean velocity profile ($\boldsymbol{U} = \big(U(z), 0, 0\big)$, as done in [18, 19]):

$$\frac{\partial \boldsymbol{u}}{\partial t} = -\left(\boldsymbol{U}\cdot\nabla\right)\boldsymbol{u} - \left(\boldsymbol{u}\cdot\nabla\right)\boldsymbol{U} - \nabla p + \nabla\cdot\left[\frac{\nu_T}{\nu}\left(\nabla\boldsymbol{u} + \nabla\boldsymbol{u}^T\right)\right] + \boldsymbol{d}, \qquad \boldsymbol{u}\cdot\nabla = 0, \qquad (1)$$

where $\nu_T(z)$ is the eddy viscosity profile and $\boldsymbol{d} = -(\boldsymbol{u}\cdot\nabla)\boldsymbol{u} + \overline{(\boldsymbol{u}\cdot\nabla)\boldsymbol{u}}$ contains all non-linear terms. An analytical approximation of the eddy viscosity profile, as used in [8, 18, 19, 24] is given by the relation:

$$\nu_T(z) = \frac{\nu}{2}\left(1 + \frac{\kappa^2\mathrm{Re}_\tau^2}{9}(2z - z^2)^2(3 - 4z + z^2)^2\left[1 - \exp\left(\frac{-\mathrm{Re}_\tau z}{A}\right)\right]^2\right)^{\frac{1}{2}} + \frac{\nu}{2}. \qquad (2)$$

The constants $\kappa = 0.426$ and $A = 25.4$ give the best fit to the mean velocity profile at $Re_\tau = 2003$ [16, 17, 29]. The mean velocity profile $\boldsymbol{U}$ is obtained by integrating $(1-z)/\nu_T(z)$ in the wall-normal direction.

### *2.2. Linear state-space model*

We require a linear time-invariant state-space model, which will allows us to design and implement a Kalman filter. We begin with the linear model introduced in 2.1, take the Fourier transforms in the homogeneous directions ($x$ and $y$), and transform it into the Orr-Sommerfeld Squire form. The wall-normal direction is then discretised using Chebyshev collocation of order 151 [30]. Convergence has been checked for the setup. The state-space model at a single wavenumber pair is thus obtained as:

$$\dot{\boldsymbol{x}}(t) = \boldsymbol{A}\boldsymbol{x}(t) + \boldsymbol{B}\boldsymbol{d}(t), \tag{3a}$$

$$\boldsymbol{y}(t) = \boldsymbol{C_y}\boldsymbol{u}(t) + \boldsymbol{n}(t), \tag{3b}$$

$$\boldsymbol{z}(t) = \boldsymbol{C_z}\boldsymbol{u}(t), \tag{3c}$$

where $\boldsymbol{x} = [\hat{\boldsymbol{w}}, \hat{\boldsymbol{\eta}}]^T$ represents the wall-normal velocity and wall-normal vorticity, $\boldsymbol{u} = [\hat{\boldsymbol{u}}, \hat{\boldsymbol{v}}, \hat{\boldsymbol{w}}]^T$ the streamwise, spanwise, and wall-normal velocity, $\boldsymbol{y}$ a sensor reading contaminated by noise $\boldsymbol{n}$, $\boldsymbol{z}$ a quantity of interest, and $\boldsymbol{d} = \left[\hat{\boldsymbol{d}}_x, \hat{\boldsymbol{d}}_y, \hat{\boldsymbol{d}}_z\right]^T$ represents the non-linear effects (see §2.1). For the purpose of the linear estimator design, the disturbances $\boldsymbol{d}$ and $\boldsymbol{n}$ are treated as unknown forcing which is white in space and time. The sensor noise is chosen to be small enough (relative to the size of disturbances) that the estimation results are insensitive to it. The state-space model (3) can be expressed as a transfer function by using the Laplace transform:

$$\begin{bmatrix} \boldsymbol{y} \\ \boldsymbol{z} \end{bmatrix} = \boldsymbol{P}(s) \begin{bmatrix} \boldsymbol{d} \\ \boldsymbol{n} \end{bmatrix}. \tag{4}$$

See Appendix A for a full definition of the state-space model (3) and the transfer function (4).

### *2.3. Linear estimator design*

The linear time-invariant model, introduced in the previous section, allows us to design an estimator using common tools from control theory. We begin the estimator design process by stating the optimal estimation problem: given sensor measurements $\boldsymbol{y}$, which are contaminated by noise $\boldsymbol{n}$, estimate a quantity of interest $\boldsymbol{z}$. We use the plant model $\boldsymbol{P}(s)$ and the $H_2$ optimal control framework to design the Kalman filter, which is denoted as $\boldsymbol{F}(s)$ in figure 1 (for example, see Seron et al. [31]). The filter generates the estimate $\hat{\boldsymbol{z}}(s)$:

$$\hat{\boldsymbol{z}}(s) = \boldsymbol{F}(s)\boldsymbol{y}(s). \tag{5}$$

Combining $\boldsymbol{F}(s)$ with $\boldsymbol{P}(s)$ forms a new transfer function, $\boldsymbol{G}(s)$, as shown in figure 1. The estimation task can be summarized as: minimize the estimation error $\boldsymbol{e}(s)$ (6) in the presence of the exogenous inputs $[\boldsymbol{d}(s) \quad \boldsymbol{n}(s)]^T$:

$$\boldsymbol{e}(s) = \boldsymbol{z}(s) - \hat{\boldsymbol{z}}(s) = \boldsymbol{G}(s) \begin{bmatrix} \boldsymbol{d}(s) \\ \boldsymbol{n}(s) \end{bmatrix}. \tag{6}$$

Now, the estimation design procedure can now be summarized as: given $\boldsymbol{P}(s)$, design $\boldsymbol{F}(s)$ such that $\boldsymbol{G}(s)$ is small. Thus, we quantify the size of $\boldsymbol{G}(s)$ via a $H_2$-norm, which is defined as:

$$||\boldsymbol{G}||_2 = \sqrt{\frac{1}{2\pi}\int_{-\infty}^{\infty} \text{Trace}[\boldsymbol{G}^*(\mathrm{j}\omega)\boldsymbol{G}(\mathrm{j}\omega)]d\omega} = \sqrt{\frac{1}{2\pi}\int_{-\infty}^{\infty}\sum_i(\sigma_i(\mathrm{j}\omega))^2 d\omega}, \tag{7}$$

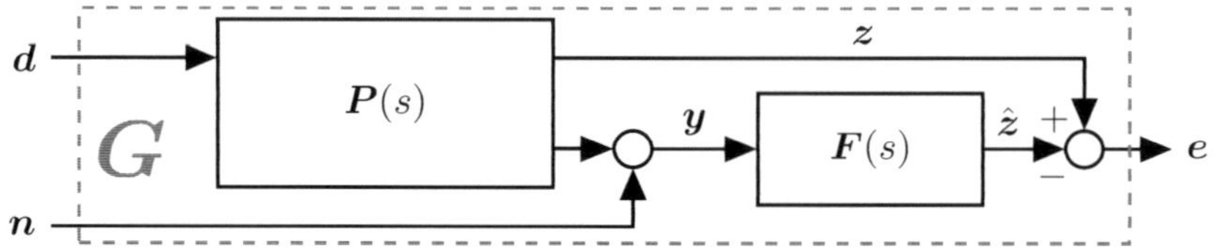


**Figure 1.** The combined Plant $\boldsymbol{P}(s)$ and Filter model $\boldsymbol{F}(s)$. The overall transfer function $\boldsymbol{G}(s)$ describes the input-output relationship of the estimation problem.

where $\sigma_i$ are the singular values of $\boldsymbol{G}$ at frequency $\omega$. The singular values represent a transfer function's gain when there are multiple inputs and multiple outputs, and hence the $H_2$-norm can be considered as an average gain over all frequencies and all directions.

### *2.4. DNS dataset*

We require two sets of time-resolved DNS data: first, the velocity or shear stress fields in the measurement plane(s) to provide the input for the estimator, and second the velocity field in the estimation plane(s) to compare the estimates with the truth. These data sets are provided by the Polytechnic University of Madrid (UPM). They were generated by direct numerical simulations (DNS) performed in a channel at $Re_\tau = 2000$. The streamwise and spanwise directions are discretised by Fourier expansion and the wall-normal direction is discretised using a compact difference scheme of 7th order. See table 1 for the details on the spatial dimensions and wavenumbers considered. Only positive spanwise wavenumbers ($k_y$) are considered, because the data is real-valued in physical space, and therefore the coefficients for modes $(+k_x, +k_y)$ are those for $(+k_x, -k_y)$ [32]. The largest temporal frequency is approximated using Taylor's hypothesis as: $\omega_{max} = \max(|k_x|)U_c = 195$, where $U_c$ is the velocity at the channel centre, and $\max(|k_x|)$ the largest streamwise wavenumber considered in this study. We therefore have $2\pi/(\omega_{max}\Delta t) = 3$ samples per period for the highest frequency, which fulfils the Nyquist criterion. Refer to table 2 for the temporal dimensions. Full details of the DNS dataset can be found in [32].

### *2.5. Estimator performance*

The filter uses the sensor measurements $\boldsymbol{y}(t)$ from the DNS data set at a chosen wall-normal height and provides estimates $\hat{\boldsymbol{z}}(t)$ at different wall-normal heights by simulating the filer in time. The error $\boldsymbol{e}(t) = \hat{\boldsymbol{z}}(t) - \boldsymbol{z}(t)$ is then formed by comparing the true flow fields $\boldsymbol{z}(t)$ at the estimated wall-normal heights with their estimates $\hat{\boldsymbol{z}}(t)$. The $H_2$-norm quantifies the performance, which

**Table 1.** The $\mathrm{Re}_\tau = 2000$ channel flow spatial dimension

| Direction | Size | Spacing | Resolved data | Smallest wavelength | Range |
|---|---|---|---|---|---|
| Streamwise ($x$): | $8\pi$ | $\Delta k_x = 1/4$ | 65 wavenumbers | $\lambda_x \geq 2\pi h/8 \approx 0.785$ | $\|k_x\| \leq 8$ |
| Spanwise ($y$): | $3\pi$ | $\Delta y_x = 2/3$ | 33 wavenumbers | $\lambda_y \geq 6\pi h/64 \approx 0.295$ | $0 \leq k_y \leq 64/3$ |

**Table 2.** The $\mathrm{Re}_\tau = 2000$ channel flow temporal dimension used for estimation

| $Re_\tau$ | $\Delta t$ | $t_{\max}$ | $U_c$ | $U_c\Delta t$ | $t_{\max}U_c/(8\pi)$ |
|---|---|---|---|---|---|
| 2000 | 0.0111 | 12.72 | 24.37 | 0.272 | 12.33 |

is obtained by integrating the error in time: $||\boldsymbol{e}||_2 = [\int_0^{t_{\max}} \boldsymbol{e}^*(t)\boldsymbol{e}(t)dt]^{1/2}$. It is normalised by $||\boldsymbol{z}||_2 = [\int_0^{t_{\max}} \boldsymbol{z}^*(t)\boldsymbol{z}(t)dt]^{1/2}$ to form the performance parameter:

$$\gamma = \frac{||\boldsymbol{e}||_2}{||\boldsymbol{z}||_2} \tag{8}$$

We can also generate performance predictions $\gamma_{\text{pred}}$ of the filter by taking $||\boldsymbol{G}||_2$ (7) and normalising it with the $H_2$ norm of $\boldsymbol{P}$. This is equivalent to taking sensor readings $\boldsymbol{y}(t)$ form the linear model itself, obtain estimates $\hat{\boldsymbol{z}}(t)$ at different wall-normal heights with the filter, and then extract the true flow fields $\boldsymbol{z}(t)$ at the estimated wall-normal heights from the linear model.

## 3. Results

We now provide the linear estimator (described in §2.3) with time-resolved DNS data at a single wall-normal height and use it to estimate the time-resolved velocity at other wall-normal heights. In practice this means forming an estimator for each of the $33 \times 65 = 2145$ wavenumber pairs of interest (see table 1).

### *3.1. Estimation in Physical space*

We begin by looking at the estimator performance in physical space by comparing the estimate of the streamwise velocity with the true field from DNS. The filter is supplied with all three velocity components at a single wall height of $z = 0.2$ ($z^+ = 400$). Figure 2 compares the estimated streamwise velocity with the true field at $z = 0.1$ ($z^+ = 200$) at a single instant in time (parts (a) and (b)); and as a function of time for a single point in the plane (part (c)). Figure 3 repeats the analysis of figure 2, this time for a wall height of $z = 0.3$ ($z^+ = 600$). In both figures we see good agreement between the model-based estimate and the true velocity field.

To show how estimation performance varies with wall-normal distance in physical space we have also included the streamwise velocity perturbations at $y = \pi/4$ in figure 4 and at $x = 2\pi/4$ in figure 5. Good agreement is seen between the model based estimated and the true velocity field, but it reduces with distance from the sensor. The average magnitude of the streamwise velocity component is over-predicted when estimating near the wall. Similarly it is under-predicted when estimating near the channel centre.

### *3.2. Estimation in wavenumber space*

We now provide a more quantitative assessment of the estimator by examining its performance across all wavenumber pairs of interest (see table 1). As in § 3.1, the estimator receives all three velocity components at $z = 0.2$ and uses them to estimate all three velocity components at $z = 0.1$ and $z = 0.3$.

Figure 6(a) plots $\gamma$ as a function of wavenumber pair $(k_x, k_y)$. In order to achieve better convergence, we calculate $\gamma$ in each of the two halves of the channel and average over the two. The colour scale is chosen to cover the range $[0, 0.7]$ to focus on the region over which the estimator performs well. (The maximum value of $\gamma$ across all wavenumber pairs is 0.94.) We see that the estimator performs best for wavenumbers satisfying approximately $|k_x| \leq 2$ and $|k_y| \leq 10$.

By assuming that the unknown disturbances $d$ are random forcing which is white in space and white in time we can also calculate the *predicted* estimator performance, which we label $\gamma_{\text{pred}}$. For this predicted estimator performance, the filter is *designed* for the linear model, but also *applied* to the linear model, and thus the DNS data is not needed at all. The predicted estimator performance $\gamma_{\text{pred}}$ is shown in figure 6(b), from which we see good agreement with the performance that is achieved in DNS. This is an encouraging result because it suggests that the physics of the largest scales are well-captured by the linear model.

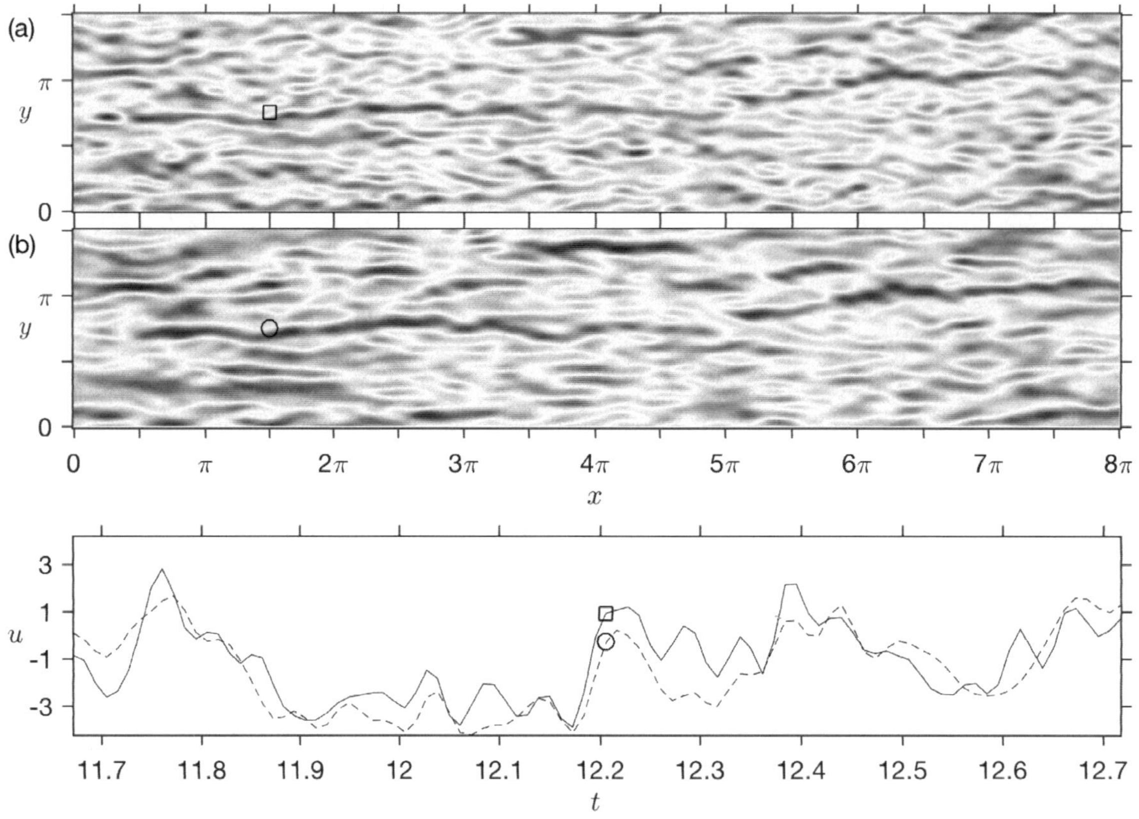


**Figure 2.** Estimation of the streamwise velocity perturbation at $z = 0.1$ ($z^+ = 200$) using the linear filter: (a) DNS data; and (b) linear estimate using measurements of all three velocity components at $z = 0.2$ ($z^+ = 400$); (c) time history at $x = 3\pi/2$, $y = 3\pi/4$ for the DNS data (—) and for the linear estimate (− −). Sixty-five contour levels are shown from $u = -5.5$ to $u = +5.5$ (red).

### *3.3. Shear stress measurements*

Velocity measurements inside the flow might not be available in an experimental setup. Therefore we now consider the performance of the estimator when its input is instead the streamwise wall shear stress, $\tau_x$ (see Appendix A).

Figure 7 is a repeat of figure 6 for streamwise wall shear stress measurements. (The quantity estimated is still all three velocity components at $z = 0.1$ and $z = 0.3$ and the maximum value of $\gamma$ across all wavenumber pairs is 1.) It is clear from figure 7 that the estimator performance is reduced significantly relative to figure 6 (i.e. relative to velocity measurements at $z = 0.2$).

To better compare results for these two measurement types, we now plot $\gamma$ and its prediction $\gamma_{\text{pred}}$ over a range of spanwise wavenumbers $k_y$ for all $|k_x| \leq 0.5$ (averaged across $k_x$ to improve convergence). This comparison is shown in figure 8(a). The predicted performance, $\gamma_{\text{pred}}$, achieves its minimum for $1.4 < k_y < 2.1$, which is consistent with the results of Illingworth et al. [8]. This corresponds well with the range of $k_y$ over which the potential for transient growth is greatest [19]; and over which the largest amplification is seen for both stochastic and harmonic forcing [21]. It appears, however, that the best performance (i.e. smallest $\gamma$) achieved by the estimator in DNS occurs at a slightly higher spanwise wavenumber of $k_y \approx 3.33$. In a similar way, in figure 8(b) we plot $\gamma$ and $\gamma_{\text{pred}}$—this time as a function of the streamwise

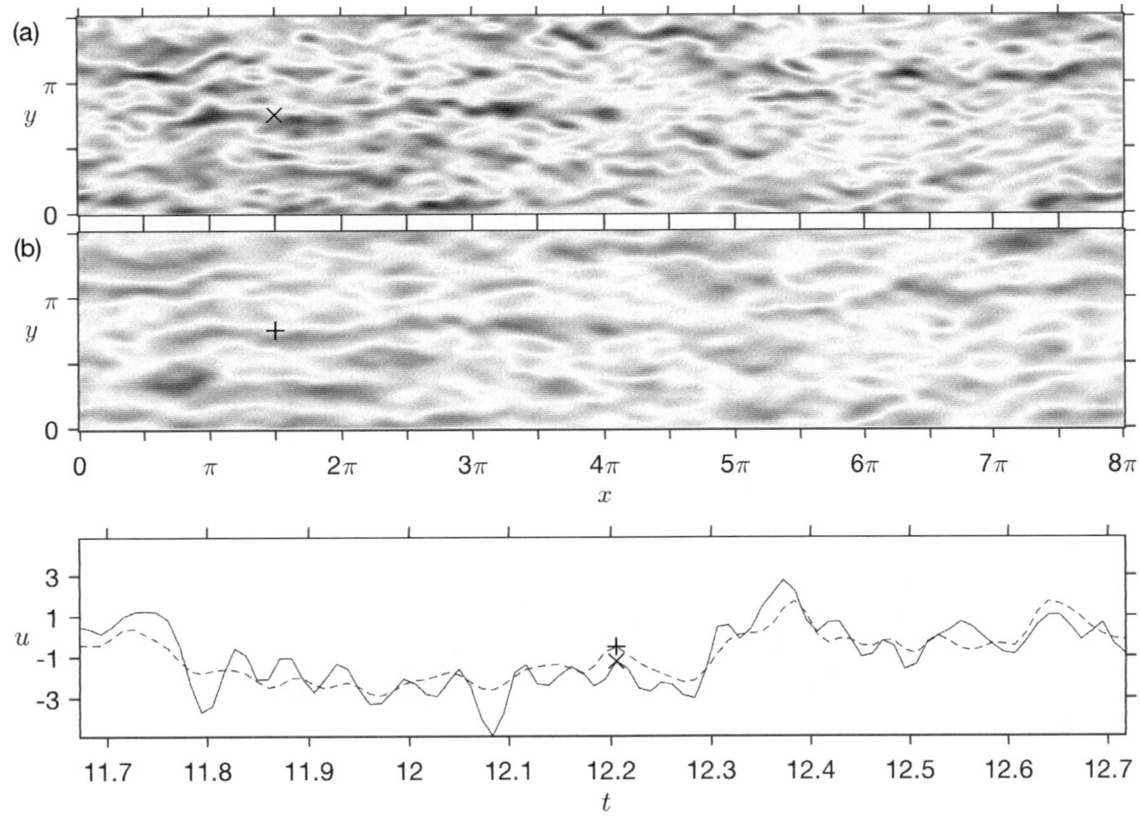


**Figure 3.** Estimation of the streamwise velocity perturbation at $z = 0.3$ ($z^+ = 600$) using the linear filter. Legend same as figure 2.

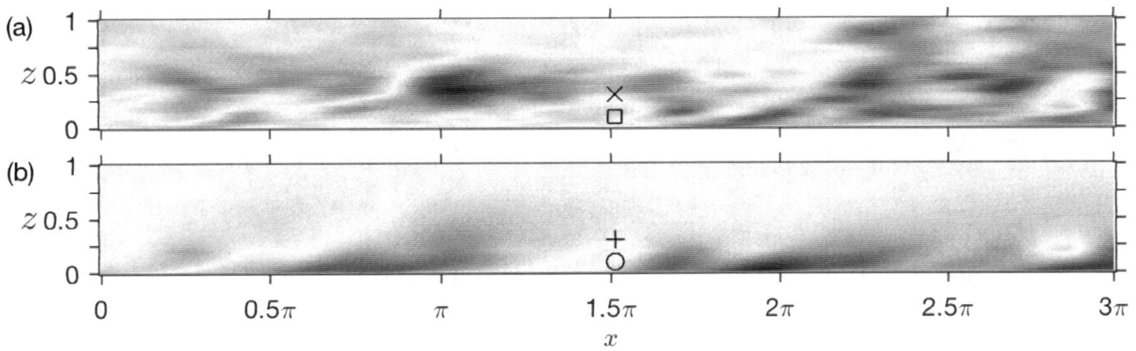


**Figure 4.** Estimation of the streamwise velocity perturbation at $y = 3\pi/4$ ($y^+ = 4712$) using the linear filter: (a) DNS data; and (b) linear estimate using measurements of all three velocity components at $z = 0.2$ ($z^+ = 400$). Sixty-five contour levels are shown from $u = -5.5$ to $u = +5.5$ (red).

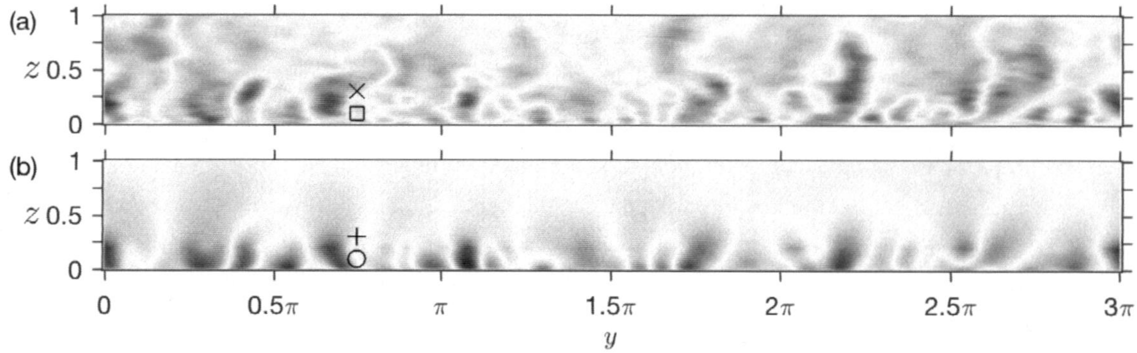


**Figure 5.** Estimation of the streamwise velocity perturbation at $x = 2\pi/4$ ($y^+ = 9425$) using the linear filter: (a) DNS data; and (b) linear estimate using measurements of all three velocity components at $z = 0.2$ ($z^+ = 400$). Sixty-five contour levels are shown from $u = -5.5$ to $u = +5.5$ (red).

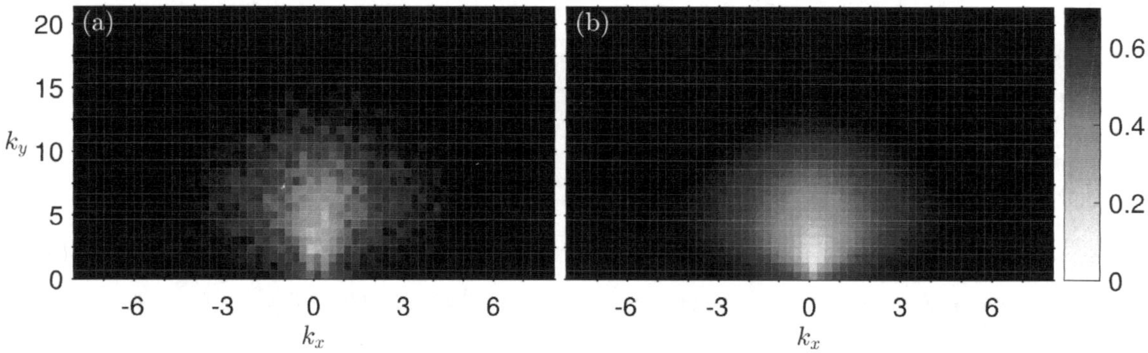


**Figure 6.** (a) Normalized estimation error $\gamma$ as a function of $(k_x, k_y)$ and (b) the prediction of the estimation error $\gamma_{\text{pred}}$. The estimator receives all three velocity components at $z = 0.2$ and uses them to estimate all three velocity components at $z = 0.1$ and $z = 0.3$. The same colour scale $\gamma \in [0, 0.7]$ is used for both plots.

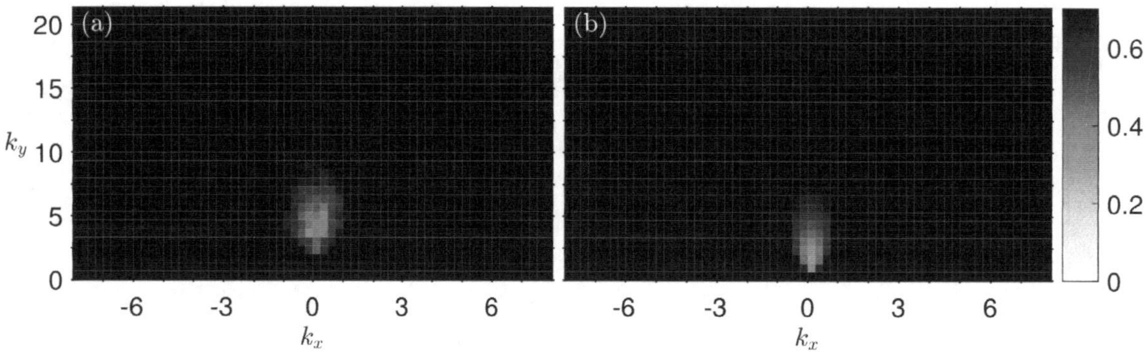


**Figure 7.** Normalized estimation error $\gamma$ as a function of $(k_x, k_y)$ and (b) the prediction of the estimation error $\gamma_{\text{pred}}$. The estimator receives the streamwise shear stress component at the wall and uses it to estimate all three velocity components at $z = 0.1$ and $z = 0.3$. The same colour scale $\gamma \in [0, 0.7]$ is used for both plots.

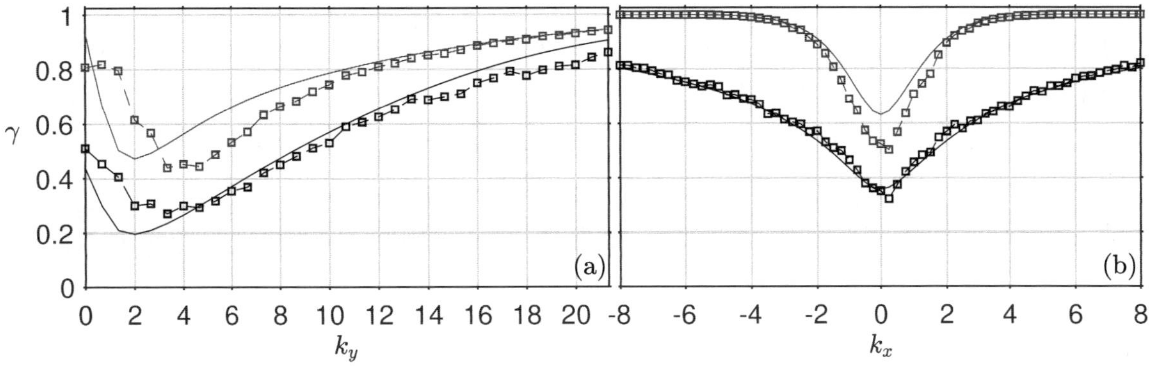


**Figure 8.** Normalized estimation error $\gamma$ (–□–) and its prediction $\gamma_{\text{pred}}$(—) : (a) as a function of $k_y$ averaged over all $k_x \leq |0.5|$; and (b) as a function of $k_x$ averaged over all $4 \leq k_y \leq 8$. Setup same as figure 6 for black results and figure 7 for blue results. The DNS results have been averaged across the two halves of the channel to improve convergence.

wavenumber $k_x$—for all $k_y$ satisfying $4 \leq k_y \leq 8$ (again, averaged across these $k_y$ to improve convergence). We see that, for both measurement types, the predicted performance agrees well with the performance actually achieved in DNS. For shear stress measurements, however, we do observe some more significant differences between the predicted and true performance (up to 22%) for the smallest values of $|k_x|$.

### *3.4. Variation of estimation performance with wall-normal height*

So far we have looked at the estimator performance when the measurement location is fixed (at $z = 0.2$ or at the wall) and when estimation is performed at $z = 0.1$ and $z = 0.3$. It is now interesting to investigate the estimator performance as these measurement and estimation locations are varied. To make the parameter space under consideration manageable, we will do this only for streamwise wavenumbers satisfying $|k_x| \leq 0.5$; and for four individual spanwise wavenumbers: $k_y = 2$, 4, 6 and 8.

We begin by fixing the measurement location at $z = 0.2$ as before, and plotting the variation of the estimator performance as the location of the (single) estimation plane is varied. This is shown in figure 9. We also show results when the measurement is instead the streamwise shear stress at the wall. The results are shown separately for the top half and bottom half of the channel to give some indication of the convergence of the results. We can make a number of observations which apply to all four spanwise wavenumbers: i) the best performance (obviously) occurs at the sensor location and increases as we move away from it; ii) the worst performance occurs at the channel centre; iii) within the proximity of the measurement location, the true performance ($\gamma$) is always better than that predicted ($\gamma_{\text{pred}}$); and (iv) a local maximum occurs in $\gamma$ in the log region at approximately $z = 0.025$ or $z^+ = 50$. An extra observation can be made for the $k_y = 2$ and $k_y = 4$ cases: (v) a local minimum occurs in $\gamma$ in the buffer layer at approximately $z = 0.0075$ or $z^+ = 15$. This could be due to the wall signature of the (relatively large) eddies that are measured at certain $z_m$ [32, 33].

Having fixed the measurement location for the velocity sensor at a single plane, we can now also vary it by adding a third dimension. This is shown in figure 10, where the estimator performance $\gamma$ is plotted as a function of *both* the measurement location $z_m$ and the estimation location $z_e$. As one would expect, the best performance is achieved when the measurement and estimation locations coincide, i.e. when $z_e = z_m$. It is interesting that, for estimation locations in the vicinity of a given measurement location, the true estimator performance ($\gamma$) is in general

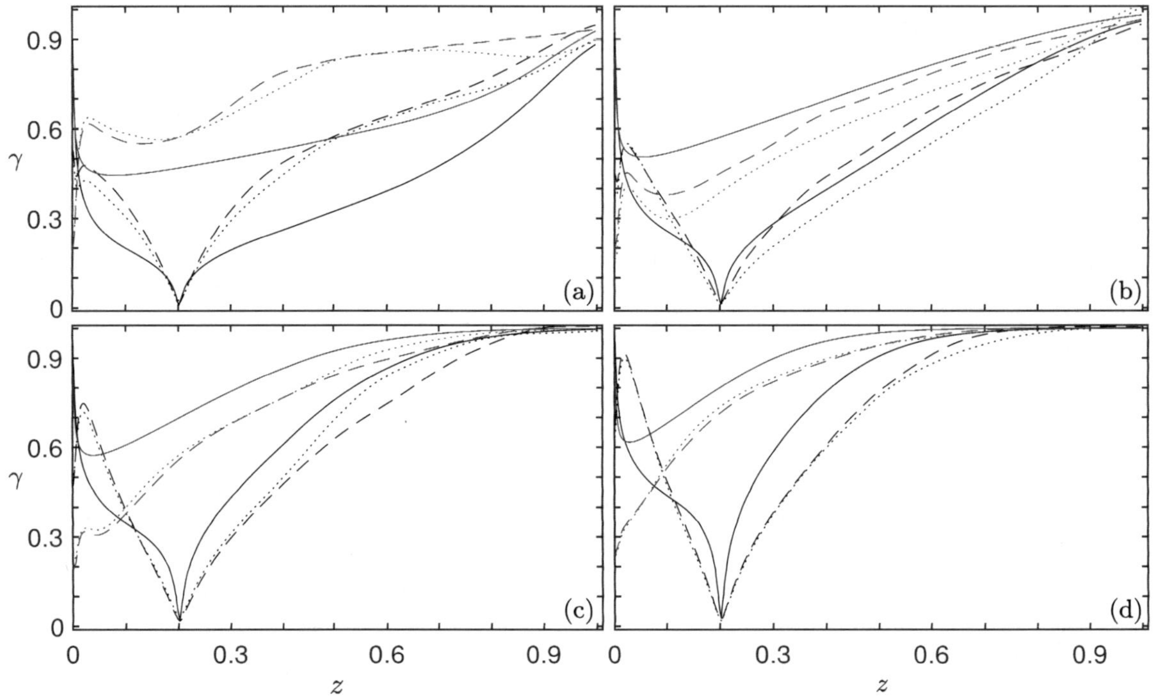


**Figure 9.** Normalized estimation error $\gamma$ for the two halves of the channel (−−, $\cdots$) and its prediction $\gamma_{\text{pred}}$ (—) as a function of wall normal location $z$ averaged over all $k_x \leq |0.5|$. This is plotted for (a) $k_y = 2$; (b) $k_y = 4$; (c) $k_y = 6$; (d) $k_y = 8$. The estimator measures all three velocity components for the black results and the streamwise shear stress component for the blue results.

smaller (i.e. better) than that predicted by the linear model ($\gamma_{\text{pred}}$).

## 4. Discussion

We now give some broader discussion on the results obtained for the two measurement types considered (i.e. velocity measurements and wall shear stress measurements).

Using velocity measurements at a single wall-normal location we are able to estimate the flow at other wall-normal locations with reasonable accuracy, as demonstrated in figures 2 - 6. This is consistent with the results observed in an earlier study [8] at $\mathrm{Re}_\tau = 1000$. The present study at $\mathrm{Re}_\tau = 2000$, however, benefits from the availability of significantly more time-resolved DNS data, and as a result we see significantly improved convergence in the normalized estimation error ($\gamma$) in figure 6. Importantly, the linear model (1) not only achieves good estimation performance, but is also able to predict the range of wavenumber pairs (figures 6 and 8) and the range of wall heights (figures 9 and 10) over which the estimator should perform well. In this respect the results in figure 6—which perhaps benefit most from the improved convergence described above—are particularly encouraging. Together these results suggest that, for the largest scales considered, linear mechanisms are important, and that they are well-represented by the linear model (1).

By measuring instead the wall shear stress it is still possible to achieve reasonable estimation performance for the largest scales. However, it is clear in figures 6 and 7 that the estimator is significantly less successful than its velocity-based counterpart. This finding is consistent with studies carried out for laminar profiles [2, 26]: the leading eigenvector of the linear system

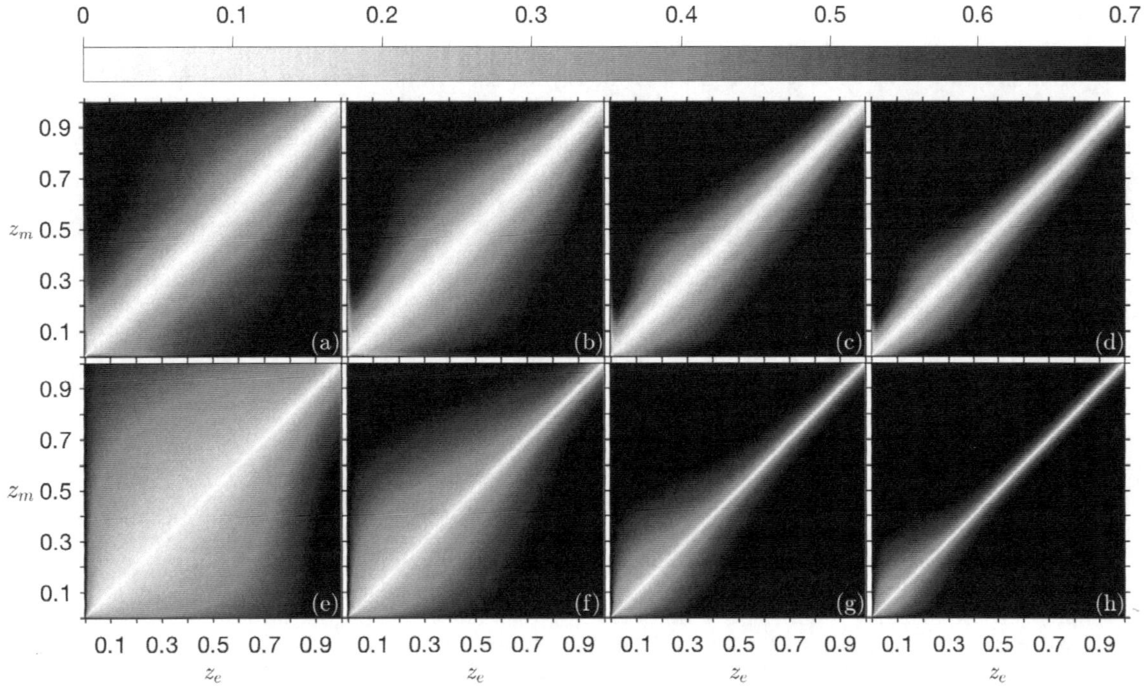


**Figure 10.** Normalized estimation error $\gamma$ (top row) and its prediction $\gamma_{\text{pred}}$ (bottom row) as function of the wall-normal measurement location $z_m$ and the wall-normal estimation location $z_e$. This is plotted for (a,e) $k_y = 2$; (b,f) $k_y = 4$; (c,g) $k_y = 6$; (d,h) $k_y = 8$. All results are averaged over all $k_x \leq |0.5|$ and the two channel halves.

includes several centred modes, with little support near the wall. (The modeshapes are only weak functions of Reynolds number.)

We also observe in figure 8 that the agreement between the true estimator performance ($\gamma$) and its prediction ($\gamma_{\text{pred}}$) is not as good as that seen for velocity measurements. This is an interesting observation, the cause of which is still under investigation. One possibility is that, for wall-based measurements, the assumption on $\boldsymbol{d}$ in (1) of delta-correlated forcing (in space and in time) is less appropriate—in other words, that second-order statistics become more important. Zare et al. [34] have recently considered coloured forcing in the linearized Navier Stokes equation to reproduce second order dynamics. It would be interesting to see if this modified linear model is able to improve i) the estimator performance and ii) the agreement between the estimator performance ($\gamma$) and its prediction ($\gamma_{\text{pred}}$)—both for velocity measurement and for wall shear stress measurements.

## 5. Conclusions

A linear Navier–Stokes-based model has been used to design a Kalman filter for fully-developed turbulent channel flow. The Kalman filter is able to estimate the largest scales with reasonable accuracy at $\text{Re}_\tau = 2000$. The results build on earlier work [8] which was performed at $\text{Re}_\tau = 1000$, and for which significantly less time-resolved data were available. Using velocity measurements as input, the filter is able to estimate the largest scales with reasonable accuracy for a range of wavenumber pairs, measurement planes, and estimation planes. Furthermore there is good agreement between the performance predicted by the model and the performance actually achieved in DNS. Using wall shear stress measurements as input, the filter is still able to estimate the largest scales with reasonable accuracy, although the filter's performance is significantly reduced relative to the velocity-based filter. It remains to be seen whether either measurement strategy is good enough for control purposes. The results warrant further research on the linear model and its applications. A further application currently under consideration is the optimal sensor placement problem: If one has access only to a single sensor, where should it be placed, and what quantity should be measured?

## Acknowledgments

The authors wish to thank the Universidad Politécnica de Madrid (UPM) for their hospitality and support during the Third Madrid Turbulence Workshop. We would also like to acknowledge the Coturb programme of the European Research Council (ERC) for the financial support that made the workshop possible. We are grateful to O. Flores for his insightful review of the manuscript.

## References

[1] Joshi S S, Speyer J L and Kim J 1997 *J. Fluid Mech.* **332** 157–184
[2] Hoepffner J, Chevalier M, Bewley T R and Henningson D S 2005 *J. Fluid Mech.* **534** 263–294
[3] Naguib A M, Morrison J F and Zaki T A 2010 *Phys. Fluids* **22** 054103
[4] Mckernan J, Whidborne J F and Papadakis G 2007 *Int. J. Control* **80** 1912–1930
[5] Jones B L, Kerrigan E C, Morrison J F and Zaki T A 2011 *Int. J. Control* **84** 1310–1325
[6] Bagheri S and Henningson D S 2011 *Phil. T. R. Soc. A.* **369** 1365–1381
[7] Chevalier M, Hoepffner J, Bewley T R and Henningson D S 2006 *J. Fluid Mech.* **552** 167–187
[8] Illingworth S J, Monty J P and Marusic I 2018 *Under consideration in J. Fluid Mech.*
[9] Farrell B F and Ioannou P J 1993 *Phys. Fluids* **5** 2600–2609
[10] Farrell B F and Ioannou P J 1996 *J. Atmos. Sci.* **53** 2025–2040
[11] Jovanović M R and Bamieh B 2001 *Am. Control Conf.* pp 1948–1953
[12] Jovanovic M 2005 *J. Fluid Mech* **534** 145–183
[13] Hårkan Gustavsson L 1991 *J. Fluid Mech.* **224** 241–260
[14] Schmid P J and Henningson D S 2012 *Stability and transition in shear flows* (Springer)
[15] Malkus W 1956 *J. Fluid Mech.* **1** 521–539
[16] Reynolds W C and Hussain A 1972 *J. Fluid Mech.* **54** 263–288
[17] Reynolds W C and Tiederman W G 1967 *J. Fluid Mech.* **27** 253–272
[18] del Alamo J C and Jimene J 2006 *J. Fluid Mech* **559** 205–213
[19] Pujals G, García-Villalba M, Cossu C and Depardon S 2009 *Phys. Fluids* **21** 015109
[20] Hwang Y and Cossu C 2010 *J. Fluid Mech.* **664** 51–73
[21] Hwang Y and Cossu C 2010 *J. Fluid Mech.* **643** 333–348
[22] McKeon B J and Sharma A S 2010 *J. Fluid Mech.* **658** 336–382
[23] McKeon B J, Sharma A S and Jacobi I 2013 *Phys. Fluids* **25** 031301
[24] Moarref R, Sharma A S, Tropp J A and McKeon B J 2013 *J. Fluid Mech.* **734** 275–316
[25] Moarref R, Jovanović M R, Tropp J A, Sharma A S and McKeon B J 2014 *Phys. Fluids* **26** 051701
[26] Bewley T R and Liu S 1998 *J. Fluid Mech.* **365** 305–349
[27] Åkervik E, Hœpffner J, Ehrenstein U and Henningson D S 2007 *J. Fluid Mech.* **579** 305–314
[28] Jones B L, Heins P H, Kerrigan E C, Morrison J F and Sharma A S 2015 *J. Fluid Mech.* **769** 687–722
[29] Cess R D 1958 *A survey of the literature on heat transfer in turbulent tube flow* (Westinghouse Research)
[30] Weideman J A and Reddy S C 2000 *ACM Trans. Math. Softw.* **26** 465–519
[31] Seron M M, Braslavsky J H and Goodwin G C 1997 *Fundamental Limitations in Filtering and Control* (Springer)
[32] Hoyas S and Jiménez J 2006 *Phys. Fluids* **18** 1–4
[33] Hutchins N and Marusic I 2007 *J. Fluid Mech.* **579** 1–28
[34] Zare A, Jovanović M R and Georgiou T T 2017 *J. Fluid Mech.* **812** 636–680
[35] Bewley T R and Protas B 2004 *Physica D* **196** 28–44

IOP Conf. Series: Journal of Physics: Conf. Series **1001** (2018) 012006 doi:10.1088/1742-6596/1001/1/012006

## Appendix A. Linear state space model of the flow

The state-space model (3) of the linear model is (1):

$$\frac{d}{dt}\begin{bmatrix}\hat{\boldsymbol{w}}\\ \hat{\boldsymbol{\eta}}\end{bmatrix} = \begin{bmatrix}\Delta^{-1}\mathcal{L}_{OS} & 0\\ -ik_yU' & \mathcal{L}_{SQ}\end{bmatrix}\begin{bmatrix}\hat{\boldsymbol{w}}\\ \hat{\boldsymbol{\eta}}\end{bmatrix} + \begin{bmatrix}-ik_x\Delta^{-1}\mathcal{D} & -ik_y\Delta^{-1}\mathcal{D} & -k^2\Delta^{-1}\\ ik_y & -ik_x & 0\end{bmatrix}\begin{bmatrix}d_x\\ d_y\\ d_z\end{bmatrix}, \tag{A.1a}$$

$$\begin{bmatrix}\hat{\boldsymbol{u}}\\ \hat{\boldsymbol{v}}\\ \hat{\boldsymbol{w}}\end{bmatrix} = \frac{1}{k^2}\begin{bmatrix}ik_x\mathcal{D} & -ik_y\\ ik_y\mathcal{D} & ik_x\\ k^2 & 0\end{bmatrix}\begin{bmatrix}\hat{\boldsymbol{w}}\\ \hat{\boldsymbol{\eta}}\end{bmatrix}, \tag{A.1b}$$

where

$$\mathcal{L}_{OS} = -jk_xU(z)\Delta + ik_xU''(z) + \nu_T\Delta^2 + 2\nu_T'\mathcal{D}\Delta + \nu_T''\left(\mathcal{D}^2 + k^2\right), \tag{A.2a}$$
$$\mathcal{L}_{SQ} = -jk_xU(z) + \nu_T\Delta + \nu_T'\mathcal{D}, \tag{A.2b}$$

$\mathcal{D} = \frac{\partial}{\partial z}$, $()' = \frac{\partial}{\partial z}()$, $k^2 = k_x^2 + k_y^2$, and $\Delta = \mathcal{D} - k^2$. The boundary conditions are: $\hat{\boldsymbol{w}}(t) = \hat{\boldsymbol{w}}'(t) = \boldsymbol{\eta}(t) = 0$.

To obtain instead the wall shear stress [2, 35] we replace (A.1b) with:

$$\tau_x = \left.\frac{1}{\mathrm{Re}}\frac{du}{dz}\right|_{\mathrm{wall}} = \left.\frac{i}{\mathrm{Re}\ k^2}\begin{bmatrix}k_x\mathcal{D}^2 & -k_y\mathcal{D}\end{bmatrix}\begin{bmatrix}\hat{\boldsymbol{w}}\\ \hat{\boldsymbol{\eta}}\end{bmatrix}\right|_{\mathrm{wall}}. \tag{A.3}$$

IOP Conf. Series: Journal of Physics: Conf. Series **1001** (2018) 012007 doi:10.1088/1742-6596/1001/1/012007

# Wall-based identification of coherent structures in wall-bounded turbulence

**C. Sanmiguel Vila and O. Flores**

Dept. Bioingeniería e Ing. Aeroespacial, Universidad Carlos III de Madrid. 28911 Leganés, Madrid. Spain.

E-mail: oflores@ing.uc3m.es

**Abstract.** During the last decades, a number of reduced order models based on coherent structures have been proposed to describe wall-bounded turbulence. Many of these models emphasize the importance of coherent wall-normal velocity eddies ($v$-eddies), which drive the generation of the very long streamwise velocity structures observed in the logarithmic and outer region. In order to use these models to improve our ability to control wall-bounded turbulence in realistic applications, these $v$-eddies need to be identified from the wall in a non-intrusive way. In this paper, the possibility of using the pressure signal at the wall to identify these $v$-eddies is explored, analyzing the cross-correlation between the wall-normal velocity component and the pressure fluctuations at the wall in a DNS of a turbulent channel flow at $Re_\tau = 939$. The results show that the cross-correlation has a region of negative correlation upstream, and a region of positive correlation backwards. In the spanwise direction the correlation decays monotonously, except very close to the wall where a change of sign of the correlation coefficient is observed. Moreover, filtering the pressure fluctuations at the wall in space results in an increase of the region where the cross-correlation is strong, both for the positively and the negatively correlated regions. The use of a time filter for the pressure fluctuations at the wall yields different results, displacing the regions of strong correlation without changing much their sizes. The results suggest that space-filtering the pressure at the wall is a feasible way to identify $v$-eddies of different sizes, which could be used to trigger turbulent control strategies.

## 1. Introduction

Wall-bounded turbulence is one of the most ubiquitous features of engineering applications involving fluids. Typically, the engineer needs to *control* these turbulent motions, either damping (i.e., to reduce drag) or promoting them (i.e., to increase mixing). Reduced-order models of wall-bounded turbulence play a key role in the still-pending task of *controlling* turbulence, since they could allow the identification of the specific times and scales where the actuation is needed [1].

During the last decades, the community has witnessed the emergence of several of these reduced order models. Some of them are semi-empirical, based on observation of coherent structures [2] on canonical wall-bounded turbulent flows [3, 4, 5]. Some are time-periodic solutions of simpler wall-bounded flows [6], which can be related to the dynamics of fully turbulent flows [7]. In most of them, wall-attached vertical velocity ($v$) eddies (reminiscent of Townsend's attached eddies [8]) are of special significance from a dynamical point of view, driving the generation of elongated streaks of streamwise velocity ($u$) through a linear amplification process [9]. The specific mechanisms by which the $v$-eddies are generated are less clear [10, 11, 12, 13], although there is agreement on the necessity of non-linearity in this part of

IOP Conf. Series: Journal of Physics: Conf. Series **1001** (2018) 012007 doi:10.1088/1742-6596/1001/1/012007

the cycle [14]. There is also reasonable agreement on that these $v$-eddies appear in all sizes, with heights ranging from the thickness of the buffer region to the thickness of the whole flow (i.e., the channel half-height or the boundary layer thickness).

From a technological point of view, the first step before using these reduced order models in any engineering application is to be able to identify these eddies with minimally-intrusive sensors at the walls. Two obvious choices are pressure taps or shear-stress sensors, where the latter should be more related to $u$-streaks and the former more related to $v$-eddies. Due to the central role of $v$-eddies in the generation of the streaks, and since it seems more reasonable to control the cause rather than the effect, the present work focuses on the detection of $v$-eddies from the pressure fluctuations at the wall. The choice is also motivated by previous works on the scaling of the spectra of the pressure fluctuations at the wall ($p_w$), which suggests that different frequency content on $p_w$ might be related to turbulent sources at different wall-normal locations [15, 16]. Since then, several other authors have analysed the spectrum of the pressure fluctuations, using DNS using numerical [17, 18] or experimental data [19, 20], corroborating the scaling arguments mentioned above. Also, models for the wall-pressure have been formulated using structures based on Townsend's attached eddy hypothesis [21], and some authors have linked specific features of the pressure fluctuations at the wall to coherent structures in the flow [22, 23].

Hence, the present paper focuses on the analysis of the cross-correlation between the pressure fluctuations at the wall and the wall-normal component of the velocity, using data from Direct Numerical Simulations (DNS) of turbulent channel flow. The specific objectives of this work are 1) to characterise the cross-correlation between the pressure fluctuations at the wall and the wall-normal velocity at different wall-distances, and 2) to evaluate if it is possible to discriminate $v$-eddies of a given size by filtering (in space or in time) the pressure fluctuations at the wall. The article is organized as follows: Section 2 presents the DNS details, as well as the definitions for the cross-correlation and the space and time filters. Section 3 presents the results for the un-filtered and filtered cases, and section 4 presents some conclusions.

## 2. Methodology

In order to evaluate the correlation between the wall-normal velocity component and the pressure fluctuations at the wall, a database with 1881 snapshots of a DNS of an incompressible turbulent channel flow is used. The turbulent channel flow has a friction Reynolds number $Re_\tau = u_\tau h/\nu = 939$, where $u_\tau$ is the friction velocity, $h$ is the half channel height and $\nu$ is the kinematic viscosity. The snapshots cover a range of approximately $20h/u_\tau$ eddy turn-over times, and they are approximately equidistant in time, at intervals of about $\Delta t^+ = 10$. Note that throughout the paper, the $+$ superscript indicates variables expressed in wall units (i.e., normalized with $u_\tau$ and $\nu$).

The algorithm employed in the DNS is the same pseudo-spectral algorithm described in [24]. The simulation is run at a constant flow rate in a computational box of size $L_x = 2\pi h$ and $L_z = \pi h$ in the streamwise ($x$) and spanwise directions ($z$), respectively. These directions are periodic, and they are discretized with $N_x = 512$ and $N_z = 512$ Fourier modes, yielding a resolution in collocation points of $\Delta x^+ = 7.7$ and $\Delta z^+ = 3.8$. The discretization of the wall-normal direction ($y$) uses $N_y = 385$ Chevychev polynomials, resulting in a maximum grid spacing at the centre of the channel of $\Delta y_c^+ = 7.7$. The wall is located at $y = 0$, and the channel centre is at $y = h$.

The cross-correlation between the vertical velocity and the pressure fluctuation at the wall is defined as

$$R_{v,p_w}(x, y, z) = \langle p_w(x', z', t')v(x' + x, y, z' + z, t')\rangle, \tag{1}$$

where $p_w$ is the pressure fluctuation at the wall, $v$ is the velocity component in the wall-normal direction, and the operator $\langle\rangle$ indicates averaging over $x', z'$ and $t'$. The correlation depends then on the horizontal separation $(x, z)$ between the points where $p_w$ and $v$ are measured, and on the wall-normal distance $y$ at which $v$ is measured. Note that no time lag between $p_w$ and $v$ is considered.

The cross-correlations between the vertical velocity and the filtered pressure fluctuations at the wall are defined analogously to equation 1. These cross-correlations are denoted $R_{v,\tilde{p}_w}$ for the space filtered pressure fluctuations ($\tilde{p}_w$), and $R_{v,\overline{p}_w}$ for the time filtered pressure fluctuations ($\overline{p}_w$).

The temporal filter is just a moving average over a characteristic time $t_c$, hence the time-filtered pressure fluctuation is

$$\overline{p}_w(x', z', t') = \frac{1}{t_c} \int_{t'-t_c}^{t'} p_w(x', z', t) \mathrm{d}t. \tag{2}$$

The spatially filtered pressure fluctuation at the wall is defined as

$$\tilde{p}_w(x', z', t') = \int_0^{L_x} \int_0^{L_z} p_w(x, z, t') K(x' - x, z' - z; l_c) \mathrm{d}z \mathrm{d}x, \tag{3}$$

where $K(x, z; l_c)$ is the kernel that defines the filter, and $l_c$ is the characteristic length scale. Two kernels are considered in the present study for the spatial filter. Most of the results presented in section 3 correspond to a Gaussian filter,

$$K(x, z; l_c) = \frac{1}{2\pi\sigma^2} \exp\left(-\frac{x^2 + z^2}{\sigma^2}\right), \tag{4}$$

with a standard deviation $\sigma = l_c/3$. For comparison, a circular top-hat filter of diameter $l_c$ has also been considered,

$$K(x, z; l_c) = \frac{4}{\pi l_c^2} \text{ if } \sqrt{x^2 + z^2} \leq l_c/2; \quad K(x, z; l_c) = 0 \text{ otherwise.} \tag{5}$$

Note that both kernels are normalized such that the integral of $K(x, z; l_c)$ on the horizontal plane is one. Also, by choosing $\sigma = l_c/3$, approximately 67% of the area of (4) lies inside the circle of diameter $l_c$ defined by (5).

As usual, the unfiltered or filtered correlations are normalized with the corresponding standard deviations of $p_w, \tilde{p}_w$ and $v$, to yield the correlation coefficients

$$\rho_{v,p_w}(x, y, z) = \frac{R_{v,p_w}(x, y, z)}{(\langle p_w^2 \rangle^{1/2} \langle v^2 \rangle^{1/2}}, \tag{6}$$

$$\rho_{v,\tilde{p}_w}(x, y, z) = \frac{R_{v,\tilde{p}_w}(x, y, z)}{\langle \tilde{p}_w^2 \rangle^{1/2} \langle v^2 \rangle^{1/2}}, \tag{7}$$

$$\rho_{v,\overline{p}_w}(x, y, z) = \frac{R_{v,\overline{p}_w}(x, y, z)}{\langle \overline{p}_w^2 \rangle^{1/2} \langle v^2 \rangle^{1/2}}, \tag{8}$$

where $\langle p_w^2 \rangle$, $\langle \overline{p}_w^2 \rangle$ and $\langle \tilde{p}_w^2 \rangle$ are constants and $\langle v^2 \rangle$ is a function of the wall normal coordinate, $y$.

## 3. Results

First, the correlation for the un-filtered wall pressure is characterized in figure 1, which shows a full three-dimensional view of $\rho_{v,p_w}(x, y, z)$ as well as three different planes: the streamwise/wall-normal plane $z = 0$ (figure 1*b*), a wall-parallel plane in the near-wall region at $y^+ = 15$ (figure 1*c*) and a wall-parallel plane further away from the wall, at $y = 0.1h$ (figure 1*d*). The solid and

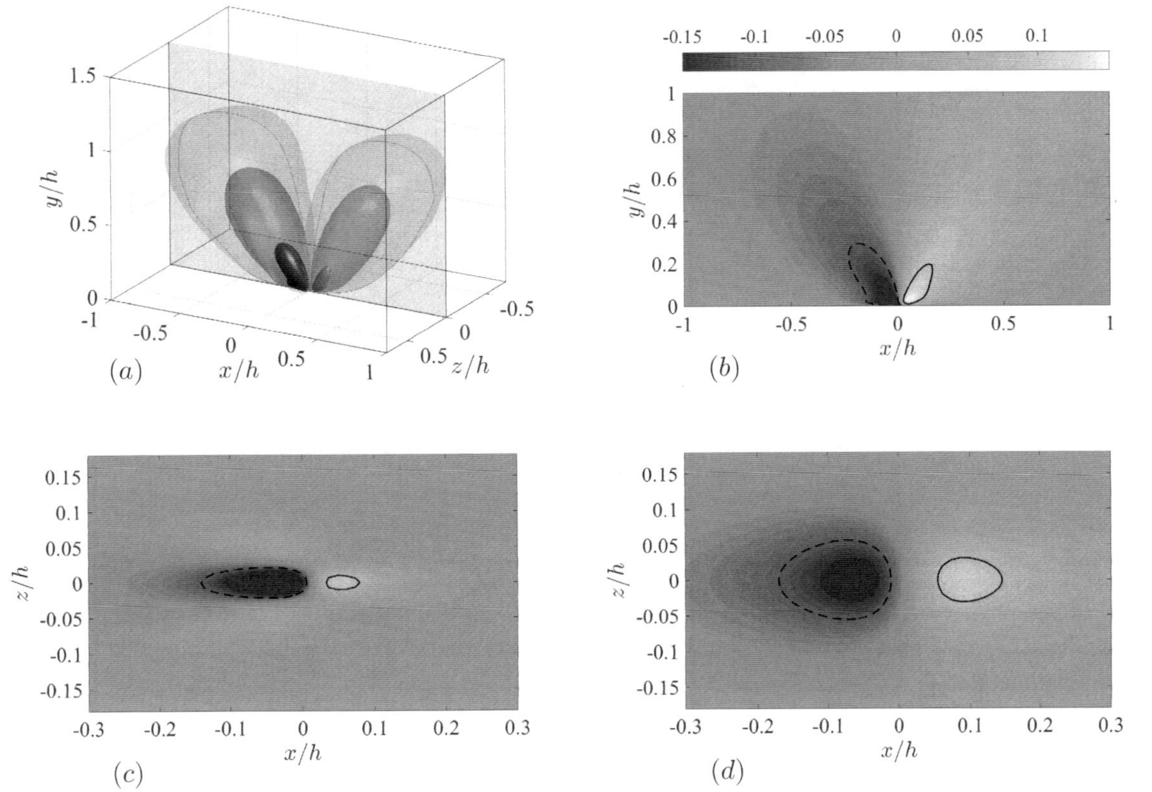


**Figure 1.** Correlation coefficient $\rho_{v,p_w}(x,y,z)$, no filter. (*a*) Three-dimensional view. The plotted surfaces correspond to $\rho_{v,p_w} = [-0.1, -0.05, -0.025, 0.025, 0.05, 0.1]$, from blue to yellow (see colorbar in part *b*). The translucent plane is $z = 0$. (*b*) $\rho_{v,p_w}(x,y,z=0)$. (*c*) $\rho_{v,p_w}(x,y^+=15,z)$. (*d*) $\rho_{v,p_w}(x,y=0.1h,z)$. The back lines in $(b-d)$ correspond to $\rho_{v,p_w} = -0.1$ (dashed) and 0.1 (solid). The colorbar in (*b*) is the same colorbar used in all panels.

dashed lines in figures 1*b*-*d* are $\rho_{v,p_w} = \pm 0.1$, which correspond also to the solid surfaces plotted in figure 1*a*. These regions ($|\rho_{v,p_w}| > 0.1$) will be referred to as the regions of strong correlation. Finally, note that the plane $z = 0$ is a symmetry plane for $\rho_{v,p_w}$, due to the statistical symmetry of the positive and negative $z$-directions in the turbulent channel flow.

Figure 1*b* shows that the correlation coefficient at $(x,y,z) = (0,0,0)$ is negative. This is the expected behaviour, since a negative (positive) wall-normal velocity should generate a positive (negative) pressure fluctuation at the wall. Figures 1*a* and *b* show that this negative correlation extends into a region that is tilted backwards, mostly upstream of the point where the pressure is measured. Although not apparent in the figure (the colorbar is saturated at $\pm 0.15$), the peak of negative correlation has a value of $(\rho_{v,p_w})_{\min} = -0.43$, occurring at the wall, slightly upstream of the point where $p_w$ is measured ($x^+_{\min} = -10, y^+_{\min} = 0.5$). On the other hand, figures 1*a* and *b* also shows a relatively strong positive correlation region, tilted forward and occurring mostly downstream of the point where $p_w$ is measured. This positive correlation is somewhat weaker than the negative correlation (i.e., the peak value is $(\rho_{v,p_w})_{\max} = 0.16$). Interestingly, the peak value is not attained at the wall, but within the near-wall region ($x^+_{\max} = 60, y^+_{\max} = 30$).

In terms of the spanwise structure of the correlation, figures 1*c* shows that the negatively

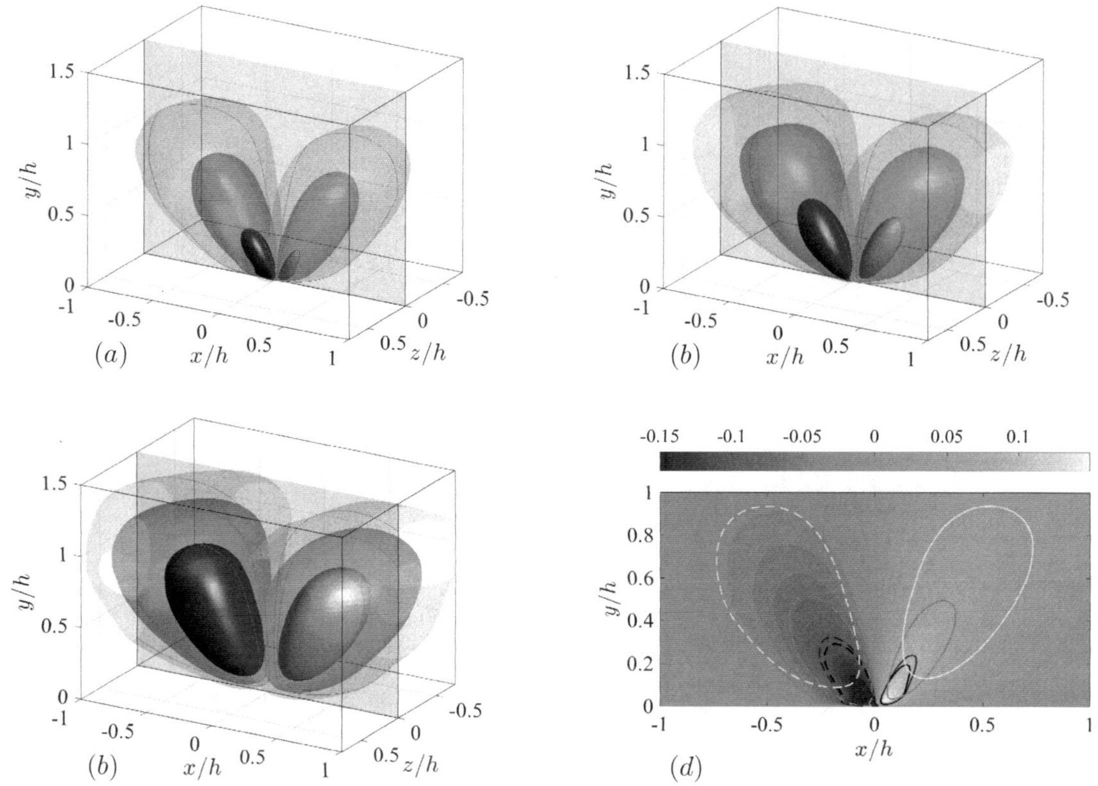


**Figure 2.** Evolution of the correlation coefficient using the space filtered pressure at the wall, $\rho_{v,\tilde{p}_w}(x,y,z)$, for different filter lengths, $l_c/h$. $(a-c)$ Three-dimensional view. The plotted surfaces correspond to $\rho_{v,p_w} = [-0.1, -0.05, -0.025, 0.025, 0.05, 0.1]$, from blue to yellow (see colorbar in part $d$). The translucent plane is $z = 0$. ($a$) $l_c = 0.025h$. ($b$) $l_c = 0.1h$. ($c$) $l_c = 0.4h$. ($d$) $\rho_{v,\tilde{p}_w}(x,y,z=0)$. The shaded contours correspond to the unfiltered cases ($l_c = 0$). The line contours correspond to $\rho_{v,\tilde{p}_w} = 0.1$ (solid) and $-0.1$ (dashed). The line colour indicate the length of the spatial filter: —, $l_c = 0$ (unfiltered); —, $l_c = 0.025h$; —, $l_c = 0.1h$; —, $l_c = 0.4h$;

correlated region (black dashed line) appears flanked by two (weaker) positively correlated regions, at spanwise separations $z \approx 0.05h \approx 50\nu/u_\tau$. This separation is consistent with the spanwise separation of the quasi-streamise vortices of the near wall region [25]. A similar structure, although weaker, is also visible around the positively correlated region downstream. Figure 1$d$ shows how this spanwise structure is lost further away from the wall.

The structure of the $\rho_{v,p_w}$ shown in figure 1 is consistent with the conditional velocity fields obtained by [22], both for positive and negative pressure peaks. Indeed, the differences in the position and intensity of the peaks corresponding to $(\rho_{v,p_w})_{\min}$ and $(\rho_{v,p_w})_{\max}$ are consistent with the results of these authors, who obtained slightly different velocity structures when performing conditional averages for positive and negative pressure fluctuations. It is also interesting to note that the pressure fluctuations at the wall do not seem to be aware of the side-by-side organization of the $v$-velocity structures observed in previous works in the logarithmic and outer region [5]. In the pressure velocity correlation, this organization is only apparent in the near-wall region.

Next, the effect of filtering the pressure fluctuations in space is evaluated, using the Gaussian filter defined in equation (4). Figure 2 shows the effect of such filter on $\rho_{v,\tilde{p}_w}$, for selected filter

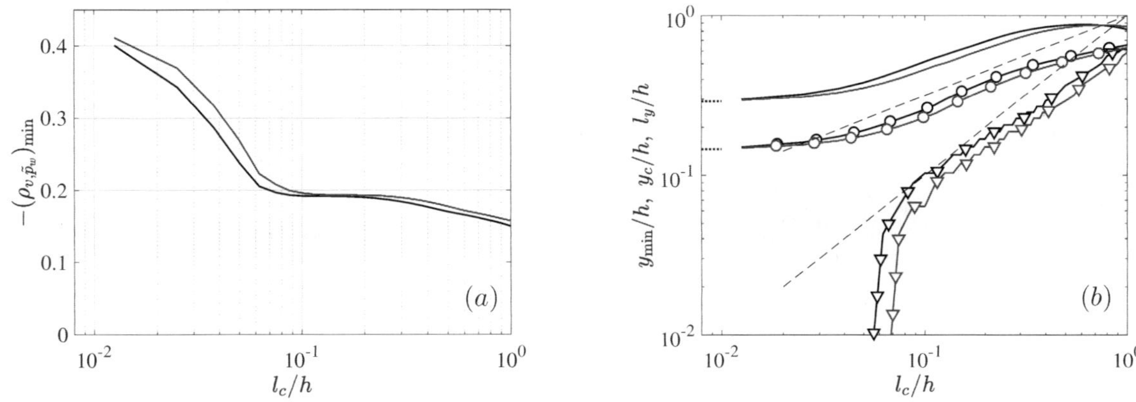


**Figure 3.** ($a$) Peak value of the negative correlation coeeficient ($-\rho_{v,p_w}$) for different filter lengths, $l_c/h$. ($b$) Characteristic lengths and wall-distances of the region of negative correlation, plotted versus the filter length, $l_c/h$. Lines and symbols are: ——, $l_y/h$; ○ , $y_c/h$; $\bigtriangledown$, $y_{\min}/h$. Colors are black for the Gaussian filter (equation 4), blue for the cylindrical top-hat filter (equation 5). The dashed lines have logarithmic slopes of 1 and 1/2.

lengths $l_c/h = 0.025, 0.1$ and $0.4$. As the filter length increases, the regions of strong correlation (namely, $|\rho_{v,\tilde{p}_w}| > 0.1$) become larger, extending further away from the wall. The inclination angle of these regions remains roughly the same, as it can be observed in the figure 2$d$. As a consequence, the streamwise separation also increases. Although not shown here, the results obtained filtering $p_w$ with the top-hat filter defined in equation (5) are qualitatively similar, suggesting that the spatially averaged pressure fluctuations can be used to identify/discriminate $v$-eddies of different sizes.

In order to quantify the changes in the size and position of the regions of intense negative correlation, the vertical size ($l_y$) and the wall-distance to the geometric centre ($y_c$) of the volume satisfying $-\rho_{v,\tilde{p}_w} > 0.1$ are computed for a number of filter lengths, in the range $l_c \in [0 - h]$. Figure 3 presents these values, together with the peak value of the negative correlation coefficient ($-(\rho_{v,\tilde{p}_w})_{\min}$) and the wall-distance where this value is attained ($y_{\min}$). The figure shows data from the two filters defined in the methodology section, equations (4) and (5), which agree remarkably well. This implies that the size of the $v$-eddies associated with different filters is roughly the same, irrespective of the details of the particular filter used. Of course, it might be necessary to adjust the cut-off length of the filters (in the present case, this is done by choosing the characteristic length of the Gaussian filter as 3 times the standard deviation).

It can be observed that for $l_c \lesssim 0.05h \approx 50\nu/u_\tau$, the filter mostly reduces the value of the peak correlation (figure 3$a$), with little changes in $y_{\min}, l_y$ or $y_c$ (figure 3$b$). In particular, $y_{\min} \approx 0$ for $l_c \lesssim 0.05h$ (i.e, the point where the peak value of negative correlation is attained is at the wall). For $0.05h \lesssim l_c \lesssim 0.3h$ the value of $(\rho_{v,\tilde{p}_w})_{\min}$ remains roughly constant, while $y_c, l_y$ and $y_{\min}$ grow with a roughly constant logarithmic slope. For larger values of the filter length ($l_c \gtrsim 0.3h$) the peak value of negative correlation starts to peel off, and for $l_c \approx h$ the growth of $y_c$ and $l_y$ seems to saturate. Similar behaviour is observed for the sizes of the positively correlated region and for the lengths in the streamwise and spanwise directions of the regions of strong (negative and positive) correlation.

It is interesting to note that in figure 3$b$ the logarithmic slope of the lengths associated the size of the region of strong correlation ($y_c \sim l_y \approx l_c^{1/2}$ ) is different than the slope of the wall-distance to the peak negative correlation ($y_{\min} \approx l_c$). At the present time, the reason for this difference is unclear. Also, the fact that $l_y$ remains roughly proportional to $y_c$ probably indicates that the

IOP Conf. Series: Journal of Physics: Conf. Series **1001** (2018) 012007 doi:10.1088/1742-6596/1001/1/012007

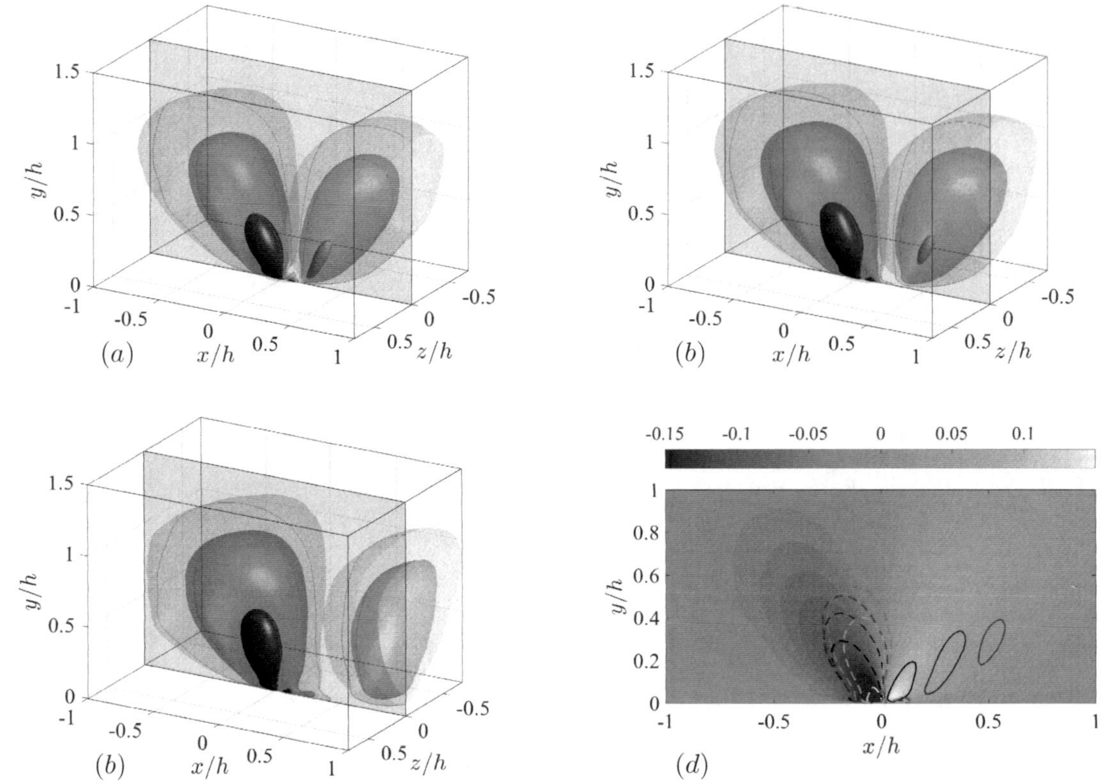


**Figure 4.** Evolution of the correlation coefficient using the time filtered pressure fluctuations at the wall, $\rho_{v,\overline{p}_w}(x, y, z)$, for different averaging times, $t_c/h$. $(a-c)$ Three-dimensional view. The plotted surfaces correspond to $\rho_{v,\overline{p}_w} = [-0.1, -0.05, -0.025, 0.025, 0.05, 0.1]$, from blue to yellow (see colorbar in part $d$). The translucent plane is $z = 0$. $(a)$ $t_c = 0.011h/u_\tau$. $(b)$ $t_c = 0.021h/u_\tau$. $(c)$ $t_c = 0.042h/u_\tau$. $(d)$ $\rho_{v,\overline{p}_w}(x, y, z = 0)$. The shaded contours correspond to the unfiltered cases ($t_c = 0$). The line contours correspond to $\rho_{v,\overline{p}_w} = 0.1$ (solid) and $-0.1$ (dashed). The line colour indicates the averaging time in (2): —, $t_c = 0$ (unfiltered); —, $t_c = 0.011h/u_\tau \approx 10\nu/u_\tau^2$; —, $t_c = 0.021h/u_\tau \approx 20\nu/u_\tau^2$; —, $t_c = 0.042h/u_\tau \approx 40\nu/u_\tau^2$; —, $t_c = 0.085h/u_\tau \approx 80\nu/u_\tau^2$.

correlation coefficient is identifying $v$-velocity eddies that are *attached* to the wall, in the sense of being able to *feel* the presence of the wall, and not necessarily to reach to the wall [3, 26, 5].

Finally, the effect of filtering in time the pressure fluctuations at the wall is evaluated in figure 4, using the moving average filter defined in equation (2). The three-dimensional structure of the time-filtered $\rho_{v,\overline{p}_w}$ is shown in figures 4$a$, $b$ and $c$, for $t_c \approx 0.01h/u_\tau, 0.02h/u_\tau$ and $0.04h/u_\tau$, respectively. In wall units, these times correspond to $t_c^+ \approx 10, 20$ and 40. It can be observed in figure 4 that the main effect of the time filter is to advect the correlation downstream, making the negatively correlated region more vertical. As shown in the representation of $\rho_{v,\overline{p}_w}$ at the $z = 0$ plane depicted in figure 4$d$, this displacement is more apparent for the region of positive correlation and occurs with a small change in the size of the strongly correlated regions. Although not shown here, the peak values of the positive and negative correlations decay monotonously with the averaging time. Moreover, averaging times $t_c \gtrsim 0.1h/u_\tau \approx 100\nu/u_\tau^2$ result in loss of the correlation.

This loss of correlation is probably not surprising. Experimental data at very high Reynolds numbers ($Re_\tau \sim 10^5 - 10^6$, see [19]) shows that the auto-correlation of $p_w$ is $\lesssim 1\%$ for $tu_\tau/h \gtrsim 0.025$. At Reynolds numbers comparable to the present $Re_\tau$, experiments [20] report that the duration of the pressure peaks at the wall is about $t^+ \gtrsim 10 - 15$, which are of the same order of magnitude as the $t_c^+$ for which loss of correlation is observed in figure 4. Note that, assuming that the Taylor hypothesis is applicable, an advection velocity of about $10 - 20u_\tau$ over a time $t \approx 0.1h/u_\tau$ results in a streamwise displacements of about $1 - 2h$.

Finally, it should be noted that the results of the time filtered $\rho_{v,\bar{p}_w}$ are probably not directly comparable to the space filtered ones: the time averaging defined in equation (2) is roughly equivalent to a spatial average in the streamwise direction only, while the spatial filters employed here are isotropic in $x$ and $z$.

## 4. Conclusions

The correlation between the wall-normal velocity and the pressure fluctuations at the wall has been characterized, using a DNS of a turbulent channel flow at moderate friction Reynolds number, $Re_\tau = 939$. The correlation coefficient shows a clear structure in the streamwise direction, with a negatively correlated region upstream of the point where the pressure is measured, and a positively correlated region downstream. In the spanwise direction, the correlation only changes sign in the near wall region. In particular, the correlation does not show any indication of spanwise pairs of $v$-eddies, as observed in other analysis [5].

Filtering in space the pressure fluctuations at the wall results in a growth of the regions of strong correlation, with sizes that are roughly proportional to $l_c^{1/2}$ (where $l_c$ is the characteristic length of the filter) for the two spatial filters considered here. This suggests that it should be possible to discriminate $v$-eddies of different times by filtering the pressure fluctuations at the wall. Further analysis at higher Reynolds numbers should be done to ascertain the limits of space filtering to detect $v$-structures of various sizes.

Finally, the effect of a simple moving time average of the pressure fluctuation at the wall has been analyzed. Contrary to the space filter, this time filter displaces the correlation in the streamwise direction, as a consequence of the advection of the eddies. The regions of strong negative correlation become more vertical, while the regions of strong positive correlation are lost. However, for the filter lengths and Reynolds numbers considered here, the time filter seems to have a smaller effect on the size of the regions of strong correlation, which suggests that these time filters are not as good candidates to discriminate $v$-eddies based on their pressure signature at the wall.

## Acknowledgments

This work has been supported by the European Research Council, under the COTURB grant ERC-2014.AdG-669505. The authors are grateful to M. Encinar, A. Ianiro and S. Discetti for their comments on a preliminar version of this manuscript.

## References

[1] C W Rowley and S T M Dawson. Model reduction for flow analysis and control. *Ann. Rev. Fluid Mech.*, 49:387–417, 2017.

[2] J Jiménez. Coherent structures in wall-bounded turbulence. *J. Fluid Mech.*, 842:P1–100, 2018.

[3] J C Del Álamo, J Jiménez, P Zandonade, and R D Moser. Self-similar vortex clusters in the turbulent logarithmic region. *J. Fluid Mech*, 561:329–358, 2006.

[4] R Adrian. Hairpin vortex organization in wall turbulence. *Phys. Fluids*, 19:041301, 2007.

[5] A Lozano-Durán, O Flores, and J Jiménez. The three-dimensional structure of momentum transfer in turbulent channels. *J. Fluid Mech.*, 694:100–130, 2012.

[6] G Kawahara, M Uhlmann, and L Van Veen. The significance of simple invariant solutions in turbulent flows. *Ann. Rev. Fluid Mech.*, 44:203–225, 2012.

[7] J Jiménez, G Kawahara, M P Simens, M Nagata, and M Shiba. Characterization of near-wall turbulence in terms of equilibrium and 'bursting' solutions. *Phys. Fluids*, 17:015105, 2005.

[8] J Jiménez. Cascades in wall-bounded turbulence. *Ann. Rev. Fluid Mech.*, 44:27–45, 2012.

[9] J C Del Álamo and J Jiménez. Linear energy amplification in turbulent channels. *J. Fluid Mech.*, 559:205–213, 2006.

[10] J Zhou, R J Adrian, S Balachandar, and T M Kendall. Mechanisms for generating coherent packets of hairpin vortices in channel flow. *J. Fluid Mech.*, 387:353–396, 1999.

[11] G Eitel-Amor, R Örlü, P Schlatter, and O Flores. Hairpin vortices in turbulent boundary layers. *Phys. Fluids*, 27:025108, 2015.

[12] O Flores and J Jiménez. Hierarchy of minimal flow units in the logarithmic layer. *Phys. Fluids*, 22:071704, 2010.

[13] A Lozano-Durán and J Jiménez. Time-resolved evolution of coherent structures in turbulent channels: characterization of eddies and cascades. *J. Fluid Mech.*, 759:432–471, 2014.

[14] N C Constantinou, A Lozano-Durán, M-A Nikolaidis, B F Farrell, P J Ioannou, and J Jiménez. Turbulence in the highly restricted dynamics of a closure at second order: comparison with DNS. *J. Phys: Conf. Ser.*, 506:012004, 2014.

[15] R L Panton and J H Linebarger. Wall pressure spectra calculations for the equilibrium boundary layer. *J. Fluid Mech.*, 65:261–287, 1974.

[16] T M Farabee and M J Casarella. Spectral features of wall pressure fluctuations beneath turbulent boundary layers. *Phys. Fluids A*, 3:2410–2420, 1991.

[17] H Abe, Y Matsuo, and H Kawamura. A DNS study of Reynolds-number dependence on pressure fluctuations in a turbulent channel flow. In *TSFP Digital Library Online*, 2005.

[18] J Jiménez and S Hoyas. Turbulent fluctuations above the buffer layer of wall-bounded flows. *J. Fluid Mech.*, 611:215–236, 2008.

[19] J C Klewicki, P J A Priyadarshana, and M M Metzger. Statistical structure of the fluctuating wall pressure and its in-plane gradients at high Reynolds number. *J. Fluid Mech.*, 609:195–220, 2008.

[20] S Ghaemi, D Ragni, and F Scarano. PIV-based pressure fluctuations in the turbulent boundary layer. *Exp. Fluids*, 53:1823–1840, 2012.

[21] B-K Ahn, W R Graham, and S A Rizzi. A structure-based model for turbulent-boundary-layer wall pressures. *J. Fluid Mech.*, 650:443–478, 2010.

[22] S Ghaemi and F Scarano. Turbulent structure of high-amplitude pressure peaks within the turbulent boundary layer. *J. Fluid Mech.*, 735:381–426, 2013.

[23] M Luhar, A S Sharma, and B J McKeon. On the structure and origin of pressure fluctuations in wall turbulence: predictions based on the resolvent analysis. *J. Fluid Mech.*, 751:38–70, 2014.

[24] J C del Álamo and J Jiménez. Spectra of the very large anisotropic scales in turbulent channels. *Phys. Fluids*, 15:L41–L44, 2003.

[25] J Jeong, F Hussain, W Schoppa, and J Kim. Coherent structures near the wall in a turbulent channel flow. *J. Fluid Mech.*, 332:185–214, 1997.

[26] O Flores, J Jiménez, and J C del Álamo. Vorticity organization in the outer layer of turbulent channels with disturbed walls. *J. Fluid Mech.*, 591:145–154, 2007.

IOP Conf. Series: Journal of Physics: Conf. Series **1001** (2018) 012008 doi:10.1088/1742-6596/1001/1/012008

# Characterization of turbulent coherent structures in square duct flow

**Marco Atzori**[1], **Ricardo Vinuesa**[1], **Adrián Lozano-Durán**[2] **and Philipp Schlatter**[1]

[1] Linné FLOW Centre, KTH Mechanics, SE-100 44 Stockholm, Sweden and Swedish e-Science Research Centre (SeRC), Stockholm, Sweden
[2] Center for Turbulence Research, Stanford, CA, USA

E-mail: atzori@mech.kth.se

**Abstract.** This work is aimed at a first characterization of coherent structures in turbulent square duct flows. Coherent structures are defined as connected components in the domain identified as places where a quantity of interest (such as Reynolds stress or vorticity) is larger than a prescribed non-uniform threshold. Firstly, we qualitatively discuss how a percolation analysis can be used to assess the effectiveness of the threshold function, and how it can be affected by statistical uncertainty. Secondly, various physical quantities that are expected to play an important role in the dynamics of the secondary flow of Prandtl's second kind are studied. Furthermore, a characterization of intense Reynolds-stress events in square duct flow, together with a comparison of their shape for analogous events in channel flow at the same Reynolds number, is presented.

## 1. Introduction

The interest in the description of duct flows originates both from practical applications (ducts are widely used in cooling systems, for instance) and from the presence of the secondary flow of Prandtl's second kind. This phenomenon, despite representing only around 2 to 3% of the bulk velocity [1], has an important impact in the wall-shear stress distribution along the perimeter of the duct. In duct flows, the simulations by Gavrilakis [2] revealed the emergence of eight counter-rotating vortices which convect momentum from the core of the duct to its bisector. More recently, Pinelli *et al.* [3] reported that the secondary flow has contributions both from the large-scale motions and from the near-wall region of turbulence.

In the present study the considered data sets are part of the direct numerical simulation (DNS) data of turbulent duct flow by Vinuesa *et al.* [4, 5] and the DNS of turbulent channel flow employed by del Álamo *et al.* [6]. We focus on square duct and channel both at a friction Reynolds number $Re_\tau$ of 180. The friction Reynolds number $Re_\tau$ is computed based on the half-height and the friction velocity $u_\tau = \sqrt{\tau_w/\rho}$ ($\tau_w$ is the wall-shear stress and $\rho$ the density of the fluid). For the duct, we will define $Re_\tau$ in terms of the centerplane friction velocity. We consider a frame of reference with the $x$, $y$ and $z$ axes aligned with the streamwise, wall-normal and spanwise directions of the channel, respectively. An analogous frame of references is adopted for the duct, but it has to be noted that in this case the $y$ and $z$ directions are equivalent given the symmetry of the problem. The instantaneous velocity components are denoted as $\tilde{u}$, $\tilde{v}$ and $\tilde{w}$, while the mean and the fluctuating components are respectively $U$, $V$ and $W$ and

$u$, $v$ and $w$. Root-mean-squares are indicated explicitly. The aim of the present study is to

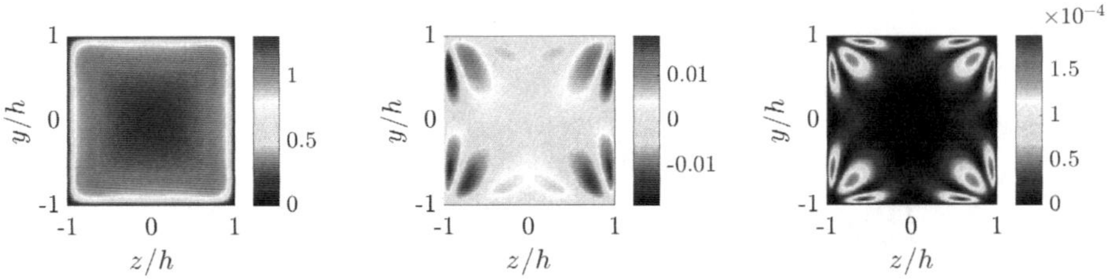


**Figure 1.** From left to right: average of the streamwise velocity component $U$, of the vertical component $V$ and of kinetic energy of the secondary flow $K = 1/2(V^2 + W^2)$ for square duct flow at $Re_\tau = 180$ at the centerplane.

extend our knowledge regarding the phenomenology of duct flows by describing the behavior of the coherent turbulent structures present in the flow. In our analysis we will follow the same procedure applied by del Álamo *et al.* [6] and Lozano-Durán *et al.* [7] for channel flows.
To identify three-dimensional structures implies finding connected components of the computational domain that fulfill a certain condition for a quantity of interest. *Vortex clusters*, first analyzed in turbulent channel flows by del Álamo *et al.* [7], denote regions of the flow where strain is small compared with the vorticity and the local pressure is low. In this case the physical quantity of interest can be, for instance, the second invariant of $\nabla \mathbf{u}$, $Q$, as proposed by Hunt *et al.* [8]

$$Q = \frac{1}{2}\left(\frac{\partial^2 u_i}{\partial x_i^2} - \frac{\partial u_i}{\partial x_j}\frac{\partial u_j}{\partial x_i}\right). \tag{1}$$

Other options are the $\lambda_2$ criterion (Jeong and Hussain [9]) and the $\Delta$ criterion (Chong *et al.* [10]). Chakraborty *et al.* [11] showed that all these methods are approximately equivalent, which strongly suggestes that the choice of a particular criterion does not affect the results. Despite the fact that for these criteria a threshold of 0 is the most natural choice, it tends to include a very large fraction of the overall volume of the domain. Therefore, a scalar value, initially prescribed *ad hoc*, is preferred (Chong *et al.* [12]). However, as firstly suggested by Nagaosa and Handler [13], in non-homogeneous flows the threshold should be such that the varying magnitude of the fluctuations in different regions of the flow is taken into account. To this end, they proposed for the identification of vortex clusters to scale $Q$ with its local root-mean-square value. Thus, the condition for belonging to a connected component becomes:

$$Q = \beta\, Q_{rms}. \tag{2}$$

The choice of $\beta$ remained relatively subjective, until Moisy and Jiménez [14] introduced the percolation analysis to obtain a quantitative measurement of the effects of different thresholds. Subsequently, del Álamo *et al.* [6] and Lozano-Durán *et al.* [7] adopted the same procedure. In the latter study the authors also considered regions with high Reynolds-stress events, by performing a three-dimensional extension of the *quadrant analysis*. The quadrant analysis was originally proposed by Wallace *et al.* [15] and employed by Willmarth and Lu [16] as an experimental technique for the detection of burst and sweep events, which are identified via a simultaneous measurement of $u$ and $v$, and Lu and Willmarth [17] introduced a threshold based on the root-mean-square of these components to identify more intense events, leading to the following condition

$$|uv| > H_{uv}\, u_{rms} v_{rms}. \tag{3}$$

Here $H$ is the so-called *hyperbolic hole*, which has the graphical interpretation of the region excluded by the hyperbola $|uv| = H$ in the Cartesian plane $u$,$v$, where different quadrants correspond to various events: $u > 0$ and $v > 0$ are outward interactions ($Q1$), $u < 0$ and $v > 0$ ejections ($Q2$), $u < 0$ and $v < 0$ inward interactions ($Q3$) and $u > 0$ and $v < 0$ are sweeps ($Q4$). The similarities between equations (2) and (3) inspires the adoption of the latter equation to detect three-dimensional structures which correspond to the four quadrants, as indeed done by Lozano-Durán *et al.* [7], who also employed the percolation analysis for the choice of $H$ (as done for $\beta$ in del Álamo *et al.* [6]). One of the main findings from these studies is the fact that the predominant structure in the logarithmic layer of wall-bounded flows is a $Q2 - Q4$ pair, which is associated with a vortex cluster [7]. Our main purpose is to extend this analysis to other kind of structures. Moreover, we clarify some details regarding the detection procedure, with the aim of extending its application and robustness for other flow configurations.

## 2. Structure identification method

### *2.1. Role of the percolation analysis*

The role of the percolation analysis in the structure identification method deserves specific attention. Here we consider a simple test case to clarify how it can be employed in order to estimate an adequate threshold for a data-set containing several snapshots of non-homogeneous turbulent flows. Firstly, let us consider a square two-dimensional lattice, where the probability of a site of the lattice being occupied or not is defined by $p$. A certain definition of connectivity among occupied sites is also provided, for instance, between nearest neighbors. If one considers increasing values of $p$ starting from 0, it will be possible to observe the existence of a critical $p_c$, above which most of the existing small connected components merge into less numerous but larger ones. This phenomenon is called *percolation crisis*. Let us introduce a scalar field $f(x, y)$ defined on a certain grid (we maintain a two-dimensional frame for the sake of simplicity), and the grid points satisfying the following condition are identified

$$f(x, y) > \alpha. \tag{4}$$

In this example, grid points are equivalent to sites of the lattice and $p$ represents the probability for each point to fulfill condition (4), thus the percolation is determined by the properties of $f$. Here we focus on three particular cases: a) $f$ is a stochastic variable uniformly distributed between 0 and 1; b) $f$ is a stochastic variable with a Gaussian distribution of zero mean and a standard deviation of 1/2; c) $f$ is defined as a uniformly distributed stochastic variable denoted as $U(a, b)$, where $a$ and $b$ are the minimum and maximum values, added to a linear function (*e.g.*: the coordinate $y$), so that $0 < f < 1$ still holds. An example could be

$$f = U(0, 1/2) + \frac{y}{2(y_{\max} - y_{\min})}. \tag{5}$$

If $f$ is homogeneously distributed, the merging process takes place simultaneously in the whole domain and it is possible to identify a short interval of values of $\alpha$ where it occurs. On the other hand, if $f$ is not homogeneous, the percolation exhibits a different behavior because the merging process cannot occur at the same time in the whole domain. Instead, it proceeds *expanding* the connected components from the region where $f$ is lower to the regions there $f$ is higher. The process is more gradual and it is not possible to clearly identify the interval of values of $\alpha$ where most of the merging processes takes place. In order to clarify these concepts, let us define $V^*(\alpha)$ and $N^*(\alpha)$ as the ratio of the volume of the largest structure divided by the total volume occupied by all the structures, and the total number of objects normalized with the largest number of identified objects over all the values of $\alpha$, respectively. In addition, we also introduce $\alpha_c$, which is the value where the derivative $\mathrm{d}V^*/\mathrm{d}\alpha$ is minimum (this would be the value of $\alpha$ for which the merging process is most intense), and $\alpha_{max}$, which is value that produces the highest number of objects. Figure 2 shows the percolation analysis for the three different test cases introduced above. Only for cases a) and b), represented by solid and dashed lines respectively, it is possible to identify unambiguously $\alpha_c$ and $\alpha_{max}$, while it is not possible for case c), represented by dotted lines. In Figure 2 it is also possible to observe the difference between cases a) and b). In the first case, since $f \in [0, 1]$, for $\alpha = 1$ no grid points fulfill condition (4), therefore $N(1) \simeq 0$. By contrast, in case c) it is possible to have points with $f > 1$ which are always above $\alpha$, thus $N(1) > 0$. These examples allow us to identify the first role of the

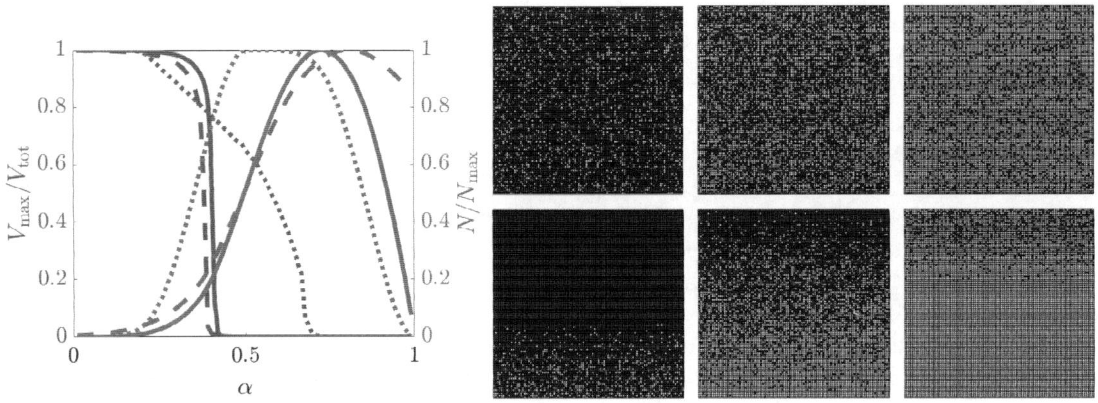


**Figure 2.** Percolation analysis for a square two-dimensional domain. (Left) The $V^*(\alpha)$ and $N^*(\alpha)$ curves are represented in blue and red, respectively. Solid lines correspond to the uniformly distributed case, dashed lines to the case with a Gaussian distribution and dotted lines denote the case with the addition of a linear function as in equation (5). (Right) Connected components (in yellow) identified in the grid corresponding to (top) solid and (bottom) dotted lines on left panel. The values of $\alpha$ are, from left to right, 0.8, 0.5 and 0.3.

percolation analysis in detecting coherent structures. Conditions such as equations (2) or (3) are equivalent to the general form $f > \alpha$ if $Q$ and $|uv|$ are scaled with factors obtained from the flow statistics of the threshold function, while $\beta$ and $H$ play the role of $\alpha$. A sharp percolation crisis obtained when gradually modifying the value of $\beta$ (or $H$) indicates that the threshold function is able to capture the non-homogeneity of the flow properly.

It is also important to take into account the influence of the size of the domain on the percolation analysis, which is related to the uncertainty of the statistics. Regarding flow statistics, if boundary effects are negligible, to employ a larger computational domain is equivalent to consider a larger data set. This is not obvious for the percolation analysis, since in a finite domain both the largest possible number of observed structures and the size of the largest structure are bounded and related to the size of the domain itself. To qualitative estimate these effects, we limit our analysis to the case of a uniformly distributed random field without superimposed bias (case a) introduced above) and we consider 50 snapshots for three different sizes of the square grid ($25 \times 25$, $100 \times 100$ and $500 \times 500$). For each of these data-sets, the percolation analysis is performed via averaging each value of $V^*$ and $N^*$ for the prescribed $\alpha$. As shown in Figure 3, smaller domains lead to a wider range of values of $\alpha$ over which the percolation crisis takes place, a fact that increases the uncertainty in the estimation of $\alpha_{max}$ and $\alpha_c$. However, the values of $\alpha_c$ and $\alpha_{max}$ resulting from the averages are consistent for all the considered sizes. This result is of great interest since the typical data-set considered in this analysis consists of a collection of instantaneous flow fields which have at least one direction of periodicity. This result therefore

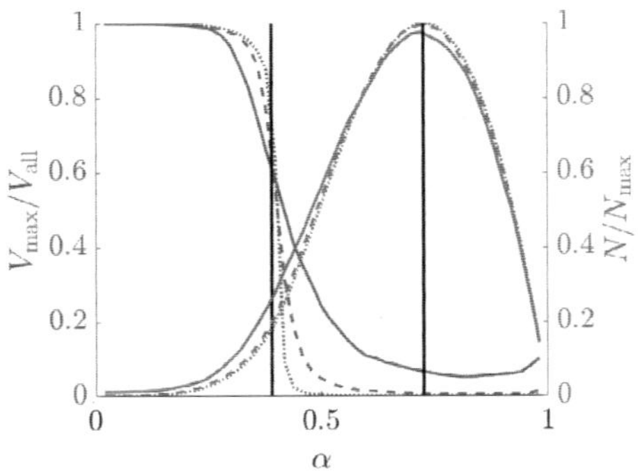


**Figure 3.** Percolation analysis averaged over 50 snapshots of square two-dimensional domains of different sizes. The $V^*(\alpha)$ and $N^*(\alpha)$ curves are represented in blue and red, respectively, and the solid, dashed and dotted lines represent the cases with $25 \times 25$, $100 \times 100$ and $500 \times 500$ grid points. The two vertical black lines represent $\alpha_c$ and $\alpha_{max}$. Note that for the averaged curves, $\alpha_c$ and $\alpha_{max}$ are identical in the three domain sizes.

justifies the employment of the percolation analysis to identify the optimal value of $\beta$ (or $H$) that maximizes the number of detected objects for a certain case, which is the second role of the percolation analysis in this context.
Based on the results of the simple test cases presented above, it is possible to summarize how the percolation analysis is employed in the present work. Firstly, it is employed to evaluate the effectiveness of the threshold functions in identifying simultaneously coherent structures in the whole domain. And secondly, it is used to estimate the value of $\beta$ (or $H$) which maximizes the number of detected objects.

### *2.2. Flow structures under consideration*

The first coherent structures under study are the intense Reynolds stresses, $uv$, which have been extensively studied by Lozano-Durán *et al.* [7] and Lozano-Durán and Jiménez [18] in turbulent channel flows. As mentioned above, their threshold function was the same as that employed by Lu and Willmarth [17], resulting in the condition given by equation (3). Inspection of the structures obtained using this condition without additional corrections revealed that the near-wall region remained occupied by large structures for higher $H$ than the rest of the domain. Considering the relatively low Reynolds number under consideration, we attribute this observation to the large outer-scaled wall-normal extent of the viscous sublayer, which results in a relatively large region where $u_{rms}v_{rms}$ is too low to work effectively as a threshold. In this case, we propose as a solution to modify the threshold function via substituting the threshold values below $y^+ = 5$ with the one at $y^+ = 5$. The effect of this correction on the percolation curve is shown in Figure 4 and we conclude that it improves the percolation behavior in the context of the present work.
Because of the geometry of the duct, $uv$ events are equivalent to $uv$ events in turbulent channels only with respect to the top and bottom walls, while they correspond to cross-flow events on the left and right walls. Moreover, in the square duct the $uw$ events on the side walls are analogous to the $uv$ on the horizontal walls due to the symmetry, and in the present work we identify them by applying the condition

$$|uw| > H_{uw}\, u_{rms} w_{rms}. \tag{6}$$

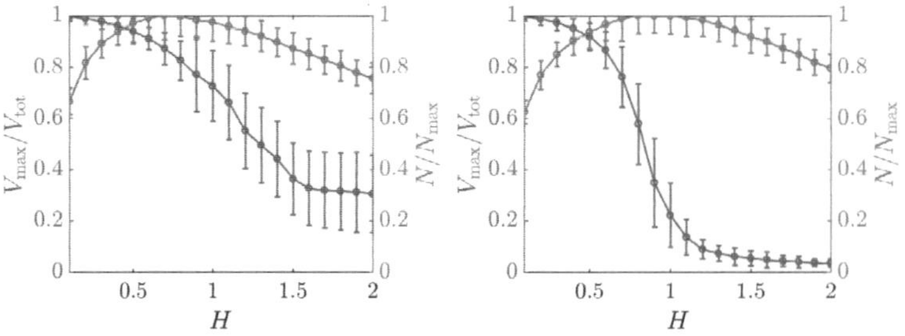


**Figure 4.** Percolation analysis for $uv$. Curves obtained (left) without and (right) with a correction for grid points below $z^+, y^+ = 5$.

Note that the percolation analysis is also analogous to the one for $uv$ structures, and therefore it is not shown here. The present analysis is extended to other structures by considering three different quantities that are related to the secondary flow of Prandtl's second kind. The first natural candidate is the Reynolds-shear stress term $vw$. As in the analysis performed for the $uv$ and $uw$ structures, we adopted the condition

$$|vw| > H_{vw}\, v_{rms} w_{rms} \tag{7}$$

for the $vw$ structures. The resulting percolation analysis is shown in Figure 5 (left), where essentially the same trend as that observed for the $uv$ structures in Figure 4 is obtained. An additional coherent structure that we take into consideration corresponds to intense regions of the normal Reynolds-stress difference $|v^2 - w^2|$. Note that this type of coherent structure is analyzed due to its role in the streamwise vorticity transport equation, see for instance Ref. [18]. The choice of the threshold in this case is less obvious. Since the values of $v^2_{rms}$ and $w^2_{rms}$ are almost equal in the core of the duct and along the diagonals, differences such as $|v^2_{rms} - w^2_{rms}|$ cannot be used as a threshold because they are close to 0 in a large portion of the domain. Considering that the difference $|v^2 - w^2|$ exhibits the highest values close the walls, where $v^2$ is larger than $w^2$ (in the case of vertical walls) – or vice-versa (when considering horizontal walls), we propose the sum $v^2_{rms} + w^2_{rms}$, which is of the order of the largest quantities near the walls and of each individual term elsewhere. The percolation condition is therefore

$$|v^2 - w^2| > H_{v^2-w^2}\,(v^2_{rms} + w^2_{rms}), \tag{8}$$

which provides similar percolation trends as the ones observed in the $uv$ and $vw$ structures, as shown in Figure 5 (center). We also consider an additional type of structure, defined by connected regions of high $K$, the instantaneous kinetic energy of the secondary flow defined by Vinuesa *et al.* [19]. Despite the fact that $K$ is calculated in terms of to the instantaneous velocity components $\tilde{v}$ and $\tilde{w}$ instead of their fluctuations, the following condition constitutes an effective threshold, as shown in Figure 5 (right)

$$K = \frac{1}{2}\left(\tilde{v}^2 + \tilde{w}^2\right) > H_K\,\frac{1}{2}(v^2_{rms} + w^2_{rms}). \tag{9}$$

Before proceeding with the analysis of the turbulent structures, it is important to note that the choice of an adequate threshold for a generic field requires further investigation. Let us consider, as an example, two alternative conditions for the detection of $uv$ events

$$|uv| > H'_{uv}\frac{1}{2}\left(u^2_{rms} + v^2_{rms}\right), \qquad |uv| > H''_{uv}\frac{1}{2}\left(u^2_{rms} + v^2_{rms} + w^2_{rms}\right). \tag{10}$$

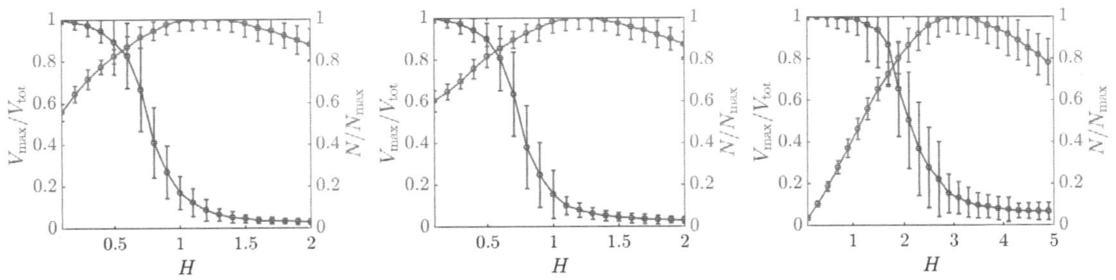


**Figure 5.** Percolation analysis for (left) $vw$, for (center) $|v^2 - w^2|$ and (right) $K = 1/2(\tilde{v}^2 + \tilde{w}^2)$.

As shown in Figure 6, in both the cases the percolation analysis gives a very similar outcome to that obtained with $|uv| > H_{uv}\, u_{rms} v_{rms}$. This shows, on the one hand, that the percolation crisis should be considered as a necessary but not sufficient condition for the acceptance of a certain threshold function, and on the other hand that a more rigorous procedure might be needed to select the most adequate threshold for a certain quantity. A possible solution, which would avoid some degree of arbitrariness, is the adoption of the root-mean-square of the considered quantities.
Nevertheless, in the present work we employed the same thresholds as those already adopted in the literature and, for the additional types of structures, we have shown that the thresholds proposed above yield adequate percolation curves.

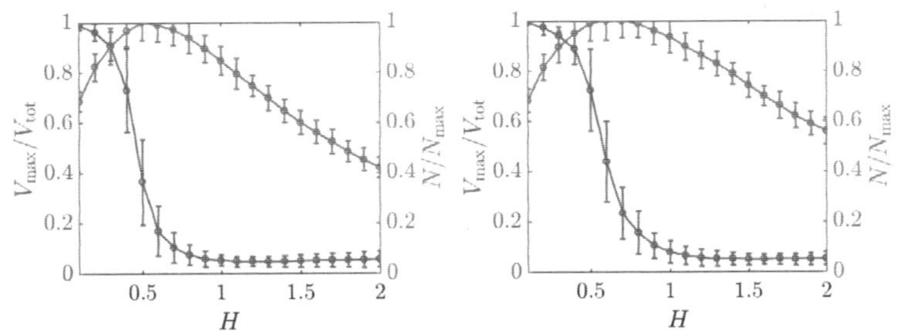


**Figure 6.** Percolation analysis for $uv$ structures obtained with two different threshold functions: (left) $|uv| > H''_{uv}/2(u^2_{rms} + v^2_{rms})$ and (right) $|uv| > H''_{uv}/2(u^2_{rms} + v^2_{rms} + w^2_{rms})$.

## 3. Results and discussion

### *3.1. Structure classification*

In their numerical studies of turbulent channel flows, del Álamo *et al.* [6] and Lozano-Durán *et al.* [7] reported the existence of different families of structures with respect with their shape and distances from the wall. More precisely, considering the joint probability density function of their minimum and maximum distances from the wall, it is possible to identify: 1) small objects located far from the walls, denoted in the following as detached (D); 2) small objects near the wall, denoted as wall-attached (WA); 3) large objects attached to the wall which extend beyond the centerline of the channel, referred to as tall-wall-attached (TWA). In both studies a "near-wall" region, defined below $y^+ = 20$, was used to classify the various structures.
In duct flows, because of the geometry, it is possible to identify at least seven analogous types

of structures: 1) detached (D), far from all the walls; 2) wall-attached (WA), close to at least one of the horizontal walls; 3) tall-wall-attached (TWA), elongated in the vertical direction and attached to the horizontal walls; 4) side-attached (SA), close to one of the vertical walls; 5) tall-side-attached (TSA), elongated in the horizontal direction and attached to at least one of the vertical wall; 6) corner-attached (CA), which are simultaneously close to both a horizontal and a vertical wall; 7) tall-corner-attached (TCA), which are large structures attached simultaneously to at least a horizontal and a vertical wall. In the present work, four near-wall regions are defined as $z^+, y^+ < 20$, evaluated based on the local value of $Re_\tau$, and the classification is based on the wall-normal distance to the closest wall of the structures, and on the location of their center of mass. The center of mass is defined as $\mathbf{r}_{\rm cm} = \sum \Delta V_i \mathbf{r}_i$, where $\mathbf{r}_i$ is the position of the $i$th grid point belonging to the structure, and $\Delta V_i$ is the volume of the hexaedron the faces of which are orthogonal to the edges departing from the $i$th node and passing through their midpoints. Tall objects considered close to the walls (WA, SA and CA) have the coordinates of the center of mass $z_{\rm cm}$, $y_{\rm cm}$ located inside the near-wall regions. Objects consider attached to the walls (TWA, TSA and TCA) have the coordinates of the center of mass located outside the near-wall regions, but at least one point in the near-wall regions. Detached objects have no points in the near-wall regions. The criteria are summarized in Table 1 and illustrated in Figure 7 (left). Since the threshold is determined regardless of the classification into different families,

**Table 1.** Conditions for the classification of structures into different families, $y^+_{\rm nw} = z^+_{\rm nw} = 20$.

| | | | | | | | |
|---|---|---|---|---|---|---|---|
| D | $y^+_{\rm min} > y^+_{\rm nw}$ | & | $y^+_{\rm cm} > y^+_{\rm nw}$ | & | $z^+_{\rm min} > z^+_{\rm nw}$ | & | $z^+_{\rm cm} > z^+_{\rm nw}$ |
| WA | $y^+_{\rm min} < y^+_{\rm nw}$ | & | $y^+_{\rm cm} < y^+_{\rm nw}$ | & | $z^+_{\rm min} > z^+_{\rm nw}$ | & | $z^+_{\rm cm} > z^+_{\rm nw}$ |
| TWA | $y^+_{\rm min} < y^+_{\rm nw}$ | & | $y^+_{\rm cm} > y^+_{\rm nw}$ | & | $z^+_{\rm min} > z^+_{\rm nw}$ | & | $z^+_{\rm cm} > z^+_{\rm nw}$ |
| SA | $y^+_{\rm min} > y^+_{\rm nw}$ | & | $y^+_{\rm cm} > y^+_{\rm nw}$ | & | $z^+_{\rm min} < z^+_{\rm nw}$ | & | $z^+_{\rm cm} < z^+_{\rm nw}$ |
| TSA | $y^+_{\rm min} > y^+_{\rm nw}$ | & | $y^+_{\rm cm} > y^+_{\rm nw}$ | & | $z^+_{\rm min} < z^+_{\rm nw}$ | & | $z^+_{\rm cm} > z^+_{\rm nw}$ |
| CA | $y^+_{\rm min} < y^+_{\rm nw}$ | & | $y^+_{\rm cm} < y^+_{\rm nw}$ | & | $z^+_{\rm min} < z^+_{\rm nw}$ | & | $z^+_{\rm cm} < z^+_{\rm nw}$ |
| TCA | $y^+_{\rm min} < y^+_{\rm nw}$ | & | $y^+_{\rm cm} > y^+_{\rm nw}$ | & | $z^+_{\rm min} < z^+_{\rm nw}$ | & | $z^+_{\rm cm} > z^+_{\rm nw}$ |

it is interesting to assess the impact of the threshold on such classification. Figure 7 (Right) shows a slice of the $uv$ field together with the connected components identified for three different values of $H$. The effect of a variation of $H$ is quite complex: structures may disappear (1); they may remarkably change their volume without splitting (2); they may split into new structures, which may or may not belong to the same family (3); they may change family (4) and they may remain almost unchanged (5). Figure 8 shows the percentage of volume for the different families as a function of $H$ for $uv$, $uw$ and $vw$ structures. It can be observed how the horizontal walls play the same role for $uv$ events as that of the vertical walls for $uw$ events, and vice-versa. Similarly, all the walls are equivalent for $vw$, $v^2 - w^2$ and $K$ structures (note that only the first ones are shown), which is consistent with the symmetry of the case. For $H \simeq 1$, which is the value that maximizes the number of objects, TCA structures occupy most of the volume. Note that the distinction between horizontal and vertical walls for the square duct may appear unnecessary, but it has been maintained because objects such as $uv$ and $uw$ have a different role depending on which the nearest wall is: SA and TSA $uv$ events are cross-flow events with respect to the vertical walls, and vice-versa for $uw$. Moreover, this subdivision allows to assess whether the sample size is large enough to reflect the statistical symmetry between $uv$ and $uw$ events, and it will be certainly needed in future comparisons with turbulent rectangular duct cases (Ref. [4]).
A comprehensive overview of how the various types of structures are subdivided into different

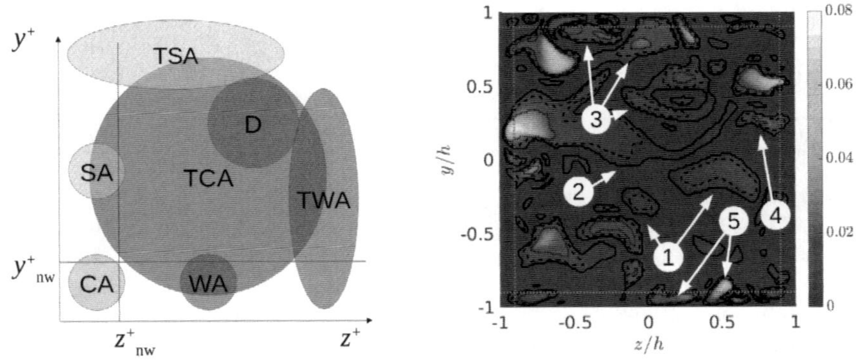


**Figure 7.** (Left) Sketch of the different families defined in Table 1 and (right) slice of a $|uv|$ field at a certain stramwise position in one of snapshots of the dataset. Solid, dashed and dotted lines represent connected components for $H = 0.5$, $H = 1$ and $H = 2$, respectively. The numbers from 1 to 5 represent the different phenomena observed when changing $H$ described in the text. The dotted red lines are $y^+_{\rm nw}$ and $z^+_{\rm nw}$ defined at the centreplane.

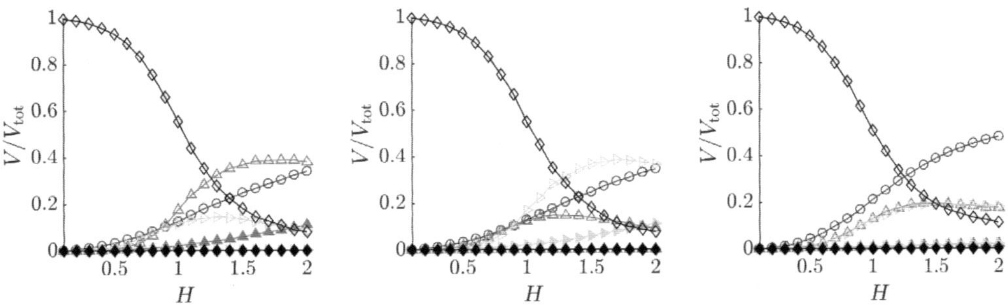


**Figure 8.** Volume occupied by each family over the total occupied volume as function of the thershold $H$, from (left) $uv$, (center) $uw$ and (right) $vw$ structures. The different families are denoted as follows: ○ detached (D), ▲ wall-attached (WA), △ tall wall-attached (TWA), ▶ side-attached (SA), ▷ tall side-attached (TSA), ♦ corner attached (CA) and ◇ tall-corner attached (TCA).

families is provided in Table 2. The corresponding fractions for $|uv|$ events in channel flow at $Re_\tau = 180$ are also reported for comparison. In all the cases, the volume of a specific structures is defined as the volume of the bounding box. In turbulent channel flows, as already reported in Ref. [7, 18] at $Re_\tau$ values from 930 to 4200, D events are the most numerous but they occupy a relatively small volume, whereas the majority of the volume is occupied by TWA events. Interestingly, in the duct most of the volume is occupied by TCA for all the considered types of structures, although the proportions among families are different depending on the structure under consideration. In particular, the fraction of the total volume occupied by the TCA is around 65% for $uv$ and $uw$ structures but below 50% for both $v^2 - w^2$ and $K$. The uncertainty of this result can be estimated by considering the discrepancy between values which are supposed to be identical due to the symmetry, and based on the results in Table 2 this uncertainty is of the order of 1%. The discrepancies among types of structures suggest the existence of qualitative differences in the shape and/or in the spatial organization of the various considered events,

**Table 2.** (Left) Fraction of the number of objects and (right) of the total volume of the objects. In both tables, $|uv|_{ch}$ are from channel flow.

| | $\|uv\|_{ch}$ | $\|uv\|$ | $\|uw\|$ | $\|vw\|$ | $\|v^2-w^2\|$ | $K$ | $\|uv\|_{ch}$ | $\|uv\|$ | $\|uw\|$ | $\|vw\|$ | $\|v^2-w^2\|$ | $K$ |
|---|---|---|---|---|---|---|---|---|---|---|---|---|
| D | 0.47 | 0.53 | 0.52 | 0.53 | 0.57 | 0.48 | 0.12 | 0.06 | 0.06 | 0.12 | 0.17 | 0.16 |
| WA | 0.23 | 0.11 | 0.05 | 0.10 | 0.08 | 0.12 | 0.01 | 0.01 | $\simeq 0$ | 0.01 | 0.01 | 0.01 |
| SA | – | 0.05 | 0.12 | 0.10 | 0.08 | 0.13 | – | $\simeq 0$ | 0.01 | 0.01 | 0.01 | 0.01 |
| TWA | 0.29 | 0.10 | 0.11 | 0.10 | 0.09 | 0.09 | 0.87 | 0.16 | 0.11 | 0.14 | 0.18 | 0.17 |
| TSA | – | 0.11 | 0.10 | 0.10 | 0.09 | 0.09 | – | 0.11 | 0.15 | 0.14 | 0.18 | 0.17 |
| CA | – | 0.03 | 0.03 | 0.02 | 0.02 | 0.02 | – | $\simeq 0$ | $\simeq 0$ | $\simeq 0$ | $\simeq 0$ | $\simeq 0$ |
| TCA | – | 0.07 | 0.07 | 0.05 | 0.06 | 0.07 | – | 0.65 | 0.66 | 0.58 | 0.44 | 0.47 |

which might be related with the different role that they have in the dynamics of the flow. A comprehensive assessment of the various types of structures will be conducted in future work and in the following we focus on the $uv$ events.

### *3.2. Characterization of uv structures*

As already mentioned, the $v$ and $w$ velocity components are equivalent (with respect to the corresponding vertical or horizontal wall) in the square duct due to symmetry, and this is also the case with the $uv$ and $uw$ structures. Thus, it is possible to employ a coordinate transformation to account for both types of structures jointly when computing statistics. Note that $uv$ structures in the duct correspond to $uv$ structures in the channel only if they are located close to the horizontal walls (WA and TWA families), while they correspond to $uw$ structures in the channel if they are close to the vertical walls (SA and TSA families). The correspondence is less obvious in the case of structures which are near the corners of the duct (CA), attached to a horizontal and a vertical wall (TCA) or far from them (D). It can be difficult to compare the size and shape of the structures in ducts and channels, even for equivalent families, because of the different geometry. In del Álamo *et al.* [6] and Lozano-Durán *et al.* [7], the structures are scaled in inner units, based on the wall-averaged friction velocity. In duct flows the presence of two orthogonal walls needs to be taken into account, and we proceed as follows: 1) the sizes of each structure is scaled with $\widetilde{Re}_\tau(y,z)$, which is defined as an average of the $Re_\tau$ values based on the closest horizontal and vertical walls weighted with the corresponding distances to those walls; 2) the structures are grouped into a total of three different *regions*, based on the distance of their respective centers of mass $\mathbf{r}_{\mathrm{cm}}$ to the closest vertical wall, as shown in Figure 9 (left). Region (a) extends from the vertical walls up to the spanwise coordinate where $Re_\tau$ is almost constant (*i.e.* $1 > |z/h| > 0.7$), regions (b) and (c) are located closer to the duct vertical centerplane (at $0.7 > |z/h| > 0.4$ and $0.4 > |z/h| > 0$, respectively). This subdivision has the rationale of distinguishing $uv$ events in the core of the duct, where structures exhibit closer similarity with the channel, from $uv$ events near the vertical walls. Although the choice of these specific regions is arbitrary, it yields reasonable results as discussed below. On the one hand, it is not possible to estimate a priori where – and even *if* – the influence of the vertical walls on the structures becomes negligible, a fact that precludes from establishing a simple subdivision into two regions based on the spanwise location. On the other hand, a finer partitioning was not practical due to the resulting low number of objects in certain regions, and such a fine subdivision is in fact not necessary for the aim of the present work. The size of the structure bounding box in viscous units are defined $\Delta x^+$, $\Delta y^+$ and $\Delta z^+$, according to the frame of reference in the duct. The probability density function (PDF) of $\Delta_x^+$ and $\Delta_y^+$ in the three different regions of the duct are shown in Figure 9 (center) and (right), respectively. In these figures we compare the PDF obtained from

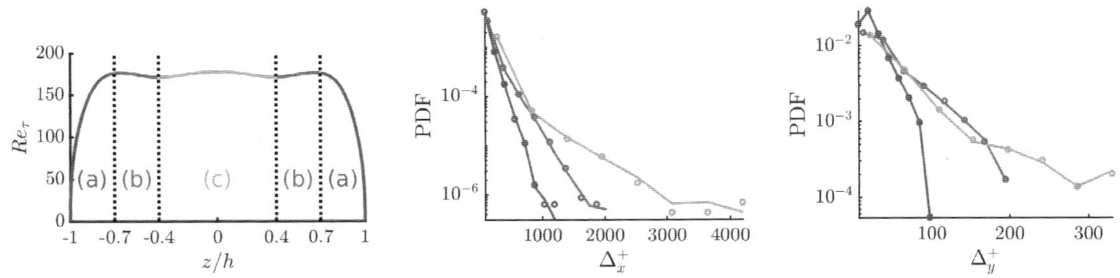


**Figure 9.** (Left) Distribution of $Re_\tau$ for the horizontal wall, where the letters and vertical dotted lines denote the three regions into which the structures are grouped).(Center) Normalized PDF of the inner-scaled structure sizes in the $x$ and (right) in the $y$ coordinates for the three regions. Solid lines and dots denotes PDFs computed based on the entire dataset and on half of the dataset, respectively.

all the instantaneous fields in the dataset (solid lines) and from half of the snapshots (dots). The good agreement between both indicates that the sample spaces under consideration in the present study ensures statistical convergence.
In order to compare the characteristics of the various families, $\Delta_y^+$ and $\Delta_z^+$ are represented as functions of $\Delta_x^+$, in the sense that a convenient binning is adopted for $\Delta_x^+$ and for each bin the averages $\langle\Delta_y^+\rangle$ and $\langle\Delta_z^+\rangle$ are computed, together with their standard deviation within the bin. Figure 10 shows a comparison of the families that are fully equivalent in channel and duct flows, *i.e.* D, WA, and TWA. Note that despite the remarkable agreement of the second-order fit with the trends observed in some of the considered cases, the fit was added only for visualization purposes, without any deeper physical implication in the context of the present work.
For the sake of brevity, in the following we will denote with a subscript the considered region, so that for instance the detached objects in region (a) of the duct are denoted as $D_{(a)}$, and the detached objects of the channel are referred to as $D_{ch}$. It is possible to observe appreciable differences among families and, occasionally, among objects which belong to the same family but are located in different regions. The family $D_{(a)}$ exhibits essentially the same behavior as that of $D_{ch}$, which is also consistent with $D_{(b)}$ (albeit this region contains structures that, for the same streamwise length, are shorter and narrower), whereas $D_{(c)}$ includes objects which are significantly shorter in both wall-normal directions. The subdivision into different regions of the duct is less relevant for WA structures, with $WA_{(a)}$, $WA_{(b)}$ and $WA_{(c)}$ exhibiting similar features reasonably consistent with those of the $WA_{ch}$ objects. Note that for the shorter structures, the channel exhibits objects with larger $\Delta_y^+$ and $\Delta_z^+$ values, a fact that could be connected with the constraining effect of the side walls in the duct. On the other hand, all the three $TWA_{(a)}$, $TWA_{(b)}$ and $TWA_{(c)}$ families of structures have statistically significant differences for the longest objects, and only $TWA_{(a)}$ exhibits a trend similar to that of $TWA_{ch}$, although with slightly smaller objects in the $y$ and $z$ directions for the same streamwise length. Interestingly, there are not any TWA structures in the duct that is large enough to reach both the horizontal walls. Regarding the duct, the identified structures are generally smaller if closer to the vertical walls, the only exception being the for $WA_{(b)}$ and $WA_{(c)}$, which are basically identical. Figure 11 shows the same size curves for of the CA and TCA families, which do not have a natural equivalent in channel flow due to the spanwise periodicity. Here, they are compared with $WA_{ch}$ and $TWA_{ch}$, which are considered to be the closest equivalents if only $uv$ structures are taken into account. CA structures are generally smaller than $WA_{ch}$, although they exhibit similar aspect ratios (which can be defined as the slopes of the $\Delta_y^+$ and the $\Delta_z^+$ curves) up to $\Delta_x^+ \simeq 200$,

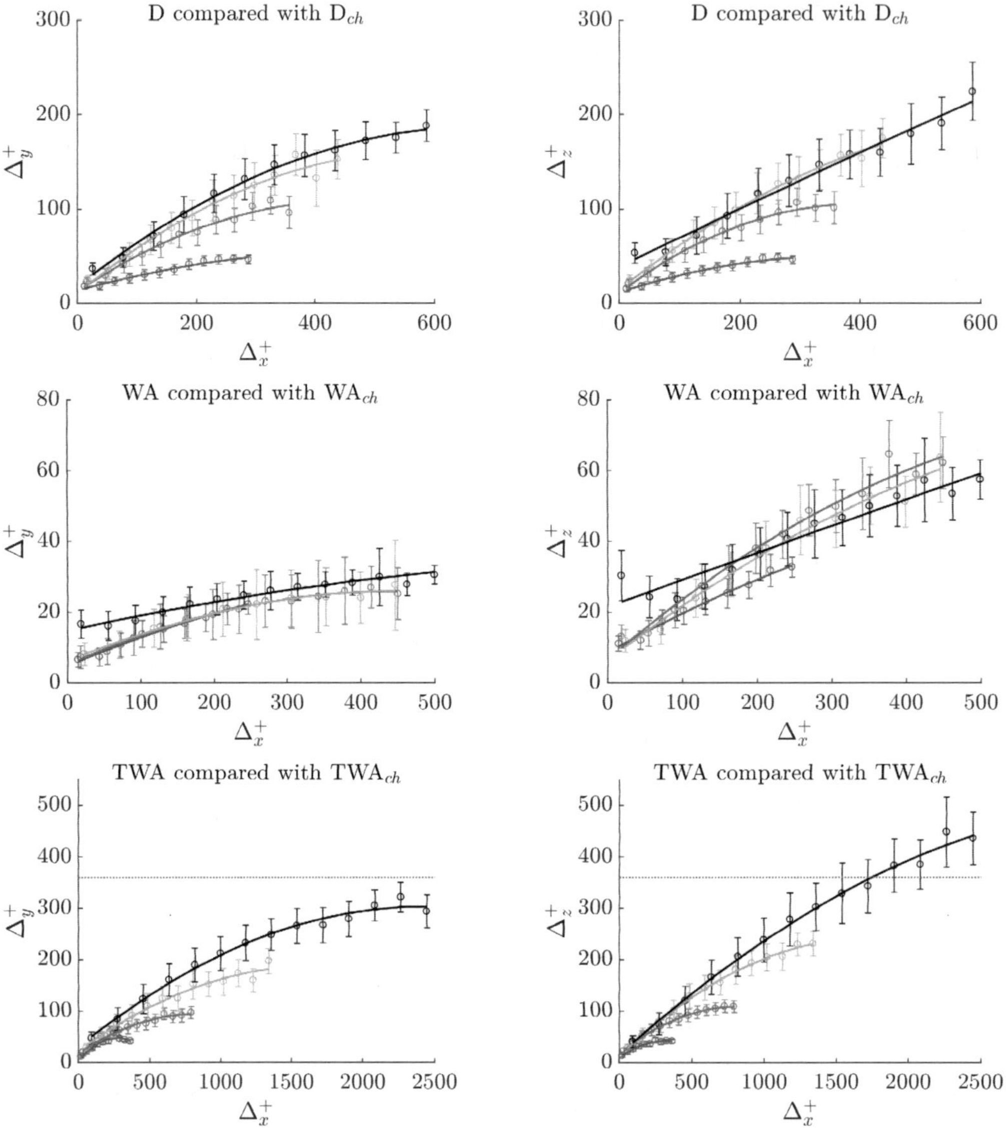


**Figure 10.** (Left) vertical size $\Delta_y^+$ and (right) horizontal size $\Delta_z^+$ as function of $\Delta_x^+$ (in viscous units), for $uv$ structures of the following families: from top to bottom, detached (D), wall-attached (WA) and tall-wall-attached (TWA). Blue, red and yellow are structures located in regions (a), (b) and (c) of the duct, consistently with Figure 9, and black is for structures of the corresponding families in channel flows. The solid lines in the figures are the results of a least-square fit with a second degree polynomial. The dotted red horizontal line represents the size of the domain for the duct.

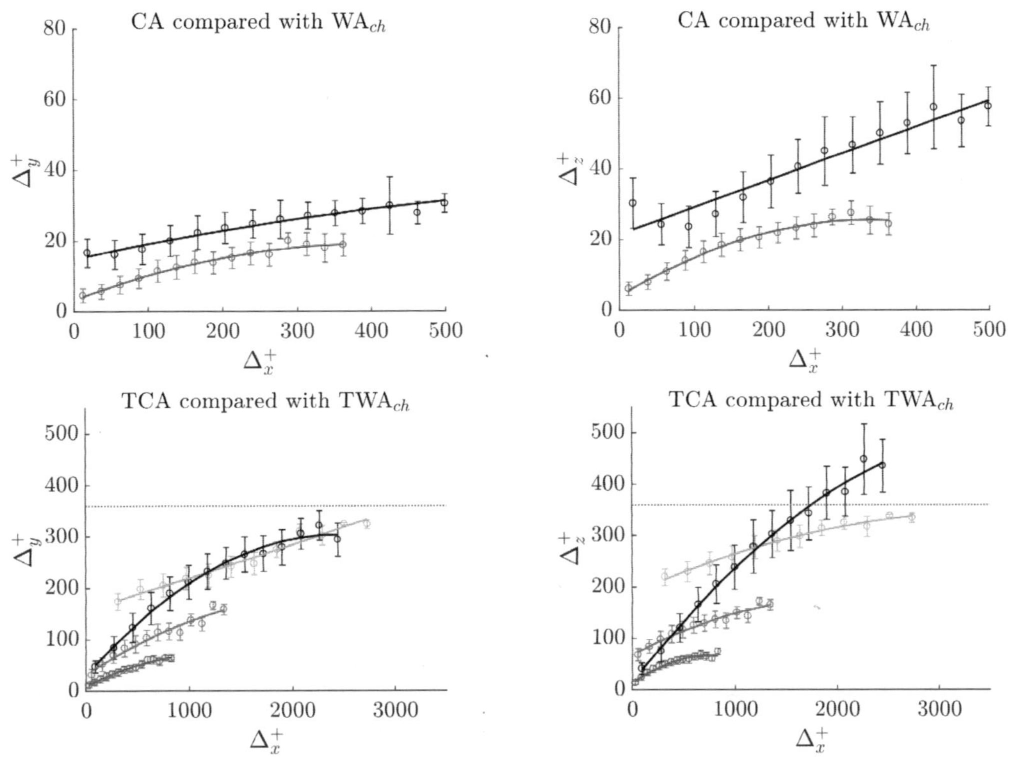


**Figure 11.** (Left) vertical size $\Delta_y^+$ and (right) horizontal size $\Delta_z^+$ as function of $\Delta_x^+$ (in viscous units), for $uv$ structures of the following families: from top to bottom, corner-attached (CA) and tall-corner-attached (TCA) in the duct, compared with wall-attached (WA) and tall-wall-attached (TWA) in the channel. Color code as in Figure 10.

while the TCA curves show a more complex behavior. The TCA$_{(a)}$ and TCA$_{(b)}$ families exhibit lower aspect ratios than the TWA$_{ch}$, and their trends are similar to the ones observed for the TWA$_{(a)}$ and TWA$_{(b)}$ families. The largest structures in the TCA$_{(c)}$ family are as tall and wide as the whole duct and when they occur they occupy most of transversal section of the duct, eventually being attached to all the walls.

The observation that the largest TWA$_{(c)}$ objects reach (and exceed) half of the width of the duct, together with the results summarized in Table 2 regarding the distribution of volume fraction for different families, suggest to consider the TCA and TWA families combined. These results are shown in Figure 12, which shows that the size in the $y$ direction of the joint family TWA$_{(c)}$+TCA$_{(c)}$ is consistent with that of the TCA$_{ch}$ family for all the lengths, and up $\Delta_z^+ \simeq 300$ and $\Delta_x^+ \simeq 2000$ in the $z$ direction.

The behavior of the remaining families, namely SA and TSA, has been studied but further analysis will be carried out, including a detailed comparison with the $uw$ structures in the channel, and it will be reported in future publications.

## 4. Conclusions and outlook

The present study is aimed at a first characterization of coherent structures in turbulent square duct flows. By applying a percolation analysis to select a threshold function that maximizes the number of detected objects, intense events associated with five different quantities have

IOP Conf. Series: Journal of Physics: Conf. Series **1001** (2018) 012008 doi:10.1088/1742-6596/1001/1/012008

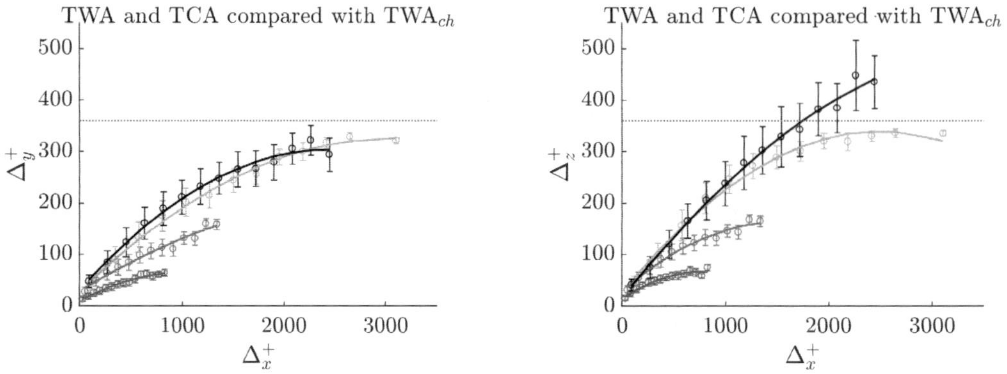


**Figure 12.** (Left) vertical size $\Delta_y^+$ and (right) horizontal size $\Delta_z^+$ as function of $\Delta_x^+$ (in viscous units), for the $uv$ structures in the duct, if the tall-wall-attached and tall-corner-attached families are considered jointly, compared with tall-wall-attached in the channel. Color code as in Figure 11.

been identified: $uv$, $uw$, $vw$, $v^2 - w^2$ and $K = (\tilde{v}^2 + \tilde{w}^2)/2$. The last three events have not been analyzed before with this percolation methodology, and our initial results show that their division into families with respect to the distance from the walls is similar but not identical to the results observed in $uv$ structures. A more detailed study regarding the shape and size of the structures is reported for $uv$ events, including a comparison with turbulent channel flow at similar a $Re_\tau$ of 180. Our main findings can be summarized as follows: 1) small $uv$ structures attached to the wall are almost identical in the two cases (channel and duct); 2) the same is true for detached objects in the spanwise core of the duct, for which the presence of the vertical walls appears to have a negligible influence; 3) both detached structures close to the side wall and tall structures attached with the horizontal wall in the duct, are narrower than the same structures of the same streamwise length in the channel as approaching the vertical walls. The longest structures from these families in the duct are shorter in the streamwise direction than the equivalent ones in the channel; 4) when considered together, tall structures attached to the horizontal wall and tall objects attached to both the horizontal and vertical walls, if located near the spanwise centerplane the duct, have statistically the same size in $y$ as that of the tall objects attached to the wall in the channel. Regarding their extent in $z$, the agreement between the duct and the channel is excellent up to $\Delta_z^+ \simeq 180$ (which is half of the total inner-scaled width of the duct), point after which the spanwise extent of the structures in the duct is smaller than that in the channel, with a maximum $\Delta_z^+$ smaller than 360 (the total inner-scaled duct width). Finally, it is interesting to note how the different behavior of the various families can be clearly observed, despite the relatively low $Re$ under consideration. These results constitute the starting point of additional studies to analyze the relations among the three-dimensional structures of different types and their role in the dynamics of flow.

Another area where additional studies are required is the percolation analysis. Firstly, it is necessary to investigate in detail whether it is possible to prescribe the threshold function with a more rigorous procedure. Indeed, if on the one hand it is possible to define different thresholds that guarantee the percolation crisis to occur for the same physical quantity, on the other hand the analogy between the threshold function and the application of a scaling suggests that it could be possible to establish the threshold based on physical arguments. Secondly, it is important to assess whether the percolation behavior is a universal feature of fully-developed turbulent flow and in this case if it is possible to extract physical insight from it. The remarkable agreement with the simple test case of the square lattice raises the question of whether the percolation behavior is only a consequence of the chaotic appearance of turbulence or not. Contrary to the example of the random lattice, for turbulent flows it is expected to have a range of the sizes of coherent structures which increases with the $Re$ due to the larger scale separation. This pictures is undoubtedly charming since it naturally fits with the concept of turbulent cascade. However, varying *ad hoc* parameters such as $H$ leads to the fact that the connected components expand through all the domain due to the definition of percolation, a fact which poses the impelling question of whether the *very* large structures observed in both duct and channel flows are really physical objects or only artifacts of a too generous detection technique. These aspects will be further investigated in the future.

**Acknowledgments**
This study was funded in part by the Coturb program of the European Research Council. MA acknowledges funding from the Swedish Foundation for Strategic Research, project "In-Situ Big Data Analysis for Flow and Climate Simulations" (ref. number BD15-0082). RV and PS acknowledge the funding provided by the Swedish Research Council (VR) and from the Knut and Alice Wallenberg Foundation. We are grateful to Davide Modesti for his careful reviewing of the manuscript.

## References

[1] Huser, A and Biringen, S 1993 *J. Fluid Mech.* **257** 65–95
[2] Gavrilakis, S 1992 *J. Fluid Mech.* **244** 101–129
[3] Pinelli, A, Uhlmann, M, Sekimoto, A and Kawahara, G 2010 *J. Fluid Mech.* **644,** 107–122
[4] Vinuesa, R Noorani, A, Lozano-Durán, A, El Khoury, G K, Schlatter, P, Fisher, P and Nagib, HM 2014 *J. Turbul.* **15** 677–706
[5] Vinuesa, R, Prus, C, Schlatter, P and Nagib, HM 2016 *Meccanica* **51** 3025–3042
[6] del Álamo, JC, Jiménez, J, Zandonade, P and Moser, RD 2006 *J. Fluid Mech.* **561** 329–358
[7] Lozano-Durán, A, Flores, O and Jiménez, J 2012 *J. Fluid Mech.* **694** 100–130
[8] Hunt, JCR, Wray, AA and Moin, P 1988 *Center Turbul. Res., Proc. Summer Prog. 1988* 193–208
[9] Jeong, J and Hussain, F 1995 *J. Fluid Mech.* **285** 69–94
[10] Chong, MS, Perry, AE and Cantwell, BJ 1990 *Phys. Fluids* **2** 765–777
[11] Chakraborty, P, Balachandar, S and Adrian, R J 2005 *J. Fluid Mech.* **535** 189–2014
[12] Chong, MS, Soria, J, Perry, AE, Chacin, J, Na, Y, and Cantwell, BJ 1996 *Center Turbul. Res., Proc. Summer Prog. 1996* 383–408
[13] Nagaosa, R and Handler, R A 2003 *Phys. Fluids* **15** 375–394
[14] Moisy, F and Jiménez, J 2004 *J. Fluid Mech.* **513** 111–133
[15] Wallace, JM, Eckelman, H and Brodkey, R S 1992 *J. Fluid Mech.* **54** 39–48
[16] Willmarth, WW and Lu, SS 1972 *J. Fluid Mech.* **55** 65–92
[17] Lu, SS and Willmarth, WW, 1973 *J. Fluid Mech.* **60** 481–511
[18] Lozano-Durán, A and Jiménez J 2014 *J. Fluid Mech.* **759** 432–471
[19] Vinuesa, R, Schlatter, P and Nagib, HM 2015 *Int. Symp. Turbul. Shear Flow Phenomena (TSFP-9)* pp 1–6

IOP Conf. Series: Journal of Physics: Conf. Series **1001** (2018) 012009 doi:10.1088/1742-6596/1001/1/012009

# Analysis of secondary motions in square duct flow

**Davide Modesti[1], Sergio Pirozzoli[2], Paolo Orlandi[2] and Francesco Grasso[1]**
[1]Cnam-Laboratoire DynFluid, 151 Boulevard de L'Hopital, 75013 Paris, France
[2]Dipartimento di Ingegneria Meccanica e Aerospaziale, Sapienza Università di Roma
Via Eudossiana 18, 00184 Roma, Italy

E-mail: davide.modesti@lecnam.net

**Abstract.** We carry out direct numerical simulations (DNS) of square duct flow spanning the friction Reynolds number range $Re_\tau^* = 150 - 1055$, to study the nature and the role of secondary motions. We preliminarily find that secondary motions are not the mere result of the time averaging procedure, but rather they are present in the instantaneous flow realizations, corresponding to large eddies persistent in both space and time. Numerical experiments have also been carried out whereby the secondary motions are suppressed, hence allowing to quantifying their effect on the mean flow field. At sufficiently high Reynolds number, secondary motions are found to increase the friction coefficient by about 3%, hence proportionally to their relative strength with respect to the bulk flow. Simulations without secondary motions are found to yield larger deviations on the mean velocity profiles from the standard law-of-the-wall, revealing that secondary motions act as a self-regulating mechanism of turbulence whereby the effect of the corners is mitigated.

## 1. Introduction

Flows within ducts with square cross section are the simplest prototype of internal flow with two-dimensional mean flow statistics, and they are common in many engineering applications, such as heat exchangers, turbomachinery, nuclear reactors, water draining and ventilation systems. These flows are characterized by the appearance of secondary flows, first observed by [1] and [2] who proposed their existence to explain deviations of the streamwise velocity isolines towards the corners. The topology of the secondary motions in square duct flow consists of eight counter-rotating vortices, bringing high momentum fluid from the core towards the corners. Early experimental studies on duct flow were carried out by [3, 4], reporting that the convection of mean streamwise vorticity plays an important contribution in the overall vorticity balance, whereas [4] observed that convection is at least one order of magnitude less than the other terms in the mean vorticity equation. Furthermore, [3] reported that secondary motions are not affected by the Reynolds number, whereas [4] found that their intensity decreases with Reynolds number, when scaled with the bulk duct velocity and it remains almost unchanged when scaled with the friction velocity. The same trend was also observed in the presence of rough walls [5]. From the above considerations it seems that no firm conclusions have been drawn from experiments, hence direct numerical simulation (DNS) may be a valuable tool to shed light on the role of secondary motions. A first DNS of incompressible duct flow was performed by [6] at $Re_b = 4410$ ($Re_b = 2hu_b/\nu$, where $h$ is the duct half side, $u_b$ the bulk velocity and $\nu$ the kinematic viscosity), corresponding to a mean friction Reynolds number of $Re_\tau^* = 150$, where $Re_\tau = hu_\tau^*/\nu$, $u_\tau^* = \sqrt{\tau_w^*/\rho_w^*}$, with $\tau_w^*$ and $\rho_w^*$ the mean viscous stress and density at the wall, respectively. The author found that convective terms in the mean vorticity equation are negligible with respect

IOP Conf. Series: Journal of Physics: Conf. Series **1001** (2018) 012009 doi:10.1088/1742-6596/1001/1/012009

**Table 1.** Flow parameters for square duct DNS. The box dimension is $6\pi h \times 2h \times 2h$ for all flow cases. $Re_b = 2hu_b/\nu$ is the bulk Reynolds number, and $Re_\tau^* = hu_\tau^*/\nu$ is the friction Reynolds number. $\Delta x$ is the mesh spacing in the streamwise direction, and $\Delta z$, $\Delta y_w$ are the maximum and minimum mesh spacings in the cross-stream direction, all given in global wall-units, $\delta_v^* = \nu/u_\tau^*$. $\Delta t_{av}^*$ is the effective averaging time interval. $\chi$ indicates the relative standard deviation of the wall shear stress, $\chi = \overline{(\tau_w - \tau_w^*)^2}^{1/2}/\tau_w^*$. Flow cases denoted with the 1 suffix are carried out by suppressing the secondary motions.

| Case | $Re_b$ | $Re_\tau^*$ | $C_f \times 10^3$ | $\chi$ | $N_x$ | $N_y$ | $N_z$ | $\Delta x^*$ | $\Delta z^*$ | $\Delta y_w^*$ | $\Delta t_{av}^* u_\tau^*/h$ |
|---|---|---|---|---|---|---|---|---|---|---|---|
| A | 4410 | 150 | 9.26 | 0.239 | 512 | 128 | 128 | 5.6 | 3.0 | 0.55 | 2120 |
| A1 | 4410 | 154 | 9.76 | 0.290 | 512 | 128 | 128 | 5.7 | 3.4 | 0.56 | 1049 |
| B | 7000 | 227 | 8.41 | 0.213 | 640 | 144 | 144 | 6.6 | 4.8 | 0.51 | 1607 |
| B1 | 7000 | 230 | 8.64 | 0.265 | 640 | 144 | 144 | 6.7 | 4.5 | 0.52 | 120 |
| C | 17800 | 519 | 6.80 | 0.161 | 1024 | 256 | 256 | 9.5 | 6.3 | 0.53 | 1387 |
| C1 | 17800 | 511 | 6.59 | 0.219 | 1024 | 256 | 256 | 9.4 | 6.1 | 0.65 | 144 |
| D | 40000 | 1055 | 5.57 | 0.126 | 2048 | 512 | 512 | 9.6 | 6.4 | 0.60 | 531 |
| D1 | 40000 | 1038 | 5.39 | 0.212 | 2048 | 512 | 512 | 9.55 | 6.37 | 0.59 | 29 |

to the other, whereas gradients of the Reynolds stresses have the same order of magnitude but opposite sign, and their difference is balanced by viscous diffusion. [7, 8] used a spectral solver to study transition in square duct flow, spanning friction Reynolds number from laminar values up to $Re_\tau^* = 150$ and observed that a very long averaging time is necessary to collect reliable mean flow statistics ($\Delta t_{av} u_b/h = 8000$). They found that, at low Reynolds number, short-time averaging gives rise to four counter-rotating eddies, rather than eight. [9] carried out DNS of square duct flow up to $Re_\tau^* = 600$, and found that below $Re_\tau^* = 300$ low-Reynolds-number effects are particularly important. Moreover, they observed a continuous trend in the position of the secondary eddies centers, which move towards the wall bisectors as the Reynolds number increases. DNS and LES of hexagonal duct flow up to $Re_\tau^* \approx 500$ showed similar trends with the Reynolds number, namely the eddies centers move towards the duct center [10]. Even though considerable efforts have been devoted to the understanding of secondary motions, a clear picture of the phenomenon is still lacking. To further clarify the role of these motions and the effect of Reynolds number, the present authors [11] carried out DNS of square duct flow up to friction Reynolds number $Re_\tau^* = 1055$, the highest value reached so far. The results show that secondary motions scale reasonably well with the bulk flow velocity, and their net effect on the friction coefficient is rather small, and it can be effectively taken into account by introducing the hydraulic diameter. In the present work we focus on the nature of secondary motions and their existence in the instantaneous flow field. We also perform numerical experiments whereby the mean cross-flow velocities are artificially suppressed, in order to quantify their effects on the mean flow statistics and on skin friction.

## 2. Methodology

The compressible Navier-Stokes equations are solved, using a fourth order co-located finite difference solver, in which the convective terms are discretized using an energy-conserving scheme, which allows to preserve total kinetic energy from convection in the inviscid limit [13]. Viscous terms are expanded to Laplacian form and discretized using standard central finite difference approximations. A novel semi-implicit algorithm is used for time advancement, in order to relax the acoustic time step limitation, thus allowing to use efficiently the same solver at all Mach numbers [14]. The streamwise momentum equation is forced in such a way to

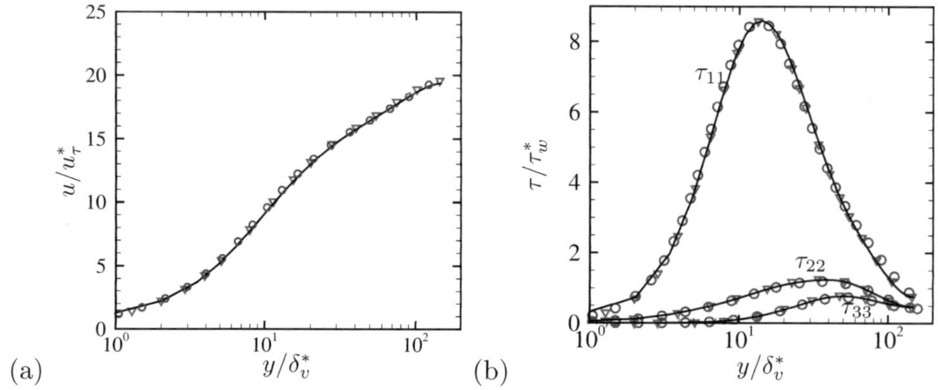


**Figure 1.** Profiles of mean velocity (a) and turbulent normal stresses (b) along wall bisector for flow case A (lines), compared to DNS data of [8] (triangles) at $Re_\tau = 150$, and [12] (circles) at $Re_\tau = 164$.

maintain a constant mass flow rate, periodicity is exploited in the streamwise direction, whereas isothermal no-slip boundary conditions are used at the walls [15]. In the following we denote the velocity components in the streamwise and wall-normal directions as $u$, $v$, $w$, respectively, whereas the overline symbol is used to indicate statistical averages in the streamwise direction and in time. We denote the local friction velocity as $u_\tau = \sqrt{\tau_w/\rho_w}$, with $\tau_w$ and $\rho_w$ the local wall shear stress and local density at the wall; $\delta_v = \nu/u_\tau$ is the local viscous length scale and $Re_\tau = hu_\tau/\nu$ the local friction Reynolds number. Quantities normalized to local and global wall-units are denoted with a + and a $*$ superscript, respectively. Four simulations at bulk Mach number $M_b = 0.2$ have been carried out in the range $Re_\tau^* = 150 - 1000$, and labeled as A-D (see table 1). The database is representative of incompressible turbulence, as the friction Mach number $M_\tau = u_\tau/c_w$ nowhere exceeds 0.01. Excellent agreement between the flow statistics of case A and previous DNS studies [8, 12] is observed (see figure 1).

Additional numerical experiments have been carried out at the same bulk Reynolds numbers as the baseline cases (and denoted with the 1 suffix), by artificially suppressing the secondary motions. For that purpose, at each Runge-Kutta sub-step we force the streamwise-averaged cross-stream velocity components and to have zero mean, by setting

$$v(x,y,z,t) \rightarrow v(x,y,z,t) - \overline{v}^x(y,z,t) \tag{1a}$$
$$w(x,y,z,t) \rightarrow w(x,y,z,t) - \overline{w}^x(y,z,t), \tag{1b}$$

where $\overline{(.)}^x$ denotes the streamwise averaging operator. Although the modified velocity field does not exactly satisfy the Navier-Stokes equations, we use the resulting statistics to understand the role of secondary motions, and to quantify their effect on the mean flow field. Previous studies [8, 12] highlighted the need for very long averaging time, thus we collected flow statistics for equivalent times $\Delta t^*_{av} = \Delta t_{av} L_x/(6h)$, which are considerably longer than in classical plane channel flow. On the other hand the numerical experiments carried out without secondary motions requires much shorter time averaging interval as a results of the suppression of the very slow dynamics associated with the secondary flows.

## 3. Secondary motions as coherent structures

Coherent structures in wall-bounded flows have been widely studied for more than a decade, but their mathematical definition is still controversial [16, 17, 18, 19]. They are usually defined

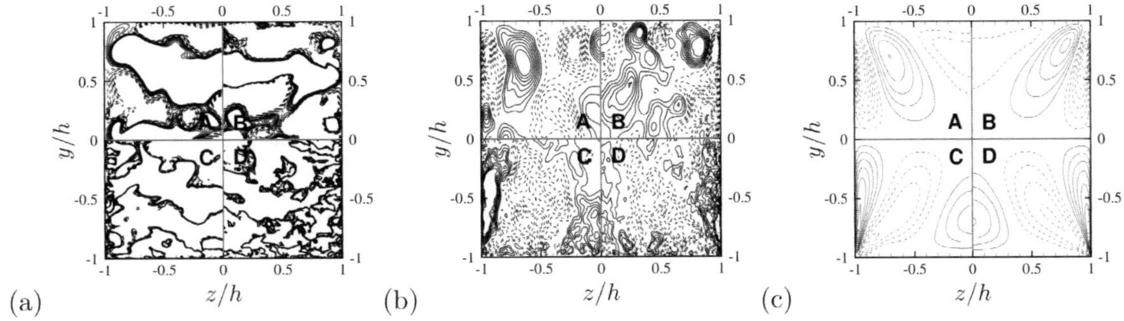


**Figure 2.** Distributions of vertical velocity $v$ in the cross-stream plane: (a) instantaneous, $v/u_b$ (b); (b) averaged in streamwise direction, $\overline{v}^x/u_b$ (c); (c) averaged in $x$ and time, $\overline{v}/u_b$. Data are reported for flow case A (top left), B (top right), C (bottom left), D (bottom right). Contour levels are shown for $-0.02 \leq (.)/(u_b) \leq 0.02$, in intervals of 0.0025 (dashed lines indicate negative values).

as flow structures which are persistent in time and space, and in this sense secondary motions in duct flow can certainly be classified as coherent, as they appear in long-time averages. On the other hand, previous studies have questioned the existence of secondary motions in the instantaneous flow field and raised the point that they may be simply the result of the averaging procedure [7, 20]. Indeed, short-time averages of low Reynolds number simulations [7] showed a topology consisting of four vortices, rather than eight, the latter only observed after long-time averaging. More recently, experiments of supersonic flow in a square duct [20] also confirmed that secondary motions are not observed in instantaneous flow visualization, but they only appear in long-time averaging or when using ad-hoc eduction techniques. figure 2 shows instantaneous, streamwise averaged and time averaged vertical velocity component. No order is found in the instantaneous cross-stream flow visualization (panel (a)), in which turbulent fluctuations high intensity comparable to $u_b$. Averaging the instantaneous three-dimensional field in the streamwise direction only (figure 2c), allows to filter out the incoherent turbulent fluctuations, thus selecting structures which are persistent in space. Indeed, panel (b) of figure 2 shows a more organized structure with a three-lobe arrangement, which is similar to that found in time-averaged fields (panel (c)). This observation suggests that secondary motions are persistent flow structures, since averaging in the streamwise direction only is sufficient to recover a velocity distribution that closely resemble the one found after long-time averaging.

## 4. The role of secondary motions

The role and the effects of secondary motions are here investigated by comparing the DNS dataset (flow cases A-D) with the numerical experiments with suppression of the secondary motions (flow cases A1-D1). In figure 3 we show the streamwise velocity in the cross-stream plane for all flow cases. By suppressing secondary motions, momentum transfer from the duct core towards the corners is inhibited, and the flow no longer exhibits the typical bulging of the velocity isolines. To further elucidate this mechanism, in figure 4 we show the various contributions to the mean streamwise momentum balance equation,

$$\underbrace{\overline{v}\frac{\partial \overline{u}}{\partial y} + \overline{w}\frac{\partial \overline{u}}{\partial z}}_{C} = \underbrace{-\frac{\partial \overline{u'v'}}{\partial y} - \frac{\partial \overline{u'w'}}{\partial z}}_{T} + \underbrace{\nu\left(\frac{\partial^2 \overline{u}}{\partial y^2} + \frac{\partial^2 \overline{u}}{\partial z^2}\right)}_{V} - \Pi, \tag{2}$$

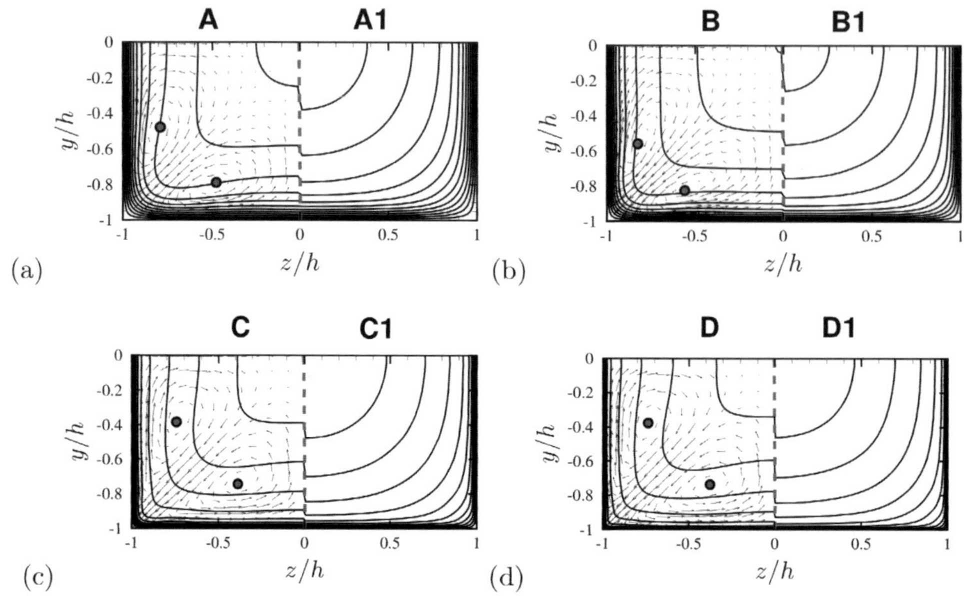


**Figure 3.** Mean streamwise velocity in the cross-stream plane with (left) and without (right) secondary motions for flow cases A-A1 (a), B-B1 (b), C-C1 (c) and D-D1 (d). 13 isolevels are shown in the range $0 \leq \overline{u}/(u_b) \leq 1.3$. Cross-stream velocity vectors are also shown and red circles indicate the position of the secondary eddies centers.

where $C$, $T$ and $V$ represent the contributions of mean convection, turbulence stresses and viscous diffusion, respectively, and $\Pi$ is the driving pressure gradient. Mean convection (figure 4a-b) is only important in the corner proximity ($y^* \sim 150$ wall units), whereas away from the corners turbulent stresses are balanced by viscous diffusion and turbulent stresses gradients (figure 4c-f). We note that diffusion is only slightly affected by suppression of the secondary motions, whereas the turbulent stresses which balance mean convection away from walls are substantially altered (see figure 4a-b).

The wall shear stress distributions are shown in figure 5. Secondary motions are found to generally increase the wall shear stress near the corners by depleting the core region, and as consequence the distributions for cases A-D are flatter than for cases A1-D1. This effect is quantified through the relative standard deviation of the wall shear stress $\chi = \overline{(\tau_w - \tau_w^*)^2}^{1/2}/\tau_w^*$ (see table 1), which is ignificantly lower in the full DNS. The likely reason for this behavior is the local thinning of the wall layer in the corner proximity owing to the inward momentum flux from the duct core, and thickening as the duct mid-point is reached because of the return flow. Because of the secondary motions a peak of $\tau_w$ forms close to the corners, at a distance which well scales in wall units (about $40\delta_v^*$). As a consequence, the profiles of $\tau_w$ become flatter and flatter ar $Re$ is increased. The duct resistance coefficient is shown in table 1, and compared with the friction law for circular pipe flow in figure 6. The results show that inhibiting momentum transfer from the core towards the corner results in drag increase for flow cases A1-B1 and in drag decrease for flow cases C1-D1. This could be anticipated from figure 5, which shows that for flow cases A and B the friction increase near corners is more than compensated by friction relief along the duct perimeter, whereas the opposite happens for flow cases C, D. We argue that this behavior may be due to the non-monotonic displacement of the center of the secondary eddies, in turn associated with post-transitional effects at the lower Reynolds numbers under

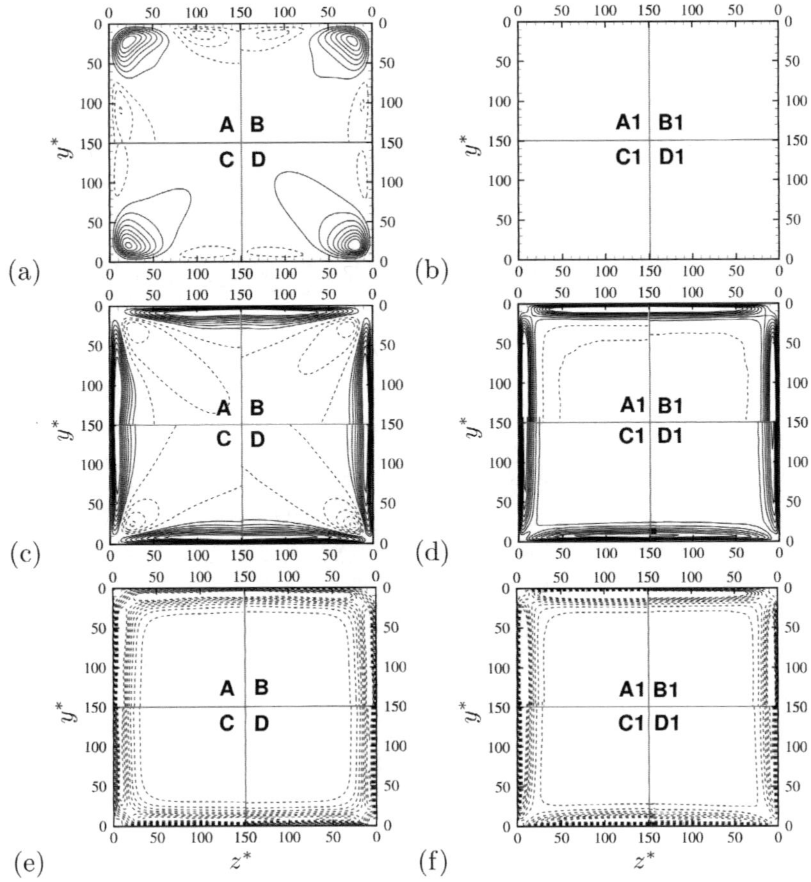


**Figure 4.** Terms in the mean streamwise momentum balance equation (2): convection (a-b), turbulence (c-d), viscous diffusion (e-f), with (left column) and without (right column) secondary motions. Contour levels are shown for $-0.05 \leq (\cdot)/(u_\tau^{*2}/\delta_v^*)^2 \leq 0.05$, in intervals of 0.005 (dashed lines denote negative values). Contour are shown for flow case A/A1 (top left), B/B1 (top right), C/C1 (bottom left), D/D1 (bottom right).

investigation (see figure 3, and consult [11]). In any case, the effect of the secondary motions on the total resistance seem to be rather small, and it may be quantified in about 3% drag increase at moderate Reynolds number, hence proportionate to their relative intensity of the secondary eddies with respect to the bulk flow.

In [11] we found that the mean velocity profiles in duct flow satisfy with good accuracy the standard law-of-the-wall up to the corner bisector. In figure 7 we show the mean streamwise velocity profiles scaled in local inner units at all $z$, up to $y = z$. The circular pipe flow data by [22] are also included for comparison. The figure shows that at low Reynolds number secondary motions affect the entire velocity distribution (figure 7a-b), whereas at higher $Re$ their influence is confined to the near-wall region (figure 7c-d). The effect of secondary motions on the mean streamwise velocity is therefore limited to a distance of about 150 viscous units from the wall, which is consistent with the range of influence of the mean convection terms in the streamwise velocity budget (see figure 4a-b). We further note that the mean velocity profile of DNS is closer to the canonical profile of pipe flow as compared to cases A1-D1. Consistent with the

IOP Conf. Series: Journal of Physics: Conf. Series **1001** (2018) 012009 doi:10.1088/1742-6596/1001/1/012009

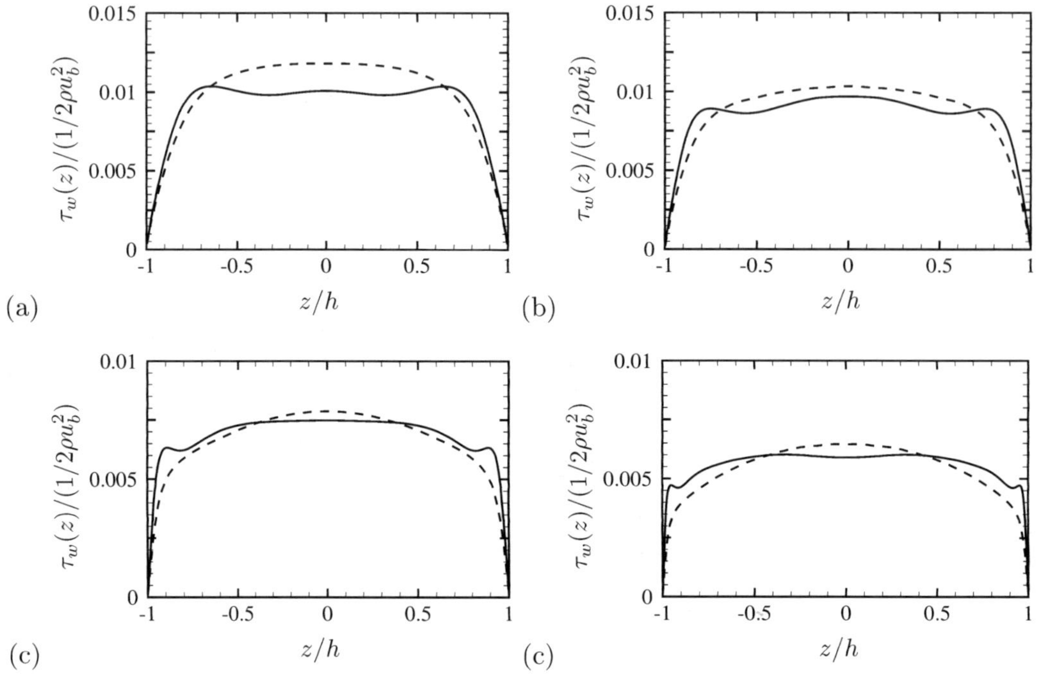


**Figure 5.** Mean wall shear stress distribution for DNS data (solid lines) and DNS with secondary motions suppressed (dashed lines). Flow cases A-A1 (a), B-B1 (b), C-C1 (c), D-D1 (d).

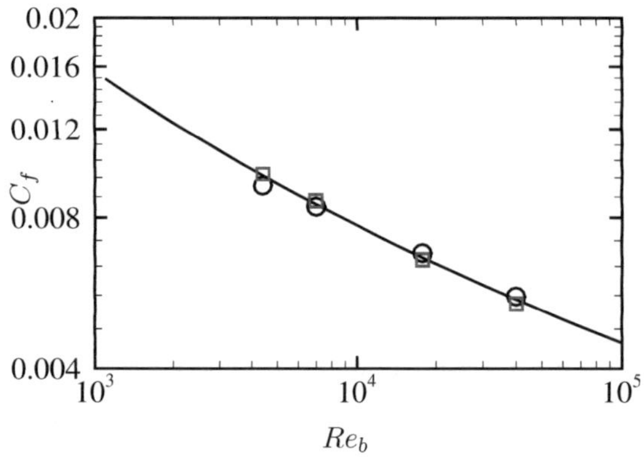


**Figure 6.** Friction coefficient as a function of bulk Reynolds number for DNS data (circles) and DNS with secondary motions suppressed (squares), compared with reference friction coefficient for circular pipe flow.

IOP Conf. Series: Journal of Physics: Conf. Series **1001** (2018) 012009 doi:10.1088/1742-6596/1001/1/012009

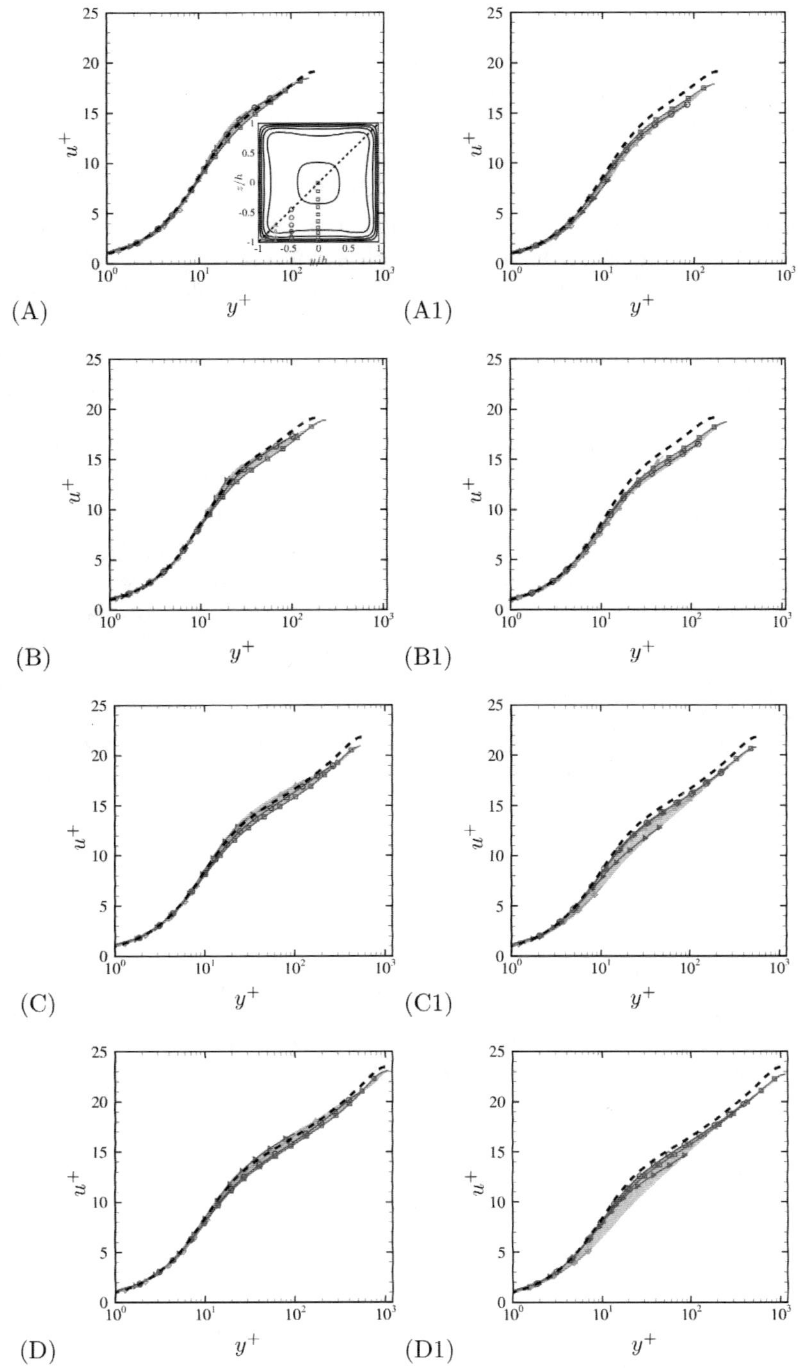


**Figure 7.** Mean streamwise velocity profiles along the $y$ direction (up to the corner bisector), given in local wall-units at all $z$. Representative stations along the bottom wall are highlighted, namely $(z+h) = 15\delta_v^*$ (diamonds), $(z+h)/h = 0.1$ (right triangles), $(z+h)/h = 0.25$ (triangles), $(z+h)/h = 0.5$ (circles), $(z+h)/h = 1$ (squares). The dashed lines denote profiles from DNS of pipe flow flow at $Re_\tau = 140$ (A-A1) [21], $Re_\tau = 180$ (B-B1), $Re_\tau = 500$ (C-C1), $Re_\tau = 1000$ (D-D1) [22]. The inset in flow case A shows the mean streamwise velocity in the cross-stream plane with symbols denoting representative sections.

previous observations related to the flatness of the wall shear stress, this observation leads to the conclusion that the effect of mean convection associated with the secondary motions is to try to restore strict wall scaling all along the duct periphery.

## 5. Conclusions

We have performed DNS of square duct flow up to friction Reynolds number $Re_\tau = 1055$ to shed light on the effect of Reynolds number on secondary motions and clarify their effect on the mean flow field. We showed that, at moderate Reynolds number, averaging in the streamwise direction is sufficient to filter out the incoherent part of the velocity fluctuation, and a secondary circulation emerges which has the same spatial organization as found from long-time averaging. This observation leads to conclude that, other than being statistical artifacts, secondary motions exist in the instantaneous flow fields, although they are difficult to visualize as their intensity is smaller than the typical turbulent fluctuations. The role of the secondary motions is quantified through comparison with numerical experiments in which secondary motions are artificially suppressed. We have found that the main effect of secondary motions is to redistribute momentum from the core towards the corner, thus causing a flattening of the mean wall shear stress, which is more evident as the Reynolds number increases. The overall effect on the duct resistance coefficient is however quite small and at moderate Reynolds number secondary motions are found to increase friction of about 3%, which is proportionate to their intensity. It is especially interesting that, in the absence of mean cross-stream flow, larger deviations of the DNS data from the law-of-the-wall are observed, hence it seems that secondary motions play the role of a self-regulatory mechanism whereby the effect of the corners is mitigated, and the velocity distributions are brought back toward a universal distribution as predicted by the wall law.

## Acknowledgments

We thank Miguel P. Encinar for being our host during the Third Madrid Turbulence Workshop, funded by the COTURB project of the European Research Council, and Prof. J. Jiménez for the invitation. We wish to thank Prof. A. Pinelli and Dr. R. Vinuesa for sharing their DNS data. We further acknowledge that most of the results reported in this paper have been achieved using the PRACE Research Infrastructure resource MARCONI based at CINECA, Casalecchio di Reno, Italy. We are finally grateful to Marco Atzori for his careful review of the manuscript.

## References

[1] Nikuradse J 1926 *Untersuchung über die Geschwindigkeitsverteilung in turbulenten Strömungen* (Vdi-verlag)
[2] Prandtl L 1927 *Reports of the Aerod. Versuchsanst, Göttingen, Germany, 3rd Series*
[3] Brundrett E and Baines W 1964 *J. Fluid Mech.* **19** 375–394
[4] Gessner F and Jones J 1965 *J. Fluid Mech.* **23** 689–713
[5] Launder B and Ying W 1972 *J. Fluid Mech.* **54** 289–295
[6] Gavrilakis S 1992 *J. Fluid Mech.* **244** 101–129
[7] Uhlmann M, Pinelli A, Kawahara G and Sekimoto A 2007 *J. Fluid Mech.* **588** 153–162
[8] Pinelli A, Uhlmann M, Sekimoto A and Kawahara G 2010 *J. Fluid Mech.* **644** 107–122
[9] Zhang H, Trias F, Gorobets A, Tan Y and Oliva A 2015 *Int. J. Heat Fluid Flow* **54** 258–267
[10] Marin O, Vinuesa R, Obabko A and Schlatter P 2016 *Phys. Fluids* **28** 125101
[11] Pirozzoli S, Modesti D, Orlandi P and Grasso F 2017 *arXiv preprint arXiv:1707.03638*
[12] Vinuesa R, Noorani A, Lozano-Durán A, El Khoury G, Schlatter P, Fischer P and Nagib H 2014 *J. Turbul.* **15** 677–706
[13] Pirozzoli S 2010 *J. Comput Phys.* **229** 7180–7190
[14] Modesti D and Pirozzoli S 2017 *J. Sci. Comput.* URL `https://doi.org/10.1007/s10915-017-0534-4`
[15] Modesti D and Pirozzoli S 2016 *Int. J. Heat Fluid Flow* **59** 33–49
[16] Townsend A 1976 *The structure of turbulent shear flow* (Cambridge U. Press)
[17] Sharma A and McKeon B 2013 *J. Fluid Mech.* **728** 196–238
[18] Adrian R and Marusic I 2012 *J. Hyd. Res.* **50** 451–464

[19] Lozano-Durán A and Jiménez J 2014 *J. Fluid Mech.* **759** 432–471
[20] Morajkar R and Gamba M 2016 *54th AIAA Aerospace Sciences Meeting* p 1590
[21] Verzicco R and Orlandi P 1996 *J. Comput. Phys.* **123** 402–414
[22] El Khoury J, Schlatter P, Noorani A, Fischer P, Brethouwer G and Johansson A 2013 *Flow Turbul. Combust.* **91** 475–495

IOP Conf. Series: Journal of Physics: Conf. Series **1001** (2018) 012010 doi:10.1088/1742-6596/1001/1/012010

# Identification of flow structures in fully developed canonical and wavy channels by means of modal decomposition techniques

**Sacha Ghebali[1], Jesús Garicano-Mena[2], Esteban Ferrer[2] and Eusebio Valero[2]**

[1]Department of Aeronautics, Imperial College London, South Kensington Campus, London SW7 2AZ, United Kingdom
[2]ETSIAE - Universidad Politécnica de Madrid, Plaza Cardenal Cisneros 3, E-28040 Madrid, Spain

E-mail: `s.ghebali14@imperial.ac.uk`

**Abstract.** A Dynamic Mode Decomposition (DMD) of Direct Numerical Simulations (DNS) of fully developed channel flows is undertaken in order to study the main differences in flow features between a plane-channel flow and a passively "controlled" flow wherein the mean friction was reduced relative to the baseline by modifying the geometry in order to generate a streamwise-periodic spanwise pressure gradient, as is the case for an oblique wavy wall. The present analysis reports POD and DMD modes for the plane channel, jointly with the application of a sparsity-promoting method, as well as a reconstruction of the Reynolds shear stress with the dynamic modes. Additionally, a dynamic link between the streamwise velocity fluctuations and the friction on the wall is sought by means of a composite approach both in the plane and wavy cases. One of the DMD modes associated with the wavy-wall friction exhibits a meandering motion which was hardly identifiable on the instantaneous friction fluctuations.

## 1. Introduction

The future of aviation relies upon more efficient and greener technologies. The reduction of the turbulent skin-friction drag, via passive or active control, can lead to major improvements in efficiency and hence has become a major target of the aeronautical industry. Turbulent-drag-reduction technologies generally aim at reducing the turbulent shear stress in the near wall region. Several DNS-based studies (Bernard *et al.* [1]) provided direct evidence of a link between the production of Reynolds stresses and the quasi-streamwise vortical structures in the buffer layer. It was shown in Kravchenko *et al.* [2] that streamwise vortices tend to be located above and to the side of high-skin-friction regions, which suggests an intimate link with the wall shear stress and consequently skin friction.

In particular, imparting transverse in-plane wall motions was shown to yield considerable levels of drag reduction (up to about 45% in Quadrio *et al.* [3]). In an effort to emulate the effect of wall motions using a passive device, Ghebali *et al.* [4] studied the effect of large-scale oblique undulations on a solid (non-moving) wall. The rationale is that although not exactly identical to the in-plane motions, the wall undulations generate a spanwise shear layer that is quantitatively similar to that arising from the imposition of an in-plane standing wave. The results showed a maximum overall-drag reduction of 0.6% at $Re_\tau \approx 360$. In this study, flow structures are extracted for a close configuration (with the parameters scaled in wall units) and compared to those found in a plane channel, at $Re_\tau \approx 180$. In order to extract the flow structures, two different modal-analysis techniques are considered, namely Proper Orthogonal Decomposition (POD) and Dynamic Mode Decomposition (DMD) (cf. Taira *et al.* [5] for a general overview).

The POD technique benefits from a long history and numerous studies of laminar and turbulent flows can be found, e.g. in [6, 7]. However, the developments in DMD techniques [8] are more recent; applications include laminar [9, 10], transitional [11] and turbulent flows [12, 13] with clear tonal frequencies, but the application of DMD to fully developed turbulent flows (with a broadband character) is still limited and is the focus of this paper. In this contribution, both POD and DMD are applied to assist in proposing links between the flow structures and friction drag.

## 2. Method

### *2.1. POD and DMD*

Dynamic Mode Decomposition (DMD) techniques [8, 14, 15] have increased in popularity over the past years and have been applied to a variety of flows, using data from numerical simulations and experiments (e.g. [10, 15, 16, 12]).

First, a brief summary of the *snapshot-based* DMD technique, as proposed by Schmid [8], is presented. Given a sequence of instantaneous flow fields numbered from 1 to $N$ (e.g. taking one or all recorded variables), the following *data matrix* can be constructed:

$$\mathbf{V}_1{}^N = \{\mathbf{v}(t_1), \mathbf{v}(t_2), .., \mathbf{v}(t_N)\}, \tag{1}$$

where the subindex and superindex identify, respectively, the first and last time instants of the sequence. The data is ordered in time, and separated by a constant sampling time interval $\Delta t^s$ such that: $t_{j+1} = t_j + \Delta t^s$ for all $j = 1, .., N-1$. In the case of linear stability analysis and within the exponential growth region, it is possible to define a linear operator $\mathbf{A}$ (i.e. a numerical approximation of the linearised Navier–Stokes operator) such that $v(t_{j+1}) = \mathbf{A}v(t_j)$. Equation (1) can then be rewritten as a Krylov sequence (Saad [17]):

$$\mathbf{V}_1{}^N = \{\mathbf{v}(t_1), \mathbf{A}\mathbf{v}(t_1), .., \mathbf{A}^{N-1}\mathbf{v}(t_1)\}. \tag{2}$$

For an ordered sequence, equation (2) can be equated to equation (1):

$$\mathbf{A}\{\mathbf{v}(t_1), \mathbf{v}(t_2), .., \mathbf{v}(t_{N-1})\} = \{\mathbf{v}(t_2), \mathbf{v}(t_3), .., \mathbf{v}(t_N)\}, \tag{3}$$

which can alternatively be written in matrix form as:

$$\mathbf{A}\mathbf{V}_1{}^{N-1} = \mathbf{V}_2{}^N. \tag{4}$$

Next, the Singular Value Decomposition (SVD) of the matrix $\mathbf{V}_1{}^{N-1} = \mathbf{U}\mathbf{\Sigma}\mathbf{W}^H$ is obtained, where the superscript $H$ denotes the conjugate transpose. The left singular vectors – the columns of $\mathbf{U}$ – correspond to the POD modes of the input data sequence. Thus, the snapshot-based DMD algorithm requires the calculation of the POD modes.

The SVD of the snapshot matrix is then inserted into equation (4), which yields $\mathbf{A}\mathbf{U}\Sigma\mathbf{W}^H = \mathbf{V}_2{}^N$. The reduced matrix $\widetilde{\mathbf{S}} = \mathbf{U}^H\mathbf{A}\mathbf{U}$ associated with the initial system described by $\mathbf{A}$, can be rewritten using the former equality as:

$$\widetilde{\mathbf{S}} = \mathbf{U}^H\mathbf{A}\mathbf{U} = \mathbf{U}^H\mathbf{V}_2{}^N\mathbf{W}\mathbf{\Sigma}^{-1}. \tag{5}$$

**Table 1.** List of simulations and discretisation parameters. Wall units are based on $h^+ = 180$.

| | $L_x/h$ | $L_z/h$ | $\Delta x^+$ | $\Delta y^+$ | $\Delta z^+$ | Num. cells | $\Delta t^+$ |
|---|---|---|---|---|---|---|---|
| Plane | 8.05 | 6.6 | 7.4 | 0.6 – 5.1 | 6.1 | 4.7 M | 0.03 |
| Wavy | 15.0 | 5.5 | 4.7 | 0.4 – 5.9 | 2.3 | 32 M | 0.01 |

The reduced matrix $\widetilde{\mathbf{S}}$ is the projection of the matrix $\mathbf{A}$ onto the POD space contained in $\mathbf{U}$, and previously obtained through a SVD [8].

Once the reduced matrix $\widetilde{\mathbf{S}}$ has been calculated, the reduced DMD modes $\mathbf{y}_i$ can be obtained, as well as the associated eigenvalues $\mu_i$ (i.e. growth rates $\Re(\mu_i)$ and frequencies $\Im(\mu_i)$ mapped to the unit circle) of the reduced system by solving the eigenvalue problem $\widetilde{\mathbf{S}}\mathbf{y}_i = \mu\mathbf{y}_i$. The approximated eigenmodes of the matrix $\mathbf{A}$ can then be recovered via a projection onto the original space, using $\boldsymbol{\phi}_i = \mathbf{U}\mathbf{y}_i$. Eventually, the growth rates and frequencies in the complex half-plane can be recovered from the eigenvalues as: $\lambda_i = \log(\mu_i)/\Delta t^s$.

Various extensions of the DMD algorithm have been proposed after the original method described by Schmid [8]. In particular, the sparsity-promoting version of Jovanović *et al.* [19] is also leveraged in this study, in an attempt to reduce the number of representative modes.

### *2.2. Numerical simulations*

The instantaneous realisations analysed in this study were generated by DNS using an in-house code solving the incompressible Navier–Stokes equations. The finite-volume discretisation of the momentum fluxes is second-order accurate, and the solution fields are integrated in time by means of a fractional-step method with a third-order gear-like scheme. The projection step is effected by solving a pressure–Poisson equation, and stable velocity–pressure coupling is ensured by the use of Rhie-and-Chow interpolation. Non-planar geometries are simulated with a body-fitted mesh. The flow is driven by a spatially homogeneous pressure gradient in the streamwise and spanwise directions, adjusted at each time step in order to enforce a unit streamwise bulk velocity and zero spanwise bulk velocity. The fluid viscosity is set so that the bulk Reynolds number is $Re_b = U_b h/\nu = 2820$, corresponding to a friction Reynolds number $Re_\tau = u_\tau h/\nu \approx 180$. Each simulation is initiated with a laminar flow, onto which random perturbations are added while the viscosity is temporarily reduced. The time advancement is continued until a fully developed turbulent state is reached, corresponding to the time instant $t = 0$.

### *2.3. Simulation set-up*

The simulation parameters are shown in table 1 for the plane and wavy channels.

The computational domain used for the wavy channel is shown in figure 1 and comprises one single undulation at an angle of 20 degrees to the main flow direction (corresponding to $\theta = 70°$ in [4]), of amplitude $A_w = 0.15h$ and wavelength $\lambda_x^+ = 15h^+ \approx 2700$. The distance between the upper and lower walls is constant, and the walls are located at

$$y_w = \pm h + A_w \sin\left(\frac{2\pi}{\lambda_x}x + \frac{2\pi}{\lambda_z}z\right). \tag{6}$$

The overall drag in this case is very close to the baseline as a consequence of the pressure drag balancing or even negating the reduction in friction, but benefits from a large amplitude which was chosen to increase the effect on the flow structure in order for the differences to be larger. As a consequence, this flow configuration may shed a more contrasted light than a small

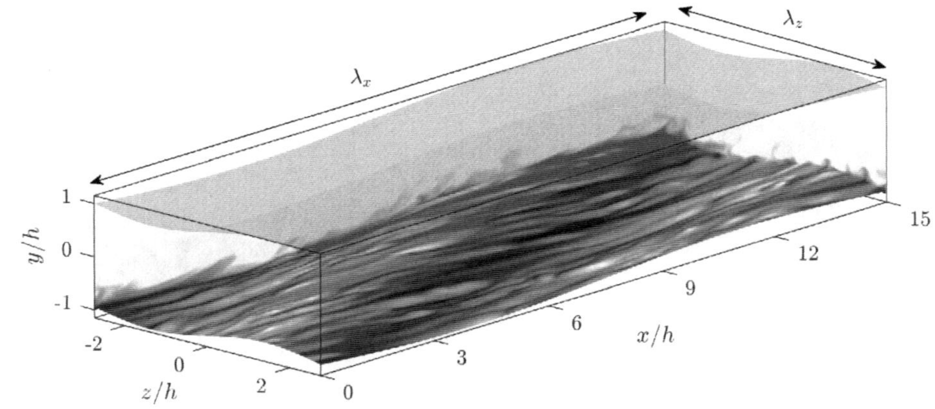


**Figure 1.** Computational set-up for the wavy channel. The contours show the instantaneous value of the streamwise velocity in a wavy plane at a constant distance from the wall $y^+ \approx 12$ and on the sides of the domain. The volume of the total computational domain is twice that saved in the instantaneous snapshots which consists of the lower half of the domain. The upper wall is represented by a green wavy surface.

**Table 2.** Content of the reduced instantaneous database used (unless stated otherwise). The solution variables are saved in single precision, and only in the lower portion on the domain. The superscript "s" stands for "snapshot" and indicates the resolution or domain length considered in the modal decomposition.

| | Total num. snapshots | $\Delta t^{s+}$ | $\Delta x^{s+}$ | $\Delta y^{s+}$ | $\Delta z^{s+}$ | $L_x^s$ | $L_z^s$ |
|---|---|---|---|---|---|---|---|
| Plane | 11635 or 5818 | 0.85 or 1.7 | 30 | 0.6 – 5.1 | 12 | $L_x$ | $L_z/4$ |
| Wavy | 4000 | 1.5 | 28 | 0.4 – 5.9 | 11 | $L_x$ | $L_z$ |

wave slope would, on two different phenomena: 1) the reduction in friction, and 2) the effects associated with the pressure drag responsible for a significant drag penalty. In view of the very fine resolution required to accurately measure the drag level, as shown in [4], this study does not aim at quantifying the precise level of drag, but rather seeks to understand the underlying physics by identifying stronger changes in the coherent structures.

### *2.4. Databases*

Extended databases consisting of instantaneous snapshots of the velocity and pressure field were generated for the two simulations. As shown in figure 1 for the wavy channel, the raw database consists of all the cells contained in the lower half of the solution domain, saved every $100\Delta t$. In order to cut the memory storage required during the post-processing, the 3D volumes were filtered by discarding data points and retaining information at regular space and time intervals $\Delta_i^s$, shown in table 2.

The consistency between the friction coefficient calculated from the driving pressure gradient and that calculated from the reduced data base is shown in figure 2 for a reduced volume of the snapshot database. The reduction in the number of statistical samples arising from the partial use of the data contained in the numerical domain results in an increased standard deviation for the friction coefficient in time with respect to that of the driving pressure gradient which accounts for the entire volume simulated. The total stress on the wall can be calculated as:

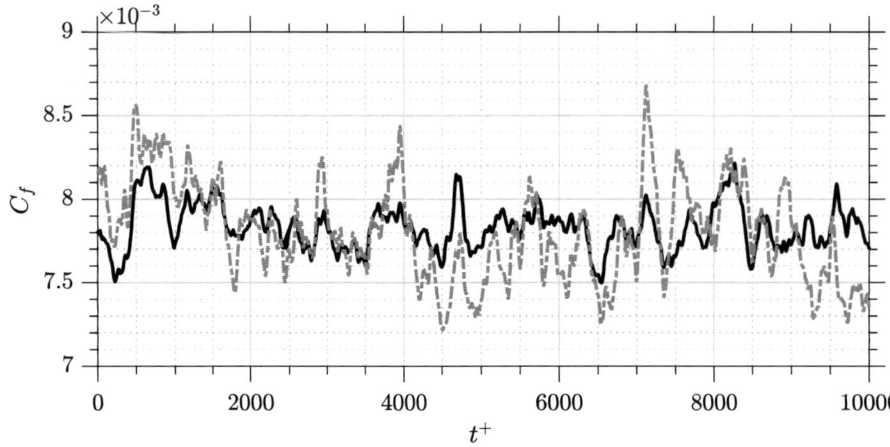


**Figure 2.** Time evolution of the friction coefficient for the plane-channel case, continuous black line: calculated in the full computational domain, dashed-dotted line: from the reduced data base.

$$\tau_{\text{tot.}} = \frac{1}{\mathcal{S}_w} \int_{\mathcal{S}_w} \left( -\mu \frac{\partial u_t}{\partial n} \boldsymbol{t} + p\boldsymbol{n} \right) \cdot \boldsymbol{e_x} \, \mathrm{d}\mathcal{S}_w, \tag{7}$$

where $\mathcal{S}_w$ is the surface of the wall, $\boldsymbol{e_x}$ is the streamwise direction unit vector, $\boldsymbol{n}$ and $\boldsymbol{t}$ are, respectively, the local unit normal and tangential vectors, and $u_t$ is the tangential velocity component. However, only the frictional part (without the pressure contribution) is considered in this study, and its calculation contains a certain degree of approximation, as the wall-normal velocity gradient is approximated by $\partial u/\partial y$. Although a more careful calculation would be required for a precise measure of the friction, it is reasonable to expect that the approximations made entail an error level lower than or comparable to the spatial-discretisation error, and make little difference in the main flow dynamics.

## 3. POD analysis

Since the DMD analysis involves a projection onto the POD modes, the latter were studied in detail for the plane-channel case. The POD modes were calculated using the method of snapshots with all three components of the velocity field. The influence of the proportion of the database used in the POD analysis was mostly evaluated by trial and error. The spatial resolution was selected by considering a small subset of snapshots and varying the resolution in each direction separately.

When every other snapshot is included in the data sequence, the POD modes are not sensitive to a shift in the data sequence by one snapshot, which suggests that the frequency of accumulation of the snapshots is sufficient. As a consequence, the snapshot sampling frequency in the wavy case was chosen to be half that of the plane case.

In all the results presented in this section, the data sequence comprises snapshots every $50\Delta t$, and $4\Delta x \times \Delta y \times 2\Delta z$, in a volume of size $L_x \times h \times L_z/4$. Although increasing the number of snapshots seemed to preferentially yield particular mode shapes, not all of them were found to be robust against a spanwise translation of the volume of snapshots considered. This may be attributed to the limited streamwise extent of the solution domain, in which structures are recirculated via the periodic boundary conditions.

As observed by Nikolaidis *et al.* [7], the convergence of the POD modes for the turbulent plane channel is very slow. To circumvent this limitation, Nikolaidis *et al.* [7] took advantage of flow symmetries and streamwise homogeneity to obtain statistical convergence. However, this operation was not possible in the present case as such symmetries are not present in the wavy case.

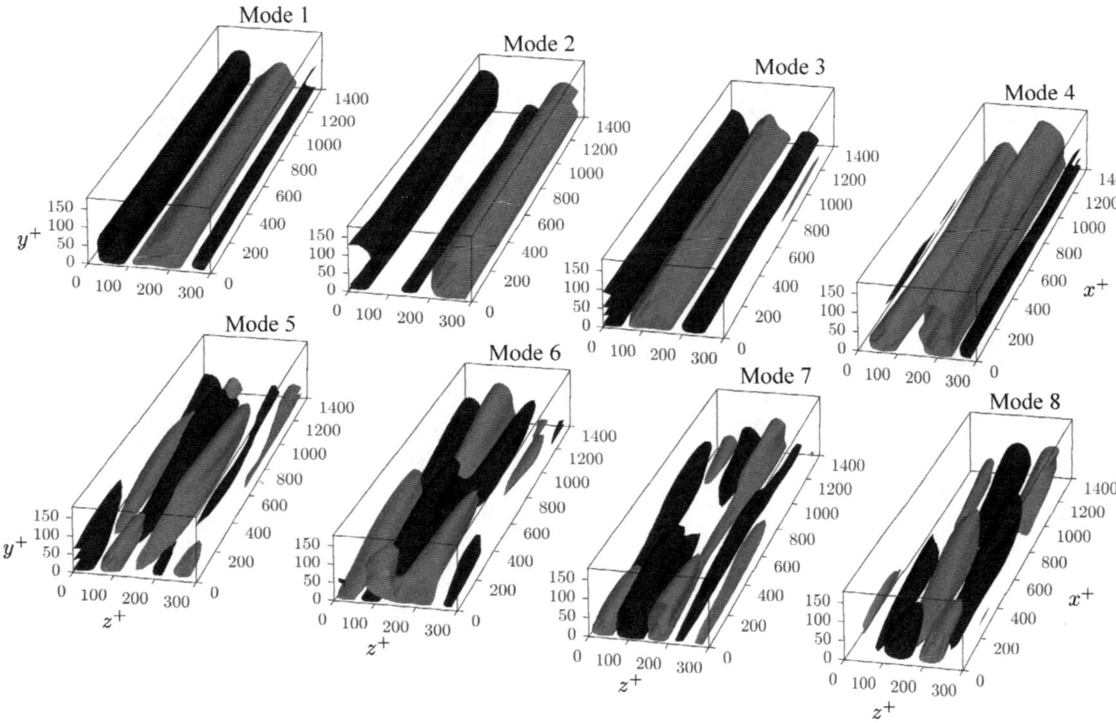


**Figure 3.** First POD modes, numbered by energy content, for the streamwise velocity using a sequence of 5817 snapshots. Low-speed regions (negative fluctuations) are shown in blue, whereas high-speed ones are shown in red.

Figures 3 to 5 show isosurfaces of the POD modes for each component of the velocity vector, for a sequence of 5817 snapshots spanning a duration of over $10,000\, u_\tau^2/\nu$, which represents about 57 eddy-turnover times. Although no statistical convergence can be claimed even for this relatively long snapshot sequence, some broad similarities are found in the spacing of the rolls, relative to the findings of [7], where the spacing of some of the first streamwise POD structures was shown to be of about $70^+$, $100^+$, and $150^+$, which is in line with, respectively, modes $\{5, 6\}$, $\{1, 3\}$, and 2. Moreover, most of the modes show a correlation between the streamwise and the wall-normal components of the velocity, with ejections correlating with low-speed streaks, and sweeps with high-speed regions, consistently with the findings of [7] who interpreted this correlation as arising from the action of the lift-up mechanism maintaining the streaks.

## 4. DMD of the velocity field (plane channel)

A DMD of the velocity field was undertaken in an attempt to identify a limited set of low-frequency modes able to reproduce the mean Reynolds shear stress from the dynamical modes. Such a procedure was studied by Sayadi *et al.* [11] for a transitional boundary layer. For brevity, the results shown are limited to the plane-channel case since the procedure is not fundamentally different in the wavy case. First, a DMD of the velocity-fluctuation vector is performed, second a sparsity-promoting technique is applied, and third, the mean Reynolds shear stress extracted

IOP Conf. Series: Journal of Physics: Conf. Series **1001** (2018) 012010 doi:10.1088/1742-6596/1001/1/012010

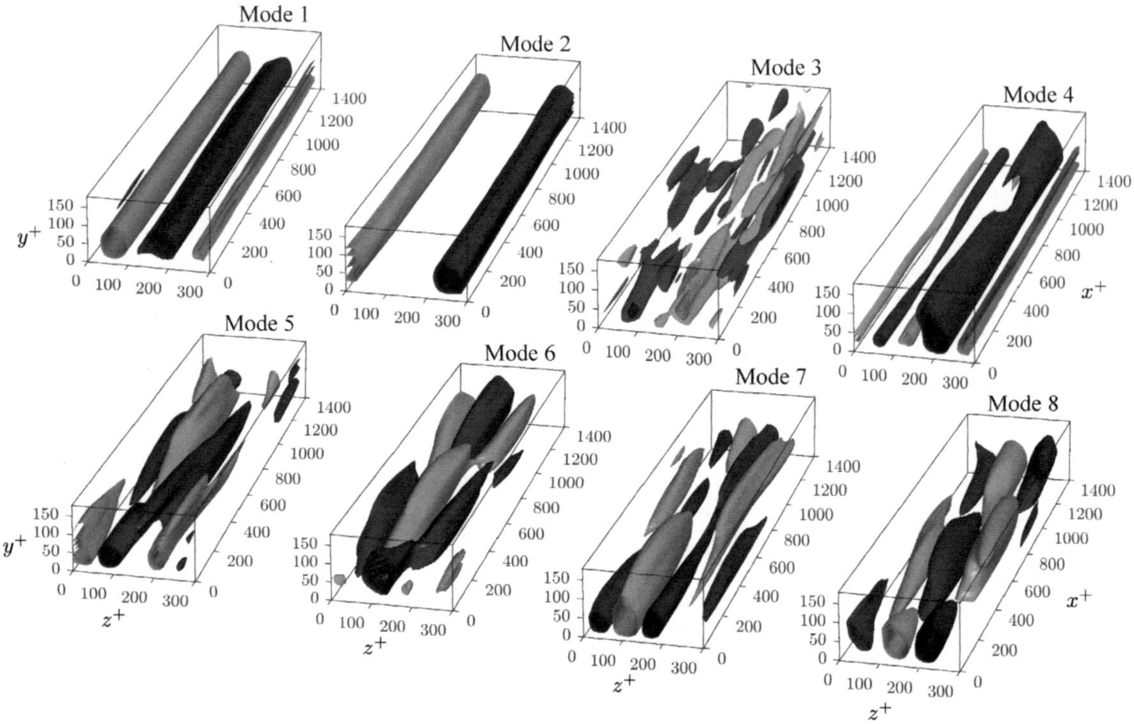


**Figure 4.** First POD modes for the wall-normal velocity using a sequence of 5817 snapshots. Same legend as figure 3.

from DNS are compared to the mean Reynolds shear stress calculated from the time average of a reduced number of the DMD modes corresponding the the streamwise and wall-normal velocity components.

Note that in all cases, although the raw snapshots are saved in single precision, performing the calculation of the DMD modes in double precision was found to be an absolute necessity to ensure that the eigenvalues were close to the unit circle.

### *4.1. DMD modes*

The frequencies and corresponding amplitudes are shown in figure 6. Some peaks were identified, and their position in the spectrum was found to be robust against a change in the length of the data sequence considered (by comparing the results with a reduced number of snapshots of 101 and 301; this observation does not apply to the lowest frequency peak which is not resolved when 301 or fewer snapshots are used). One peak appears to be located at $T^+ = O(100)$, which may be related near the wall to a spatial dimension of $\lambda^+ = O(1000)$ using Taylor's hypothesis and the convection velocity at the wall $u^+ \approx 10$ (see e.g. del Álamo *et al.* [18]). This suggests that the modes surrounding this peak might be linked to the near-wall streaks, and it is remarkably close to the optimal actuation period for oscillating walls $T^+ = 125$ in Quadrio *et al.* [3]. Additionally, the large amplitudes of the low-frequency modes are brought to light, and may suggest the presence of long structures or reflect the occurrence of structure recirculation arising from the finite domain length.

The DMD modes corresponding to the peaks identified in figure 6 are shown in figure 7. Although the mode numbers do not bear any physical meaning, unlike with the POD where they are ranked by energy content, the mode number is used as a more convenient label to identify

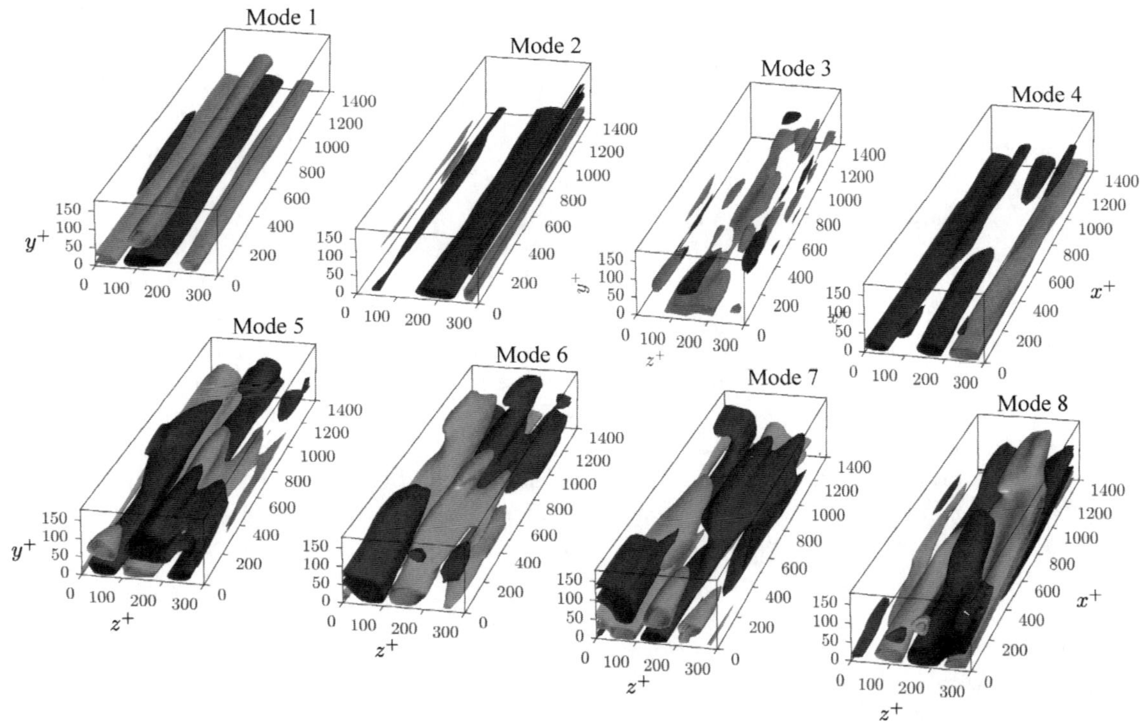


**Figure 5.** First POD modes for the spanwise velocity using a sequence of 5817 snapshots. Same legend as figure 3.

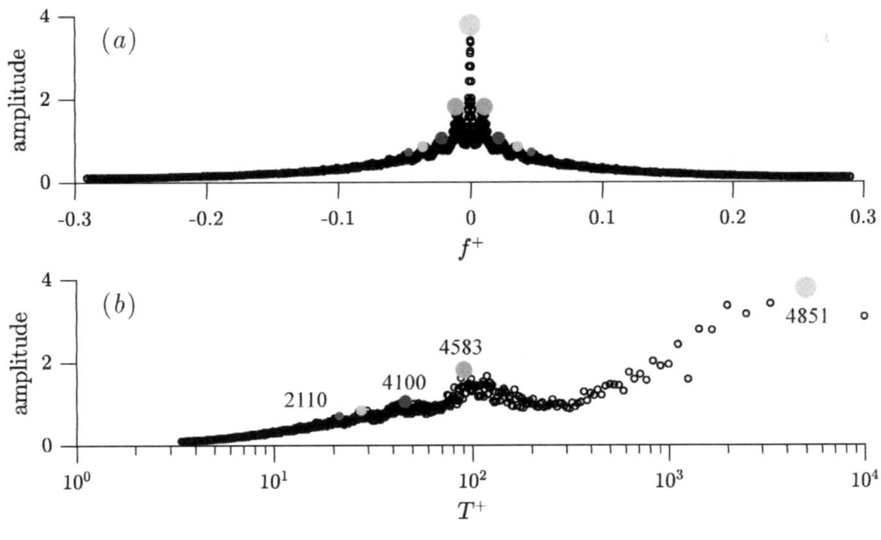


**Figure 6.** Spectrum of the DMD modes for the plane channel, using 5818 snapshots. A set of low-frequency modes is identified by larger coloured circles. (*a*) Amplitude $a_i$ of the DMD modes versus frequency $f_i$, (*b*) corresponding time period and labels (mode numbers) used to identify each mode.

the different frequencies. Interestingly, the mode number 4583, corresponding to $T^+ \approx 100$, displays elongated structures, with low-speed and high-speed regions aligned at an angle to the flow. The $u$-structures are inclined, starting from the wall and reaching the outer region. Again, the low-speed $u$-structures correlate with positive wall-normal velocities (ejections) whilst high-speed ones correlate with negative values (sweeps). The spanwise structures are observed to be much flatter and wider, and are tilted in both the $x$- and $y$-directions.

### *4.2. Sparsity promotion*

The sparsity-promoting technique by Jovanović *et al.* [19] was applied to the DMD for the plane channel using a set of 301 snapshots with the spatial and temporal spacings shown in table 2.

The Matlab source code available from `http://people.ece.umn.edu/~mihailo/software/dmdsp/` was employed. As the penalty $\gamma$ is increased, the number of modes with non-zero amplitude decreases, and so does the fidelity of the modal reconstruction, as quantified in figure 8($a$) using the same norm as Jovanović *et al.* [19]. Figures 8($b$) and 9 show the spectrum corresponding to the highest value of $\gamma$ (smallest number of non-zero-amplitude modes) compared to the full spectrum. The broadband nature of the fully developed turbulent flow makes the sparsity-promoting algorithm fail to isolate few modes, and the performance quickly decays with the level of sparsity.

### *4.3. Reconstruction of the Reynolds shear stress*

Using the large set of 5818 snapshots, the DMD modes of the streamwise and wall-normal velocities are used in order to estimate the corresponding mean Reynolds shear stress. Figure 10 shows a comparison of the Reynolds shear stress obtained from DNS and that obtained from the sum over a range of modes of the contributions from the product of the DMD modes of the streamwise and wall-normal velocities (cf. Sayadi *et al.* [11]). The good statistical convergence of the Reynolds stress from DNS is demonstrated by the fact that the dash-dotted line – corresponding to the DNS snapshot sequence – is smooth and does not feature any high-frequency oscillation away from the wall. Unlike in the transitional flow of [11] where only a few low-frequency modes sufficed to estimate the shear stress, it was found that a significant proportion of the modes is necessary in the fully-turbulent plane channel. This is a known limitation of the DMD algorithm associated with the absence of a clearly defined frequency in the dynamics, so that important structures are spread across a large range of frequencies, both temporally and spatially. This also echoes the difficulty faced to isolate a small number of modes with the sparsity-promoting technique in section 4.2.

## 5. Composite DMD

A composite DMD was performed in order to seek a dynamical link between the friction at the wall and other quantities such as velocity, pressure, or the Reynolds shear stress by appending the wall friction in the snapshot matrix. Thus, the snapshot matrix in the results shown in this section contained the streamwise fluctuations scaled with the bulk velocity, together with the fluctuations of the non-dimensional friction coefficient. Owing to resource limitations, only a small sequence of 301 snapshots is considered in the present analysis. However, as a consequence of the convergence issues evidenced in section 3, the modes obtained may be only relevant to the dynamics of the snapshot sequence selected, and not necessarily to that of the turbulent flow associated with the flow configuration.

Figures 11 and 13 show the DMD spectra obtained for the plane and the wavy case using a sequence with similar spacing (cf. table 2) and composed of 301 snapshots. The amplitude of the lowest-frequency mode is found to be higher in the wavy case, and the frequencies are more evenly distributed than in the plane case. Some low-frequency modes are shown in figures 12 and 14 for the plane channel and the wavy channel, respectively. In general, low-speed regions

IOP Conf. Series: Journal of Physics: Conf. Series **1001** (2018) 012010 doi:10.1088/1742-6596/1001/1/012010

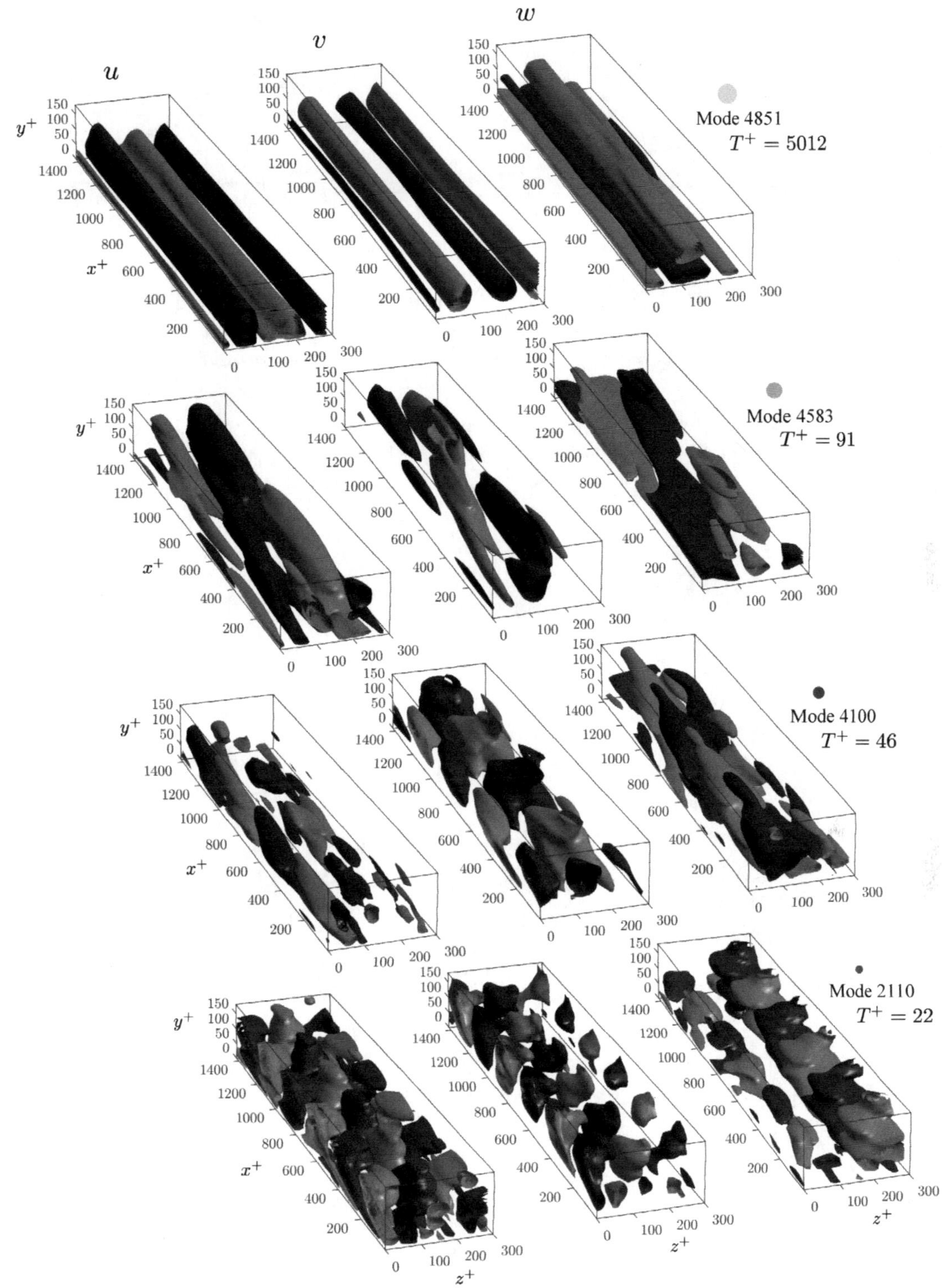


**Figure 7.** Real part of the DMD modes of the three velocity-fluctuation components (displayed in columns and ordered by amplification level from bottom to top) for the four lower-frequency modes labelled in figure 6 (rows), for the plane channel using 5818 snapshots.

IOP Conf. Series: Journal of Physics: Conf. Series **1001** (2018) 012010 doi:10.1088/1742-6596/1001/1/012010

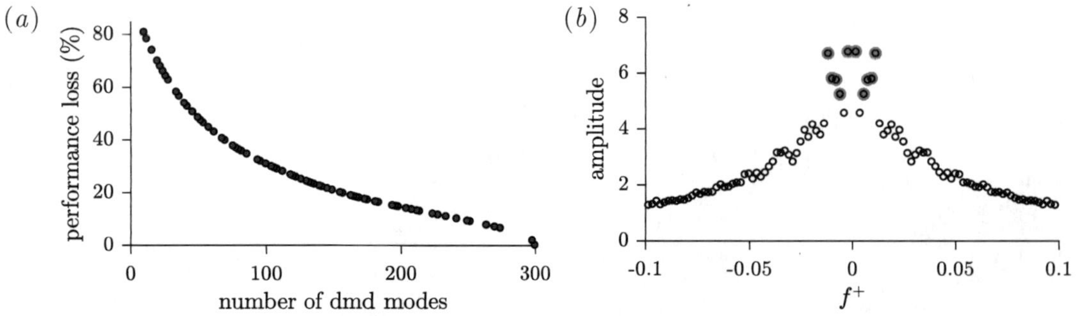


**Figure 8.** Sparsity-promoting DMD analysis for the plane channel, using 301 snapshots. ($a$) Performance loss with the number of non-zero amplitudes, ($b$) DMD spectrum, open circles: full DMD, filled circles: modes isolated from the sparsity-promoting optimisation for $\gamma = 3000$.

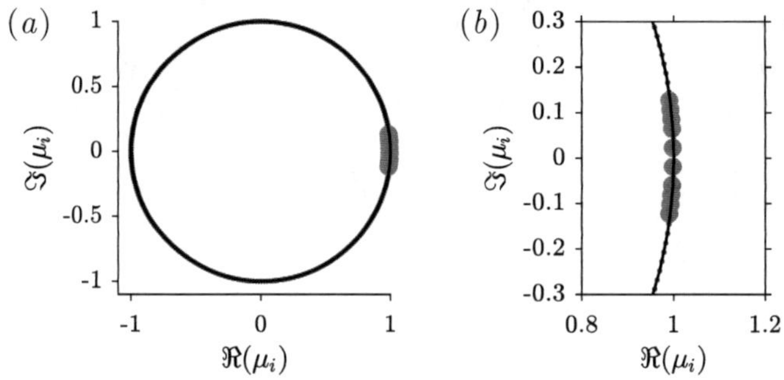


**Figure 9.** Spectrum corresponding to the most reduced set of modes. Black line: unit circle, blue dots: full DMD spectrum, red filled circles: modes isolated from the sparsity-promoting optimisation for $\gamma = 3000$. ($a$) Full spectrum, ($b$) zoom on the isolated modes.

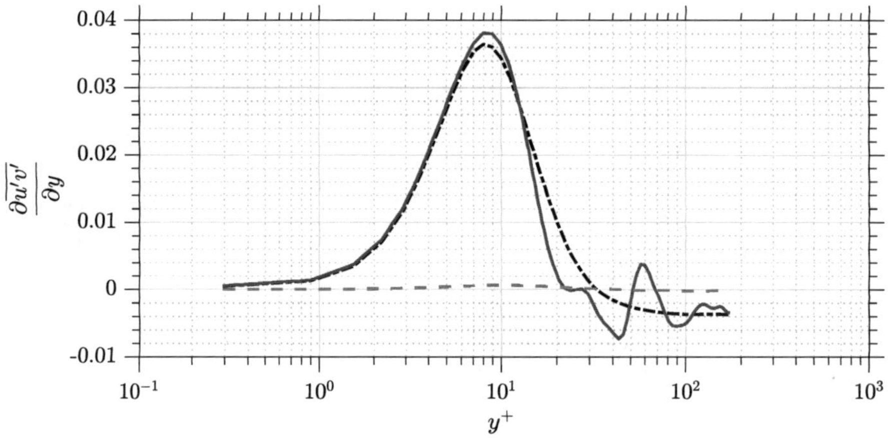


**Figure 10.** Wall-normal derivative of the Reynolds shear stress evaluated from DNS (dash-dotted line), the set of modes highlighted in figure 6 (red dashed line), and modes 2000 to 5000 (blue line).

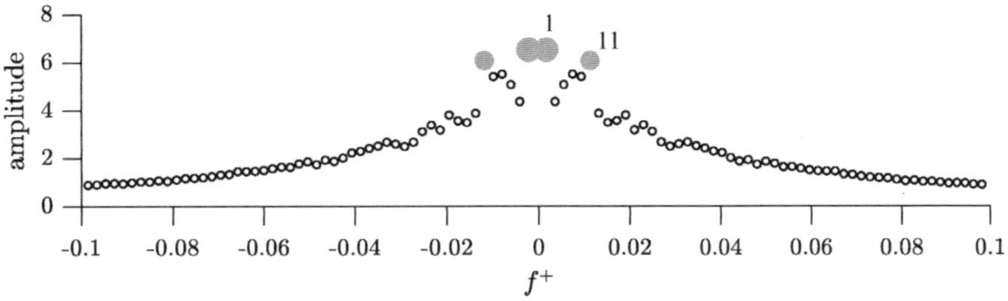


**Figure 11.** DMD spectrum for the plane channel, using the first 301 snapshots. Wall units are based on the actual friction velocity.

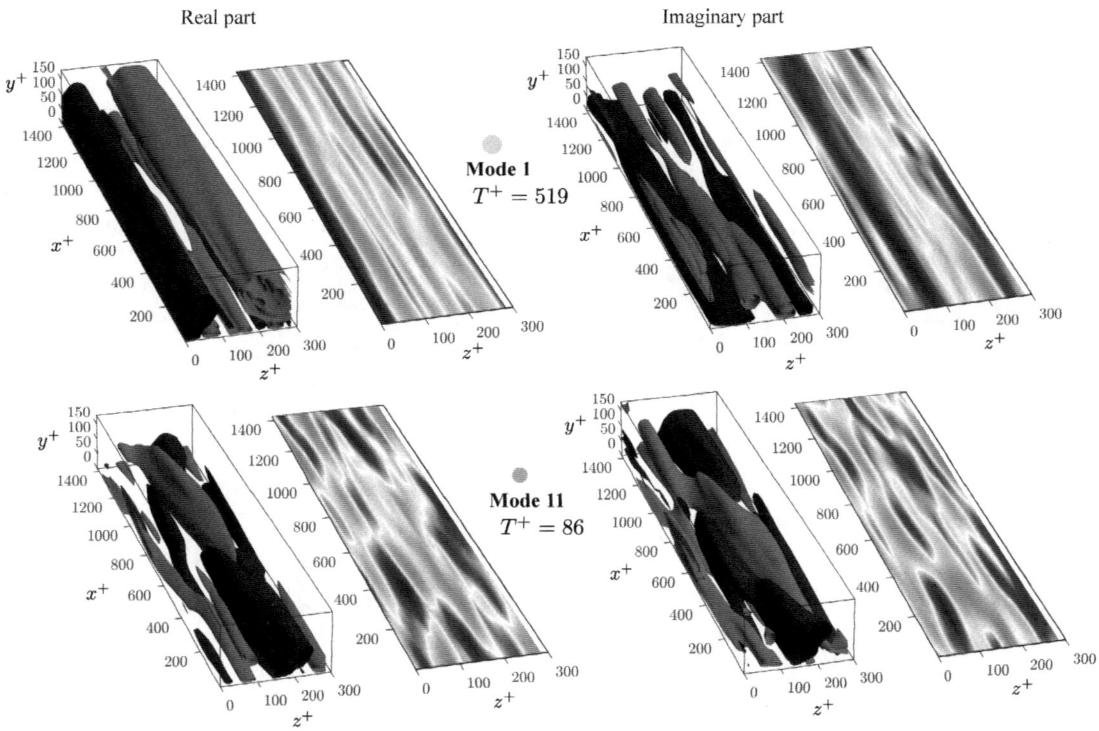


**Figure 12.** DMD modes, highlighted in figure 11, for the streamwise fluctuations and wall friction side by side for the plane channel, using 301 snapshots. The three-dimensional isosurfaces correspond to the DMD modes of the streamwise fluctuations, while the contours correspond to the DMD mode of the friction fluctuations.

correlate with low-friction regions, and conversely high-speed regions correlate with high-friction levels. Again, there seems to be a preference in the alignment of the low- and high-speed regions, making an angle of about 10 degrees with the main flow direction. Moreover, the three-dimensional arrangement of the streamwise-fluctuation modes presents tilted structures (around the $z$-axis) with low- and high-speed regions overlapping each other.

The DMD mode number 3 is shown for the wall friction in figure 15($b$), and compared to an instantaneous view of the wall friction from DNS in figure 15($a$). Although the instantaneous contours mostly show elongated structures aligned with the main flow direction, the DMD mode brings to light a meandering motion resulting from the presence of the oblique wavy boundary.

IOP Conf. Series: Journal of Physics: Conf. Series **1001** (2018) 012010 doi:10.1088/1742-6596/1001/1/012010

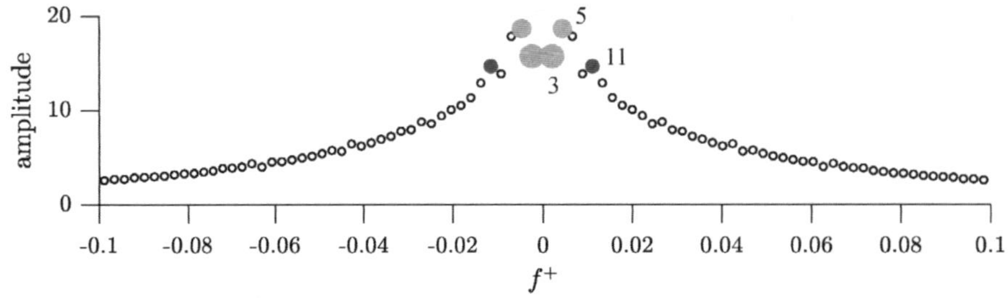


**Figure 13.** DMD spectrum for the wavy wall, using the first 301 snapshots. Wall units are based on the baseline friction.

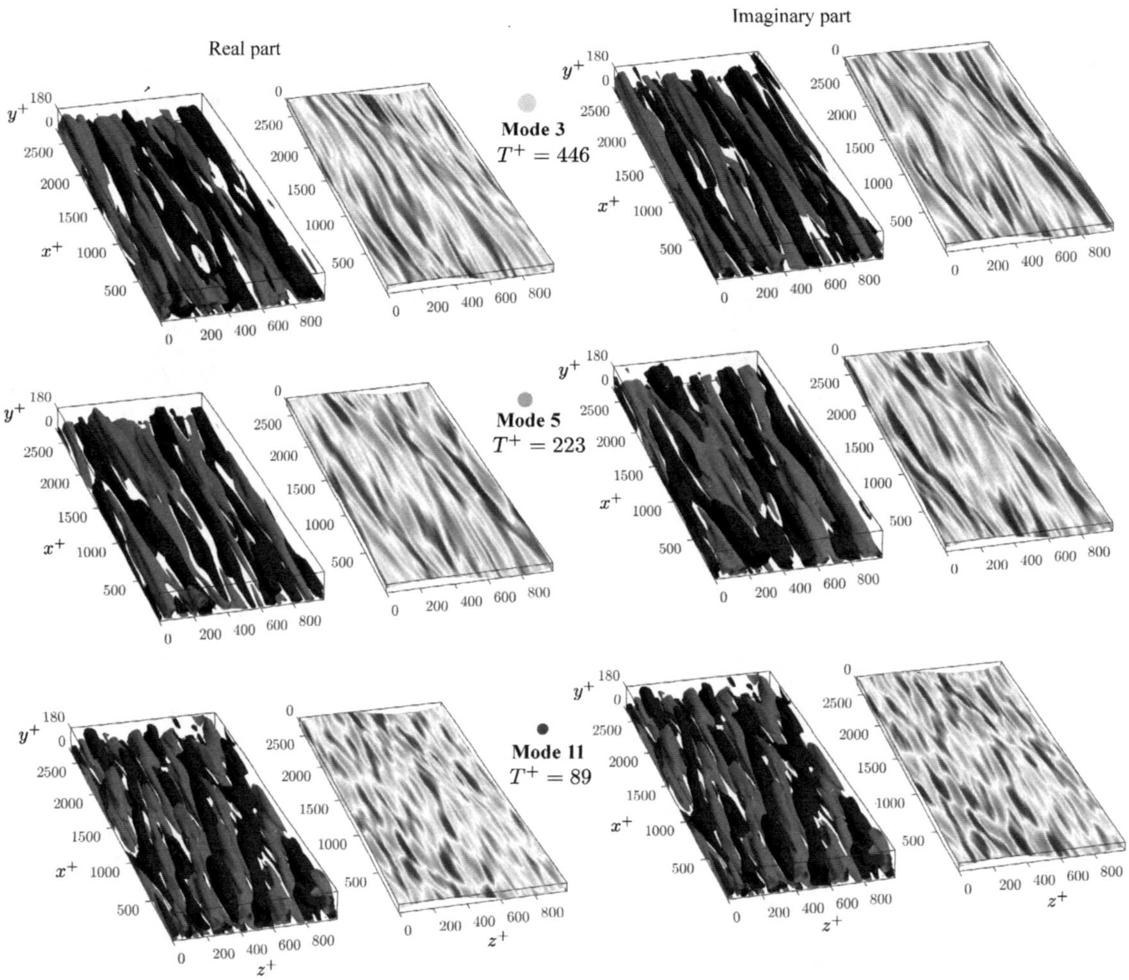


**Figure 14.** DMD modes, highlighted in figure 13, ordered from lowest (top row) to highest frequency (bottom row), of the streamwise fluctuations and wall friction side by side for the wavy channel, using the first 301 snapshots. Wall units are based on $h^+ = 180$.

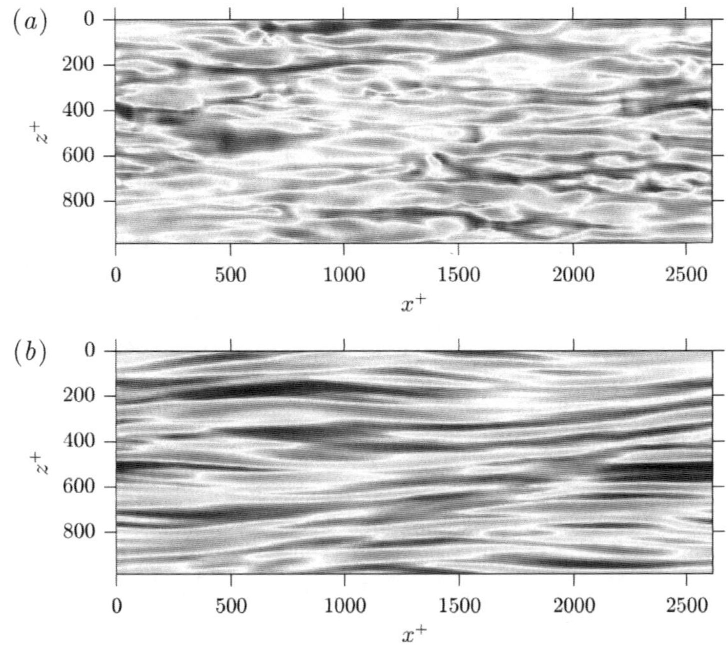


**Figure 15.** Comparison of the instantaneous friction fluctuations with the real part of the third DMD mode. Wall units are based on $h^+ = 180$. $(a)$ Instantaneous friction fluctuations at the wall for the 110th snapshot, $(b)$ third DMD mode.

## 6. Summary and outlook

Simulations of fully developed turbulent channel flows were performed, at $Re_\tau \approx 180$, with plane and wavy walls. Instantaneous realisations of these flows were analysed by means of POD and DMD. The DMD revealed that most of the modes are necessary to represent the flow accurately, as a consequence of the broadband spectrum of the flow dynamics. Eventually, a composite DMD was undertaken in order to find dynamic correlations between the wall friction and the three-dimensional streamwise-fluctuation field. It was observed for the wavy wall that one of the wall-friction modes could enable an alternative representation of the effect of the wavy wall, highlighting a spanwise-wavy pattern that was not identifiable from the instantaneous wall-friction. Future work will include a deeper assessment of the selection of the snapshot sequence, gathering more data by leveraging parallel map-reduce algorithms. This would enable higher Reynolds numbers to be processed, and potentially shed light on the inner-outer interactions and their consequences on the turbulent skin friction. The composite DMD may be studied in more detail for the pressure fluctuations, or kinetic-energy production or dissipation, especially in the case of the wavy wall. Eventually, it may be interesting to consider very small regions in a Lagrangian framework [20] in order to reduce the dilution of the travelling structures across a broad range of modes.

## Acknowledgements

This work was funded in part by the Coturb program of the European Research Council, and by the EU–China project DRAGY (Grant Agreement 690623). Access to Imperial College High Performance Computing Service, doi: 10.14469/hpc/2232, is also acknowledged. The authors are grateful to R. García-Mayoral for careful refereeing comments.

## References


[1] Bernard P S, Thomas J M and Handler R A 1993 "Vortex dynamics and the production of Reynolds stress" *J. Fluid Mech.* **253** 385419

[2] Kravchenko A G, Choi H and Moin P 1993 "On the relation of near-wall streamwise vortices to wall skin friction in turbulent boundary layers" *Phys. Fluids A* **5** 3307–3309

[3] Quadrio M and Ricco P 2004 "Critical assessment of turbulent drag reduction through spanwise wall oscillations" *J. Fluid Mech.* **521** 251271

[4] Ghebali S, Chernyshenko S I and Leschziner M A 2017 "Can large-scale oblique undulations on a solid wall reduce the turbulent drag?" *Phys. Fluids* **29** 105102

[5] Taira K, Brunton S L, Dawson S T M, Rowley C W, Colonius T, McKeon B J, Schmidt O T, Gordeyev S, Theofilis V and Ukeiley L S 2017 "Modal analysis of fluid flows: an overview" *AIAA J.* **55** 12

[6] Berkooz G, Holmes P and Lumley J L 1993 "The proper orthogonal decomposition in the analysis of turbulent flows" 1993 *Annu. Rev. Fluid Mech.* **25** 539–575

[7] Nikolaidis M-A, Farrell B F, Ioannou P J, Gayme D F, Lozano-Durán A and Jiménez J 2016 "A POD-based analysis of turbulence in the reduced nonlinear dynamics system" *J. of Phys.: Conf. Series* **708** 012002

[8] Schmid P J 2010 "Dynamic mode decomposition of numerical and experimental data" *J. Fluid Mech.* **656** 5–28

[9] Ferrer E, De Vincente J and Valero E 2014 "Low cost 3D global instability analysis and flow sensitivity based on dynamic mode decomposition and high-order numerical tools" *Int. J. Numer. Meth. Fluids* **76** 169–184

[10] Bagheri S 2013 "Koopman-mode decomposition of the cylinder wake" *J. Fluid Mech.* **726** 596–623

[11] Sayadi T, Schmid P J, Nichols J W and Moin P 2014 "Reduced-order representation of near-wall structures in the late transitional boundary layer" *J. Fluid Mech.* **748** 278–301

[12] Duke D, Soria J and Honnery D 2012 "An error analysis of the dynamic mode decomposition" *Exp. Fluids* **52** 529–542

[13] Le Clainche S, Vega J M and Soria J 2017 "Higher order dynamic mode decomposition of noisy experimental data: the flow structure of a zero-net-mass-flux jet" *Exp. Therm. Fluid Science* **88** 336–353

[14] Chen K K, Tu J H and Rowley C W 2012 "Variants of dynamic mode decomposition: boundary condition, Koopman, and Fourier analyses" *J. Nonlinear Sci.* **22** 887–915

[15] Mezić I 2013 "Analysis of fluid flows via spectral properties of the Koopman operator" *Annu. Rev. Fluid Mech.* **45** 357–378

[16] Rowley C W, Mezia I, Bagheri S, Schlatter P and Henningson D S 2009 "Spectral analysis of nonlinear flows" *J. Fluid Mech.* **641** 115–127

[17] Saad Y 1992 *Numerical Methods for Large Eigenvalue Problems* Manchester University Press

[18] Del Álamo J C and Jiménez J 2009 "Estimation of turbulent convection velocities and corrections to Taylor's approximation" *J. Fluid Mech.* **640** 5–26

[19] Jovanović M R, Schmid P J and J W Nichols 2014 "Sparsity-promoting dynamic mode decomposition" *Phys. Fluids* **26** 024103

[20] Sesterhenn J and Shahirpour A 2016 "A Lagrangian dynamic mode decomposition" *ArXiv e-prints* 1603.02539

IOP Conf. Series: Journal of Physics: Conf. Series **1001** (2018) 012011 doi:10.1088/1742-6596/1001/1/012011

# Manipulation of near-wall turbulence by surface slip and permeability

**G. Gómez-de-Segura, C. T. Fairhall, M. MacDonald, D. Chung and R. García-Mayoral**

Department of Engineering, University of Cambridge, Trumpington St, Cambridge CB2 1PZ, UK

E-mail: r.gmayoral@eng.cam.ac.uk

**Abstract.** We study the effect on near-wall turbulence of tangential slip and wall-normal transpiration, typically produced by textured surfaces and other surface manipulations. For this, we conduct direct numerical simulations (DNSs) with different virtual origins for the different velocity components. The different origins result in a relative wall-normal displacement of the near-wall, quasi-streamwise vortices with respect to the mean flow, which in turn produces a change in drag. The objective of this work is to extend the existing understanding on how these virtual origins affect the flow. In the literature, the virtual origins for the tangential velocities are typically characterised by slip boundary conditions, while the wall-normal velocity is assumed to be zero at the boundary plane. Here we explore different techniques to define and implement the three virtual origins, with special emphasis on the wall-normal one. We investigate impedance conditions relating the wall-normal velocity to the pressure, and linear relations between the velocity components and their wall-normal gradients, as is typically done to impose slip conditions. These models are first tested to represent a smooth wall below the boundary plane, with all virtual origins equal, and later for different tangential and wall-normal origins. Our results confirm that the change in drag is determined by the offset between the origins perceived by mean flow and the quasi-streamwise vortices or, more generally, the near-wall turbulent cycle. The origin for the latter, however, is not set by the spanwise virtual origin alone, as previously proposed, but by a combination of the spanwise and wall-normal origins, and mainly determined by the shallowest of the two. These observations allow us to extend the existing expression to predict the change in drag, accounting for the wall-normal effect when the transpiration is not negligible.

## 1. Introduction

The use of complex surfaces to reduce turbulent skin friction is an area of active scientific and practical interest. Some of the most popular techniques include riblets [1], superhydrophobic surfaces [2] or permeable substrates [3, 4]. It is generally accepted that the drag reduction produced by these surfaces is due to the cross flow being more severely hindered than the streamwise mean flow. Luchini *et al.* [5, 6] and Jiménez [7] provided an expression to estimate quantitatively the change in drag produced by these surfaces. If the surface texture is small compared to the near-wall turbulent structures, the overlying flow perceives the texture as a uniform surface, inducing a homogeneous shear in the streamwise and spanwise directions. However, the virtual wall from which the streamwise and spanwise velocities originate can differ. The effect of the texture can then be reduced to a reference plane with non-zero velocities,

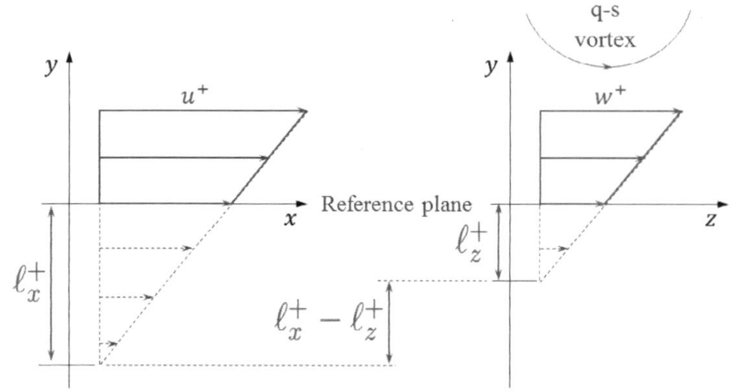


Figure 1: Sketch of streamwise and spanwise slip lengths. The term 'q-s vortex' refers to quasi-streamwise vortex.

contrary to conventional smooth walls, where all velocity components are zero at the surface. This reference plane is generally taken at the top plane of the surface geometry, such as at the riblet tips or at the plane above the gas pockets for superhydrophobic surfaces. The effect of the texture is typically modelled by slip boundary conditions at this reference plane $y = 0$, $u|_{slip} = \ell_x \, \partial u/\partial y|_{y=0}$ and $w|_{slip} = \ell_z \, \partial u/\partial y|_{y=0}$, where $\ell_x$ and $\ell_z$ are the streamwise and spanwise slip lengths, or alternatively, virtual origins. The slip lengths, $\ell_x$ and $\ell_z$, are defined as the depth below the reference plane where the streamwise and spanwise velocities induced by the overlying shear become zero when extrapolated. These concepts are portrayed in Figure 1. In the superhydrophobic community the term 'slip length' is used to denote virtual origins, while in the riblet community the term 'protrusion height' is used instead. Note that throughout this paper we will consider $x, y, z$ for the streamwise, wall-normal and spanwise coordinates and $u, v, w$ for the velocity components in $x$, $y$ and $z$ directions.

The classical theory of wall turbulence suggests that, sufficiently far away from the wall, the only appreciable effect of surface manipulations is to modify the intercept of the logarithmic velocity profile, while the Kármán constant remains unaltered [8]. Therefore, the change in drag with respect to the smooth wall is directly related to the change of the intercept of the velocity profile, $\Delta U^+$, where the superscript '+' denotes viscous units. An upward shift of the logarithmic region, $\Delta U^+ > 0$, translates into a decrease in drag, and vice-versa. Following the theory of Luchini *et al.* [5] and Jiménez [7], for sufficiently small surface textures the change in drag or $\Delta U^+$ is given by the difference between the streamwise and the spanwise slip lengths scaled in viscous units,

$$\Delta D \propto \Delta U^+ = \mu_0 \left(\ell_x^+ - \ell_z^+\right), \tag{1}$$

where $\mu_0$ is a universal constant, estimated as $\mu_0 \approx 0.785$ in [9] and $\mu_0 \approx 0.66$ in [7]. From a physical point of view, Eq. (1) indicates that the change in drag is due to a relative $y$-displacement of the quasi-streamwise vortices with respect to the mean flow. If the overlying spanwise flow, which is essentially induced by quasi-streamwise vortices, is hindered more than the mean streamwise flow, the virtual origin seen by quasi-streamwise vortices is above that seen by the mean flow, i.e. $\ell_z^+ < \ell_x^+$. Compared to a smooth wall, the vortices are 'pushed' further away from the streamwise virtual origin, even if they approach the reference plane. As a result, the local momentum flux close to the streamwise virtual origin decreases, thereby reducing the friction drag. Conversely, if $\ell_x^+ < \ell_z^+$, the vortices move closer to the origin of the mean flow, and friction drag increases.

This linear theory is only valid as long as the texture length scales are small compared to the

IOP Conf. Series: Journal of Physics: Conf. Series **1001** (2018) 012011 doi:10.1088/1742-6596/1001/1/012011

characteristic length scales of near wall turbulence. In the vicinity of the wall, the smallest typical length scale is the characteristic diameter of quasi-streamwise vortices, which is approximately 15 wall units [7]. As the texture size increases, the turbulent structures begin to perceive the surface texture. They no longer perceive the presence of the texture as a homogeneous slip, and the linear slip-length model in Eq. (1) breaks down [10, 11].

However, Fukagata *et al.* [12] and Busse & Sandham [13] observed that the linear dependence in Eq. (1) breaks down for spanwise slip lengths smaller than expected, of the order of $\sim 1$ wall unit. Beyond $\ell_z^+ \approx 1$, the spanwise slip length saturates. This saturation effect can be observed in Figure 2a, where the change in drag in terms of $\Delta U^+$ obtained by Busse & Sandham [13] for a wide range of $\ell_x^+$ and $\ell_z^+$ is portrayed. Analysing their results, the change in drag can still be represented by an expression similar to Eq. (1), if an effective spanwise slip length, $\ell_{z,eff}^+$, is used instead of $\ell_z^+$. The shift in the velocity profile is then $\Delta U^+ \propto \left(\ell_x^+ - \ell_{z,eff}^+\right)$, where $\ell_{z,eff}^+$ is empirically found from their data to be

$$\ell_{z,eff}^+ = \frac{\ell_z^+}{1+\ell_z^+/4}. \tag{2}$$

The two-dimensional parametric space can then be fitted into a single curve, as shown in Figure 2b. This results would be consistent with $\mu_0 \approx 1$ in Eq.(1). The data shows that isotropic textures with equal streamwise and spanwise slip lengths can also reduce drag, provided that the slip they induce is large enough, as the adverse effect of the spanwise slip saturates for $\ell_z^+ \gtrsim 1$. This is in agreement with the DNS results of Min & Kim [14], who studied the effect of streamwise and spanwise slip lengths independently as well as combined.

In this work, we will study the effect of introducing virtual origins for the tangential velocities, $u$ and $w$, and also for the wall-normal velocity, $v$. Although imposing a virtual origin for $u$ and $w$ in terms of slip lengths is very intuitive, defining a virtual origin for $v$ is not as simple, and the origin of $v$ has never been included in this framework. However, in the literature the effect of wall-normal velocity has been investigated separately [15], showing that turbulence is highly sensitive to transpiration, and it is, for instance, responsible for the eventual degradation of performance for drag-reducing technologies such as riblets [1]. The objective of this project is to study the effect of imposing different virtual origins for the three velocity components independently. In this paper, we conduct preliminary work with that aim, restricting ourselves to cases where all three virtual origins are the same, or cases where the origins for the tangential velocities are the same and that for the wall-normal velocity is zero. The paper is organised as follows. Firstly, the saturation of the spanwise slip length is discussed in Section 2. The numerical method, together with the models for defining virtual origins is presented in Section 3. Results are then presented in Section 4, followed by a discussion in Section 5. Finally, conclusions are summarised in Section 6.

## 2. Saturation of spanwise slip

As mentioned above, when the spanwise slip length increases beyond $\ell_z^+ \gtrsim 1$, its effect on the overlying flow saturates [13]. Consequently, equal slip lengths, $\ell_x^+ = \ell_z^+$, also lead to a decrease of drag, contrary to what the linear theory of slip lengths in Eq. (1) predicts. Note that both the simulations of [13] and the linear theory of slip lengths [5, 7] assume zero wall-normal velocity fluctuations at the reference plane. However, real surfaces that produce non-zero slips, such as riblets [1], porous substrates [16] or superhydrophobic surfaces [17], induce additionally a non-zero wall-normal velocity at the reference plane. The absence of transpiration in slip-only simulations leads to the saturation of the spanwise slip observed in Figure 2.

According to [5], the change in drag is due to a relative displacement of the streamwise mean flow and the quasi-streamwise vortices, but these vortices induce both spanwise and wall-normal

IOP Conf. Series: Journal of Physics: Conf. Series **1001** (2018) 012011 doi:10.1088/1742-6596/1001/1/012011

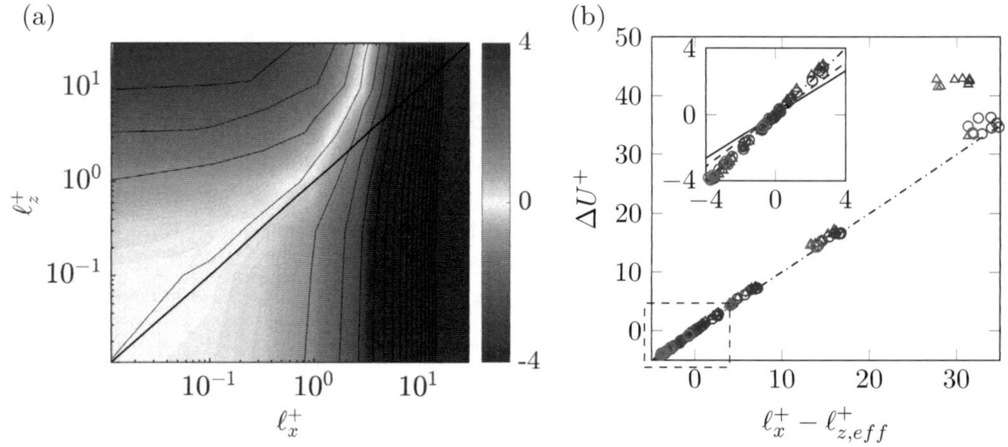


Figure 2: (a) Map of $\Delta U^+$ for different slip lengths, $\ell_x^+$ and $\ell_z^+$ from Busse & Sandham [13] for $Re_\tau = 180$. The black solid line corresponds to $\ell_x^+ = \ell_z^+$. (b) $\Delta U^+$ as a function of $\ell_x^+ - \ell_{z,eff}^+$, where $\ell_{z,eff}^+ = \ell_z^+/(1+\ell_z^+/4)$, using the same data as in (a). Triangles correspond to simulations at $Re_\tau = 180$ and circles at $Re_\tau = 360$. From blue to red the spanwise slip increases. The dash-dotted line represents Eq. (1) with the slope $\mu_0 = 1.0$. The inset shows a zoom for the small values, where the solid line represents Eq. (1) with the slope from [7], $\mu_0 = 0.66$; the dashed line that from [9, 6], $\mu_0 = 0.785$; and the dash-dotted line $\mu_0 = 1.0$.

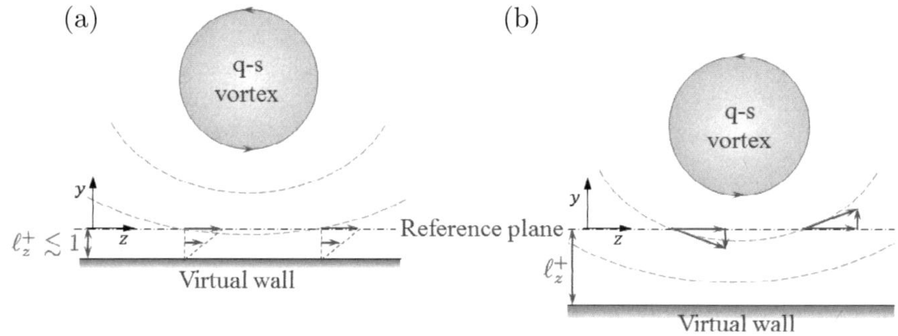


Figure 3: Schematic of the spanwise and wall-normal velocities induced by quasi-streamwise vortices at the reference plane for (a) virtual origins $\lesssim 1$ wall unit and (b) larger virtual origins.

velocities near the wall. Their distance to the wall would therefore not only be affected by the virtual origin for $w$, but also by the virtual origin for $v$. The sketch of quasi-streamwise vortices in Figure 3 helps to visualise this idea. For $\ell_z^+ \lesssim 1$, the wall-normal velocity induced by the vortices at the reference plane is small compared to the spanwise velocity (Figure 3a). Consequently, the effect of $v$ can be neglected and the virtual origin for the vortices is essentially that for the spanwise velocity. Therefore, for $\ell_z^+ \lesssim 1$ the change in drag is properly captured by the linear expression in Eq. 1, where the origin of quasi-streamwise vortices is set by that of the spanwise velocity alone, $\ell_z^+$.

On the other hand, if the reference plane is closer to the mean height of vortex cores, as in Figure 3b, the wall-normal velocity induced by the vortices at the reference plane is no longer negligible. When impeding $v$ at the reference plane, such as in the study by Busse & Sandham [13], the virtual origin that the vortices perceive is no longer set by $\ell_z^+$ alone, as the downward translation of the vortices is now restricted by $v$ being zero at the reference plane. For spanwise

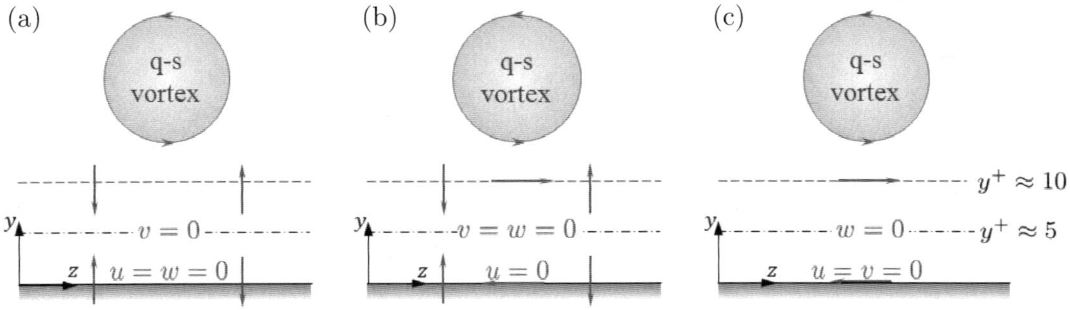


Figure 4: Schematics of the different control strategies studied by Choi *et al.* [18]. (a) *v-control*, (b) *v-w-control* and (c) *w-control*.

Table 1: Different control strategies studied by Choi *et al.* [18] and the equivalency with the displacement of virtual origins. The slip lengths are measured from the intermediate plane, i.e. the dash-dotted plane in Figure 4. Drag reduction (DR) is measured in terms of the change in the mean pressure gradient for a fixed mass flow rate.

| | *v-control* | *v-w-control* | *w-control* |
|---|---|---|---|
| $\ell_x^+$ | 5 | 5 | 5 |
| $\ell_z^+$ | 5 | 0 | 0 |
| $\ell_y^+$ | 0 | 0 | 5 |
| *DR* | 25% | 30% | 30% |

slip lengths of a few wall units, for which the saturation has already been reached, a further increase of $\ell_z^+$ has little or no effect on the origin of the overlying turbulence, since the wall-normal velocity is zero at the reference plane and the quasi-streamwise vortices cannot approach it any further. For the vortices to move closer to the reference plane, the wall-normal velocity would also need to perceive a non-zero virtual origin, i.e. some form of transpiration would need to exist at the reference plane. In this paper, in addition to the streamwise and spanwise virtual origins traditionally defined, we will also include a virtual origin for the wall-normal velocity.

In retrospect, the opposition control studies of Choi *et al.* [18] can be interpreted as an alternative method of imposing different virtual origins for the different velocity components but, in contrast with [13, 7, 5, 6], including a non-zero virtual origin for $v$. This would suggest that some active control techniques are based on the same drag-reduction mechanism as passive controls. Choi *et al.* explored several active control strategies for the wall-normal and spanwise velocities, including opposition control of the wall-normal velocity alone (*v-control*), control of the spanwise velocity alone (*w-control*), and combined control of $v$ and $w$ (*v-w-control*). In each case, the velocity components imposed at the wall were opposite to those measured at $y^+ \approx 10$. The authors concluded that the reduction of skin friction in the three cases was a result to the outward shift of the origin of turbulence with respect to the mean flow. These simulations can be interpreted as imposing different origins for the different velocities, since imposing at $y^+ = 0$ a velocity equal and opposite to that at $y^+ = 10$ is roughly equivalent to imposing a zero velocity at $y^+ \approx 5$. Schematics for the three cases considered are shown in Figure 4. The drag reduction achieved, together with the equivalent virtual origins, taking as reference plane that at $y^+ = 5$, can be found in Table 1. Note that these values only allow for qualitative comparison, since the

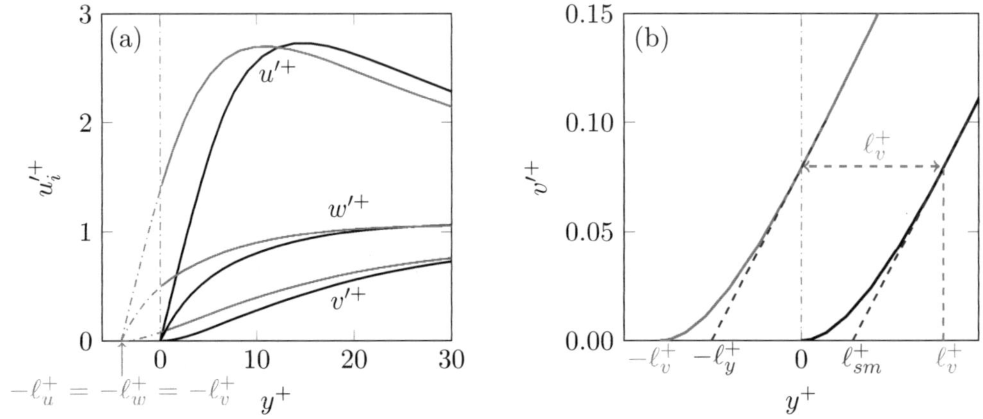


Figure 5: Schematics showing (a) the definition of the virtual origins $\ell_u^+$, $\ell_w^+$ and $\ell_v^+$ as the shift of the rms velocity fluctuations with respect to a smooth channel; (b) the difference between $\ell_v^+$ and $\ell_y^+$. Black, smooth channel; red, virtual origins equal and below the reference plane.

above analogy only produces a rough estimate of the effective virtual origins. The *v-control* case would be analogous to setting virtual origins $\ell_x^+ = \ell_z^+ = 5$ and $\ell_y^+ \approx 0$. This case corresponds to one of the cases from Busse & Sandham [13] portrayed in Figure 2, and a similar drag reduction is obtained, which is well captured by the modified drag-reduction expression introduced in Section 1, $\Delta U^+ \propto \left(\ell_x^+ - \ell_{z,eff}^+\right)$. The *v-w-control* case would be analogous to virtual origins $\ell_x^+ = 5$ and $\ell_y^+ = \ell_z^+ \approx 0$, which is also one of the cases from [13] included in Figure 2. The last case, *w-control*, is equivalent to imposing a virtual origin for $u$ and $v$, but not for $w$, which has not been considered in the previous studies, and corresponds to $\ell_x^+ = \ell_y^+ = 5$ and $\ell_z^+ \approx 0$. The obtained drag reduction is practically equal to that of the *v-w-control* case. By comparing *w-control* to *v-control*, it may be inferred that the effect of the spanwise slip length, $\ell_z^+$, is more restrictive than that of the wall-normal component, $\ell_y^+$.

Before moving to the modelling of virtual origins, some terms need to be defined. In the literature, the concepts of slip lengths and virtual origins have been used interchangeably, but here we will distinguish them. Hereinafter we will denote $\ell_x^+$, $\ell_z^+$ and $\ell_y^+$ as the slip lengths in the streamwise, spanwise and wall-normal directions. These slip lengths relate the velocities with their wall-normal gradients,

$$u|_{slip} = \ell_x \left.\frac{\partial u}{\partial y}\right|_{y=0}, \quad w|_{slip} = \ell_z \left.\frac{\partial w}{\partial y}\right|_{y=0} \quad \text{and} \quad v|_{slip} = \ell_y \left.\frac{\partial v}{\partial y}\right|_{y=0}, \tag{3}$$

and can be used to set Robin boundary conditions. Note that for $y$, $\ell_y$ does not convey a slip effect. The slip lengths simply correspond to the $y$-locations where the velocity components become zero if linearly extrapolated from the reference plane.

The virtual origins of $u$, $w$ and $v$, on the other hand, will be denoted by $\ell_u^+$, $\ell_w^+$ and $\ell_v^+$, respectively. These are obtained from the shifts of the rms velocity fluctuations, $u'^+$, $w'^+$ and $v'^+$ in the wall-normal direction with respect to the smooth channel profiles, as illustrated in Figure 5a. For small slip lengths, we expect the tangential slip lengths, $\ell_x^+$ and $\ell_z^+$, to be essentially equal to the virtual origins, $\ell_u^+$ and $\ell_w^+$. The case of $v$, however, is slightly more delicate. Due to the parabolic nature of the $v'$ profile at the wall, the virtual origin of $v$, $\ell_v^+$, differs from the slip length $\ell_y^+$ even for small values. From Figure 5b, we can observe that $\ell_y^+$ and $\ell_v^+$ relate to each other by $\ell_y^+ = \ell_v^+ - \ell_{sm}^+$, where $\ell_{sm}^+$ is obtained from extrapolating the slope of $v'^+$ at $y^+ = \ell_v^+$ for a smooth channel. Note that the value of $\ell_{sm}^+$ is a function of $\ell_v^+$, as it depends on the local slope at which the extrapolation of the curve is calculated. Equivalently, the origins of the mean flow, $\ell_U^+$, and that of the quasi-streamwise vortices, $\ell_{\omega_x}^+$, are defined as the wall-normal displacement of the mean flow and the streamwise vorticity $\omega_x'^+$ with respect to the smooth channel case. The streamwise vorticity $\omega_x'^+$ is mainly due to quasi-streamwise vortices. It therefore provides a good measure to identify the translation of these vortices [18], as it will be observed in the results.

## 3. Numerical setup

In this section, the numerical setup for the simulations is presented together with the models used to define the virtual origins. We conduct simulations of channel flows with off-wall boundary conditions to represent the virtual origins for the three velocity components.

The DNS code is adapted from that of García-Mayoral & Jiménez [1]. The incompressible Navier-Stokes equations are solved using a fractional step method combined with a three-step Runge-Kutta [19] in a doubly-periodic channel of height $2\delta$. The spatial discretisation is spectral in $x$ and $z$ using Fourier series, and the wall-normal direction is discretised using a second-order centred finite difference scheme. Simulations are conducted at constant pressure gradient, starting from a smooth wall flow at $Re_\tau \simeq 180$, which is the Reynolds number based on the half-height of the channel and the friction velocity $u_\tau$. The size of the computational domain is $2\pi \times \pi \times 2$ in the streamwise, spanwise and wall-normal directions, respectively, and simulations are performed for a grid of $128 \times 128 \times 153$ points, which corresponds to a resolution of $\Delta x^+ \approx 8$, $\Delta z^+ \approx 4$ and $\Delta y^+ \simeq 0.3 - 3$.

The virtual origins below the simulation boundary are modelled by means of off-wall conditions for the three velocities, which have previously been used to model the presence of smooth walls below the simulation boundary [20, 21, 22]. Although traditionally Robin boundary conditions have been used to impose the tangential virtual origins in terms of slip lengths [14, 13], the origin for the wall-normal velocity has not been introduced in that framework. Therefore, before being able to set the virtual origins for the three velocities independently, we explore different techniques to represent $\ell_v^+$. Two different techniques are investigated:

a) an impedance condition [15], relating $v$ at the simulation boundary to the pressure, which is obtained assuming that the flow between the virtual wall and the simulation boundary is dominantly viscous and governed by Stokes equations. The off-wall conditions for this technique are detailed in subsection 3.1.
b) Robin off-wall conditions, defined in Eq. (3), which linearly relate the velocity components with their wall-normal gradients.

To validate the above models, we will first consider cases with the same virtual origin for all the velocities, which we will denote by a simple $\ell^+$ for simplicity ($\ell_u^+ = \ell_v^+ = \ell_w^+ = \ell^+$). If the technique is successful at mimicking a virtual wall at $y^+ = -\ell^+$, flow for a smooth channel with an actual wall at the height of the virtual origin should be recovered. Compared to a smooth channel, the turbulence statistics should then be shifted by an amount equal to the virtual origin, as in Figure 5a.

### *3.1. Stokes model*

To derive the off-wall conditions with the Stokes model we will consider cases for which the virtual origins for all the components are at the same $y$-location, $y^+ = -\ell^+$. If we consider a small $\ell^+$, the advective terms in the region delimited by $y^+ = -\ell^+$ and $y^+ = 0$ would be negligible compared to the viscous terms. The flow can thus be modelled as a Stokes flow. To obtain the off-wall conditions at $y^+ = 0$ for the DNSs, we solve analytically the Stokes equation,

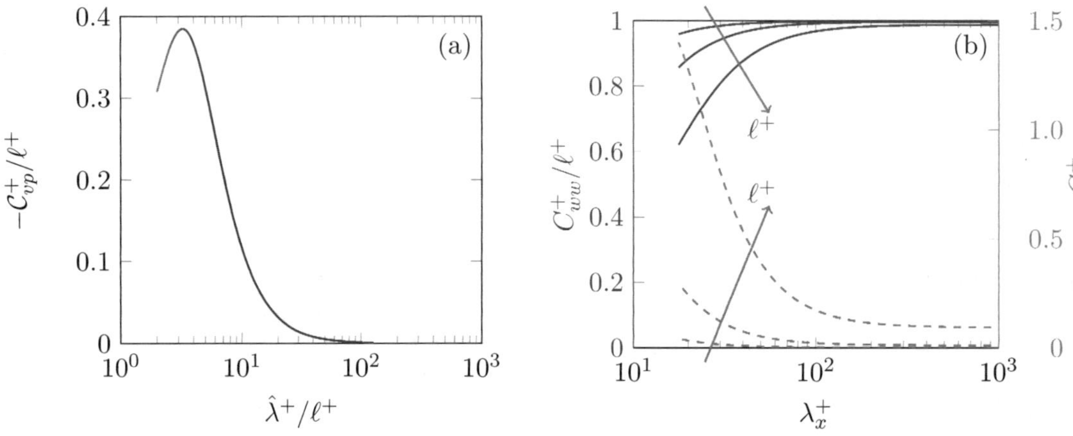


Figure 6: (a) Dimensionless transpiration coefficient $\mathcal{C}_{vp}/\ell^{+}$ as a function of the spherical wavelength $\hat{\lambda} = 2\pi/\sqrt{\alpha_x^2 + \alpha_z^2}$. (b) $\mathcal{C}_{vp}^{+}$, red dashed lines, and $\mathcal{C}_{ww}^{+}$, blue solid lines, for $\lambda_z^{+} \approx 100$ as a function of $\lambda_x^{+}$ for three different virtual origins $\ell^{+} = 1, 2$ and 4. Arrows indicate increasing $\ell^{+}$.

$-\nabla p + \nu \nabla^2 \mathbf{u} = 0$, where $\mathbf{u} = (u, v, w)$ is the velocity vector in $(x, y, z)$ coordinates, and $\nu$ is the kinematic viscosity. We assume the density $\rho$ to be unity for convenience. The problem can be expanded using Fourier series in $x$ and $z$. The Stokes equation is then solved as a response to waves of tangential shear, $d\hat{u}/dy$ and $d\hat{w}/dy$, and pressure, $\hat{p}$, at $y^{+} = 0$, where the hat indicates variables in Fourier space. At $y^{+} = -\ell^{+}$, impermeability and no-slip conditions are applied. Considering these boundary conditions, the analytical solution of the problem provides the following relationships between the velocities and the pressure at $y^{+} = 0$,

$$\hat{u}|_{y=0} = \mathcal{C}_{uu} \left.\frac{d\hat{u}}{dy}\right|_{y=0} + \mathcal{C}_{uw} \left.\frac{d\hat{w}}{dy}\right|_{y=0} + \mathcal{C}_{up}\hat{p}|_{y=0^{+}}\,, \tag{4a}$$

$$\hat{w}|_{y=0} = \mathcal{C}_{wu} \left.\frac{d\hat{u}}{dy}\right|_{y=0} + \mathcal{C}_{ww} \left.\frac{d\hat{w}}{dy}\right|_{y=0} + \mathcal{C}_{wp}\hat{p}|_{y=0^{+}}\,, \tag{4b}$$

$$\hat{v}|_{y=0} = \mathcal{C}_{vu} \left.\frac{d\hat{u}}{dy}\right|_{y=0} + \mathcal{C}_{vw} \left.\frac{d\hat{w}}{dy}\right|_{y=0} + \mathcal{C}_{vp}\hat{p}|_{y=0}\,. \tag{4c}$$

The constants $\mathcal{C}_{ij}$ depend on the virtual origin $\ell^{+}$, but also on the streamwise and spanwise wavelengths, $\lambda_x^{+}$ and $\lambda_z^{+}$. The corresponding expressions for the upper wall can be obtained by symmetry, and together they provide off-wall conditions for the flow within the channel. The dominant terms are those in the diagonal, i.e. $\mathcal{C}_{uu}$ and $\mathcal{C}_{ww}$ for $u$ and $w$, which would be equivalent to the slip lengths in the streamwise and spanwise directions, and $\mathcal{C}_{vp}$ for $v$, which would correspond to the impedance of [15].

To illustrate how the apparent permeability of the wall varies for different structures, the transpiration coefficient $\mathcal{C}_{vp}^{+}$ for different wavelengths is portrayed in Figure 6a. Note that the Stokes equation is isotropic in the $x$ and $z$ directions, meaning that $\mathcal{C}_{vp}^{+}$ follows a single curve when plotted against a spherical wavelength, $\hat{\lambda} = 2\pi/\sqrt{\alpha_x^2 + \alpha_z^2}$. Values in Figure 6a are scaled with the distance to the virtual wall, $\ell^{+}$. As observed in the figure, the reference plane is essentially impermeable to large structures, while the impermeability relaxes as the structures become smaller. Actual correlations of the pressure with the wall-normal velocity from direct observations in DNS are discussed in this volume by Sanmiguel Vila & Flores [23].

Table 2: Boundary conditions for the three velocity components for each case simulated. $Re_\tau$ is the Reynolds number measured at the virtual origin of the mean flow for each case, i.e. at $y^+ = -\ell_x^+$. The wavelength-dependent coefficients are obtained from solving the Stokes flow in $-\ell^+ < y^+ < 0$ (Figure 6)

.

| Case | $\hat{u}^+\|_{y^+=0} =$ | $\hat{w}^+\|_{y^+=0} =$ | $\hat{v}^+\|_{y^+=0} =$ | $Re_\tau$ |
|---|---|---|---|---|
| $FP222$ | Eq. (4a)$= f(\alpha_x, \alpha_z)$ | Eq. (4b)$= f(\alpha_x, \alpha_z)$ | Eq. (4c)$= f(\alpha_x, \alpha_z)$ | 183 |
| $DIP222$ | $C_{uu}^+(\alpha_x, \alpha_z)\frac{d\hat{u}}{dy}^+$ | $C_{ww}^+(\alpha_x, \alpha_z)\frac{d\hat{w}}{dy}^+$ | $C_{vp}^+(\alpha_x, \alpha_z)\hat{p}^+$ | 183 |
| $DHP222$ | $\ell_x^+\frac{d\hat{u}}{dy}^+$ | $\ell_z^+\frac{d\hat{w}}{dy}^+$ | $C_{vp}^+(\alpha_x, \alpha_z)\hat{p}^+$ | 183 |
| $DHV222$ | $\ell_x^+\frac{d\hat{u}}{dy}^+$ | $\ell_z^+\frac{d\hat{w}}{dy}^+$ | $\ell_y^+\frac{d\hat{v}}{dy}^+$ | 184 |
| $DHV220$ | $\ell_x^+\frac{d\hat{u}}{dy}^+$ | $\ell_z^+\frac{d\hat{w}}{dy}^+$ | 0 | 183 |
| $FP444$ | Eq. (4a)$= f(\alpha_x, \alpha_z)$ | Eq. (4b)$= f(\alpha_x, \alpha_z)$ | Eq. (4c)$= f(\alpha_x, \alpha_z)$ | 181 |
| $DHP444$ | $\ell_x^+\frac{d\hat{u}}{dy}^+$ | $\ell_z^+\frac{d\hat{w}}{dy}^+$ | $C_{vp}^+(\alpha_x, \alpha_z)\hat{p}^+$ | 185 |
| $DHV444$ | $\ell_x^+\frac{d\hat{u}}{dy}^+$ | $\ell_z^+\frac{d\hat{w}}{dy}^+$ | $\ell_y^+\frac{d\hat{v}}{dy}^+$ | 185 |
| $DHV440$ | $\ell_x^+\frac{d\hat{u}}{dy}^+$ | $\ell_z^+\frac{d\hat{w}}{dy}^+$ | 0 | 185 |

Additionally, in Figure 6b the dependence on the streamwise wavelength of the transpiration coefficient $C_{vp}^+$ is compared to that of the slip coefficient, $C_{ww}^+$, at $\lambda_z^+ = 100$ for a range of $\lambda_x^+$ and for different $\ell^+$. Structures of the size of streaks ($\lambda_x^+ \approx 1000$) and quasi-streamwise vortices ($\lambda_x^+ \approx 100$) are subject to essentially the same slip length as the mean, $C_{ww}^+ = \ell^+$. Hence, the slip coefficient $C_{ww}^+$ is practically homogeneous and it could be treated as such, as it will be shown in the next section. This, although not shown, is also the case for $C_{uu}^+$. The transpiration $C_{vp}^+$, however, is more sensitive to the wavelengths. Although $C_{vp}^+$ is zero for very large $\hat{\lambda}^+$, its value increases as the wavelengths decrease. This can be observed for $\ell^+ = 4$ in Figure 6b, where the surface is already quite permeable for $\lambda_z^+ = 100$. Hence, while the dependence on the wavelength of $C_{uu}^+$ and $C_{ww}^+$ can be neglected, the wavelength-dependent nature of the transpiration coefficient $C_{vp}^+$ needs to be accounted for.

### *3.2. Cases under study*

As previously mentioned, our first objective is to find appropriate slip and impedance boundary conditions at $y^+ = 0$ to mimic the effect of a wall at $y^+ = -\ell^+$. Once we have identified a suitable technique, we will extend the method to different virtual origins for the different velocities.

We explore the following alternatives:

a) First, we will consider the full conditions from Eq. (4), which are derived from solving the Stokes flow between $y^+ = -\ell^+$ and $y^+ = 0$. Three cases will be analysed, $\ell^+ = 1, 2$ and 4.

b) To later define different virtual origins for each velocity component, we eliminate the coupling between different components from Eq. (4), by retaining only the dominant terms for each component, i.e. the slip coefficients $C_{uu}$ and $C_{ww}$ for $u$ and $w$, and the transpiration coefficient $C_{vp}$ for $v$. The idea is to later define these separately, as if due to different virtual origins.

c) In an attempt to define the virtual origins in a simpler manner, we will further simplify the boundary conditions from case b). Following previous studies on slip lengths [14, 13]

Table 3: Virtual origins for the velocities, $\ell_u^+$, $\ell_w^+$ and $\ell_v^+$, the mean flow, $\ell_U^+$, and quasi-streamwise-vortices, $\ell_{\omega_x}^+$, obtained from DNS for the different cases conducted. $\ell_{sm}^+$ corresponds is obtained from extrapolating the v-profile for a smooth channel, as defined in Figure 5b.

| | $FP222$ | $DIP222$ | $DHP222$ | $DHV222$ | $DHV220$ | $FP444$ | $DHP444$ | $DHV444$ | $DHV440$ |
|---|---|---|---|---|---|---|---|---|---|
| $\ell_u^+$ | 2.0 | 2.0 | 2.0 | 2.0 | 2.0 | 3.9 | 4.0 | 4.0 | 4.0 |
| $\ell_w^+$ | 2.1 | 1.7 | 1.7 | 2.0 | 2.0 | 5.7 | 4.0 | 3.0 | 3.1 |
| $\ell_v^+$ | 2.1 | 1.5 | 1.5 | 2.3 | 1.3 | 7.4 | 9.5 | 3.5 | 1.7 |
| $\ell_{sm}^+$ | – | – | – | 0.8 | 0.0 | – | – | 1.5 | 0.0 |
| $\ell_{mean}^+$ | 2.0 | 2.0 | 2.0 | 2.0 | 2.0 | 3.7 | 3.5 | 3.9 | 4.0 |
| $\ell_{\omega_x}^+$ | 2.1 | 1.5 | 1.5 | 2.0 | 1.3 | – | – | 3.0 | 1.7 |

and the discussion at the end of subsection 3.1 on wavelength dependence, we will impose homogeneous tangential slip lengths, $\ell_x^+$ and $\ell_z^+$, to model the virtual origins of $u$ and $w$, while keeping a wavelength-dependent transpiration, $\mathcal{C}_{vp}^+$, as before.

d) Alternatively, instead of conditions derived from Eq. (4), we implement Robin conditions as defined in Eq. 3, even for the wall-normal velocity. In a first approach to introducing different virtual origins for the different velocities, we will study cases with $\ell_x^+ = \ell_z^+ = 1, 2$ and 4 and $\ell_y^+ = 0$, as in [13], and compare them with the simulations $\ell_x^+ = \ell_z^+ = \ell_y^+ = 1, 2$ and 4.

The different simulations and their numerical parameters are summarised in Table 2. Each case is labelled with 2 or 3 letters followed by 3 numbers. The letters refer to the equations used as boundary conditions and the three numbers correspond to the desired values for the virtual origins for $u$, $w$ and $v$, respectively, i.e. $\ell_u^+$, $\ell_w^+$ and $\ell_v^+$. For the first letter, $F(ull)$ means that the full conditions from Eq. (4) are used, whereas $D(iagonal)$ means that only the three dominant terms, $\mathcal{C}_{uu}$, $\mathcal{C}_{ww}$ and $\mathcal{C}_{vp}$ are used, and the velocities are then uncoupled. The second letter denotes the type of boundary condition used for $u$ and $w$: $H(omogeneous)$ when homogeneous slip lengths are considered, and $I(nhomogeneous)$ when wavelength-dependent coefficients, resulting from the analytical solution of the Stokes equation, are considered. The last letter denotes the type of boundary condition used for $v$: $P$ when an impedance between $v$ and $p$ is imposed (Stokes model), and $v$ when $v = \ell_y dv/dy$ is used instead.

## 4. Results and discussion

From the three virtual origins under study, i.e. $\ell^+ = 1, 2$ and 4, the first two exhibit proportionately the same behaviour, and hence only results for $\ell^+ = 2$ and 4 are presented here. Throughout this section, results are scaled with magnitudes measured at the virtual origin of the mean flow, $\tilde{\delta} = \delta + \ell_U$, so that $\tilde{\delta}$, the half-channel height perceived by the mean flow, is greater than for the smooth case, and $u_\tau = u_{\tau_{y=0}}(\tilde{\delta}/\delta)^{1/2}$. This is essentially equivalent to ascribing the position of a reference smooth wall for comparison where the mean flow perceives it, i.e. at $y^+ = -\ell_U^+$.

We find the different virtual origins by measuring the wall-normal displacement of the mean flow and rms velocity profiles relative to smooth wall data, as was illustrated in Figure 5. The different virtual origins for each case, i.e. $\ell_U^+$, $\ell_u^+$, $\ell_w^+$, $\ell_v^+$, and $\ell_{\omega_x}^+$, are assembled in Table 3. To assess the validity of the off-wall conditions, in Figures 7-13 the profiles are shifted by their

IOP Conf. Series: Journal of Physics: Conf. Series **1001** (2018) 012011 doi:10.1088/1742-6596/1001/1/012011

corresponding virtual origins. The degree of resemblance of the shifted profiles to the smooth wall data allows us to determine the degree of success of the technique.

### *4.1. Fully coupled boundary conditions*

We first impose the full boundary conditions provided by Eq. (4), with the aim of modelling a smooth wall at $y^+ = -2$ and at $y^+ = -4$. The mean flow and rms fluctuations are portrayed in Figure 7. At a first glance, we can already conclude that case $FP444$ fails to mimic a wall at $y^+ = -4$, while case $FP222$ seems to succeed, as all the shifted curves fall on top of the smooth channel data.

For $FP222$ the mean velocity profile in Figure 7a shows a virtual origin of $\ell_U^+ \approx 2$. Similarly, all the three velocity fluctuations as well as the streamwise vorticity have been shifted towards the wall by 2 wall units with respect to the smooth wall, thereby implying that the virtual origin of the quasi-streamwise vortices is $\ell_{\omega_x}^+ \approx 2$, same as the mean flow. Thus, the boundary conditions have successfully reproduced the effect of having a virtual origin at $y^+ = -2$. The validity of the model for case $FP222$ is confirmed by the spectral density distributions in Figures 8a-c, which show that the distribution of the velocity fluctuations among different length scales agrees reasonably well with that of the smooth channel case. In these figures, the data for case $FP222$ is compared to the corresponding data for a smooth wall at $y_{smooth}^+ = y^+ + \ell^+$.

While the mean profile for $FP222$ collapses with the profile for the corresponding smooth channel, that for $FP444$ exhibits a downward shift of the logarithmic region. This suggests that the model does not merely shift the velocities by the virtual origin, as observed in the rms velocity and $\omega_x^+$ fluctuations shown in Figures 7b-c. For case $FP444$ the near-wall turbulence is completely disrupted compared to a smooth wall case. The peak of the streamwise fluctuations in Figure 7 decays, while that of the streamwise vorticity increases, suggesting a weakening of the streaks and a strengthening of the vortices. Inspecting the spectral energy distributions of the three velocities, Figures 8d-f, the wall-normal velocity exhibits the most dramatic change. The spectrum shows the appearance of new energetic structures, short in $x$ and wide in $z$. The source of the energy can be attributed to the large values of the impedance coefficient $\mathcal{C}_{vp}^+$ assigned to small length scales, as observed in Figure 6a. This will be further discussed in the following subsections.

These results suggest that the conditions derived from considering a Stokes flow underneath are a good model for $\ell^+ \lesssim 2$, but are not valid for $\ell^+ = 4$.

### *4.2. Uncoupled boundary conditions*

The implementation of the full conditions from Eq. (4) involves the full coupling between the three velocities and the pressure, which makes setting different virtual origins for each component difficult. In what follows we neglect certain terms while still maintaining a common virtual origin, aiming to decouple the three velocity conditions to later impose different origins. For that, two different approaches are tested: case $DIP222$, where only the dominant term for each velocity component has been retained, i.e. the slip coefficients $\mathcal{C}_{uu}^+$ and $\mathcal{C}_{ww}^+$ for $u$ and $w$, and the transpiration coefficient $\mathcal{C}_{vp}^+$ for $v$; and case $DHP222$ and $DHP444$, similar to the previous case but where homogeneous slip lengths in $x$ and $z$ have been used, removing any dependencies of the slip coefficients on the length scales. The latter is supported by previous studies on slip lengths [14, 13], where they considered homogeneous slip lengths for $u$ and $w$ defined by Robin conditions.

For the case of a virtual origin at $y^+ = -2$, similar results are obtained with the two approaches, $DIP222$ and $DHP222$, as shown in red in Figures 9a-c and 9d-f, respectively. This agrees well with the observations in Section 3.1, where we concluded that $\mathcal{C}_{ww}^+$ and $\mathcal{C}_{uu}^+$ are essentially homogeneous for the energetically significant flow length scales. When comparing the current two cases with the previous case with full off-wall conditions, $FP222$, little difference is

IOP Conf. Series: Journal of Physics: Conf. Series **1001** (2018) 012011 doi:10.1088/1742-6596/1001/1/012011

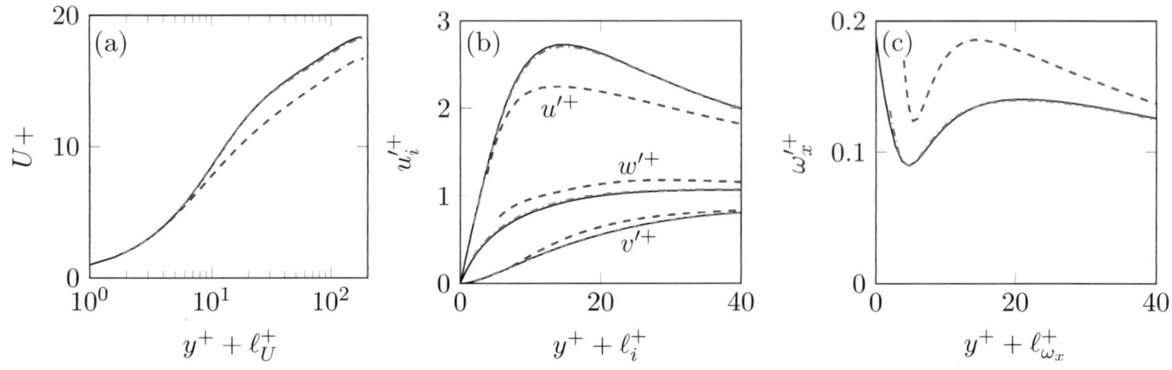


Figure 7: (a) Mean velocity profile shifted by the virtual origin of the mean flow, $\ell_U^+$; (b) rms velocity fluctuations shifted by their corresponding virtual origins, $\ell_u^+$, $\ell_v^+$ and $\ell_w^+$; (c) streamwise vorticity fluctuations shifted by the virtual origin of the vortices. In the case $FP444$ it is shifted by $\ell^+ = 4$. Solid lines, smooth channel; dash-dotted, $FP222$; and dashed, $FP444$.

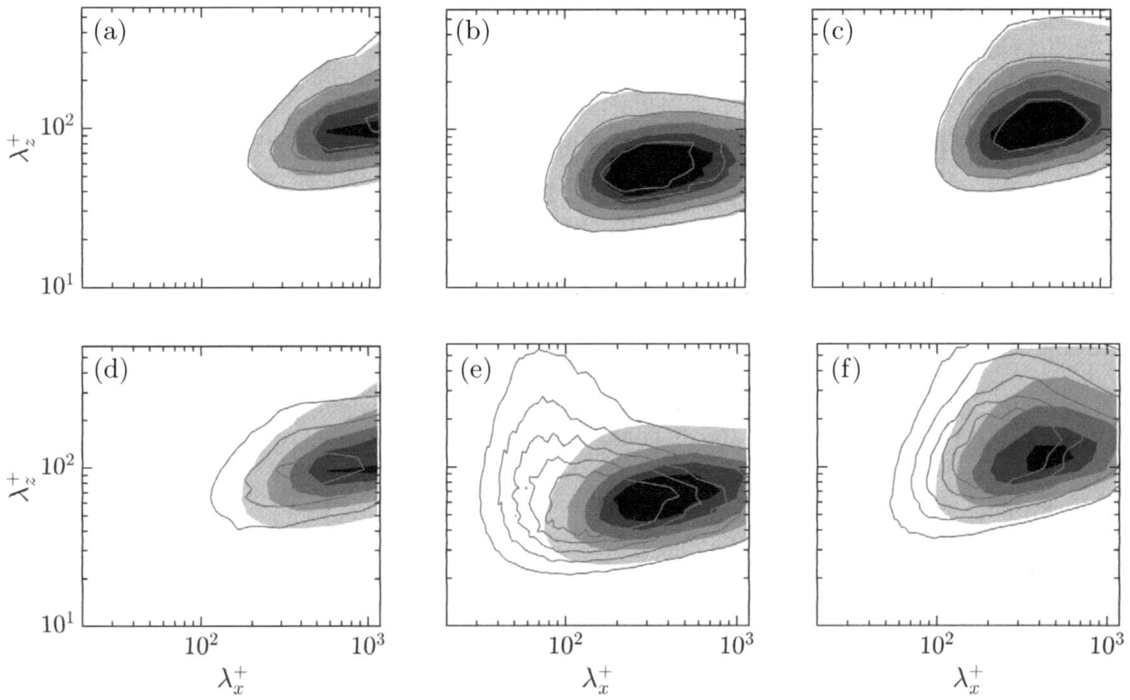


Figure 8: Premultiplied spectral energy densities of the velocity components. (a-c) Case $FP222$. Shaded contours represent the smooth channel at $y^+ \approx 7$ and the solid lines the present DNSs shifted by their corresponding virtual origin, at $y^+ \approx 7 - \ell_i^+$, $\ell_i^+$ being $\ell_u^+$, $\ell_v^+$ and $\ell_w^+$. (a) $k_x k_z E_{uu}$, (b) $k_x k_z E_{vv}$ and (c) $k_x k_z E_{ww}$. (d-f) Case $FP444$. Shaded contours represent the smooth channel at $y^+ \approx 13$ and the solid lines the present DNSs shifted by their corresponding virtual origin. (d) $k_x k_z E_{uu}$, (e) $k_x k_z E_{vv}$ and (f) $k_x k_z E_{ww}$.

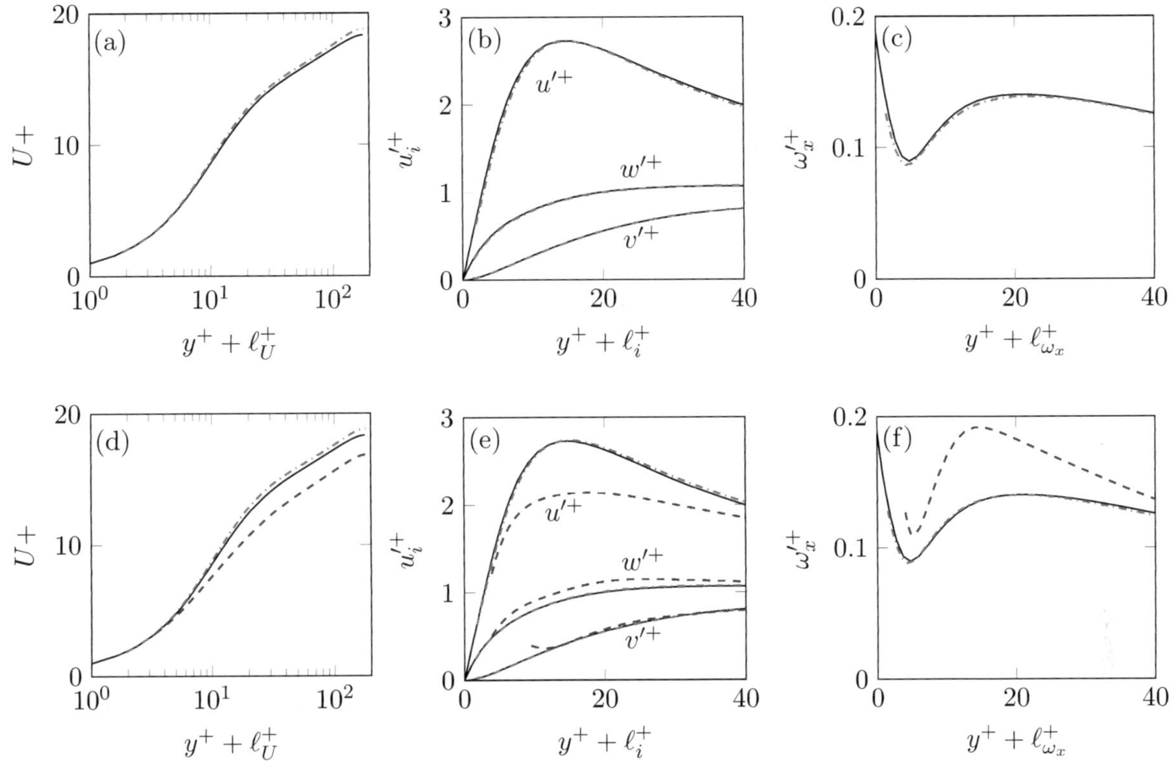


Figure 9: (a,d) Mean velocity profile; (b,e) rms velocity fluctuations where $\ell_i^+ = \ell_u^+$, $\ell_v^+$ and $\ell_w^+$; (c,f) streamwise vorticity fluctuations. (a-c) Solid lines, smooth channel; dash-dotted, case $DIP222$. (d-f) Solid, smooth channel; dash-dotted, case $DHP222$; dashed, case $DHP444$. All results are shifted by their corresponding virtual origins, except for case $DHP444$ in (f), for which the shift is $\ell^+ = 4$.

observed. The mean flows have virtual origins of 2 wall units, same as for $FP222$. The velocity fluctuations, however, are slightly shifted from the $FP222$ case, although it is indiscernible in the figures. In particular, the virtual origin of the wall-normal velocity is slightly less than 2 wall-units, $\ell_v^+ \approx 1.5$. The virtual origin of $\omega_x'$, which relates to the displacement of quasi-streamwise vortices, is also $\ell_{\omega_x}^+ \approx 1.5$, same as $\ell_v^+$. The difference in virtual origins for the mean velocity, $\ell_U^+ \approx 2$, and the streamwise vortices, $\ell_{\omega_x}^+ \approx 1.5$, results in a small upward shift of the log law, as observed in Figure 9a.

For a virtual origin at $y^+ = -4$, only the second approach ($DHP444$) has been tested, obtaining similar results to case $FP444$. This approach failed to model a virtual origin at $y^+ = -4$, as we could expect from the previous case. The $v'^+$ profile in this case diverges even more than before. Although not shown here, in this case the new energetic region observed in the v-spectrum of $FP444$ (see Figure 8e) intensifies, indicating that the disparity between the current case and the corresponding smooth channel is indeed due to the large permeability values at small length scales. In order to eliminate the appearance of the new structures observed in this case, brute-force filtering was also tested in a separate simulation. The boundary condition $v^+ = \mathcal{C}_{vp}^+ p^+$ was applied to large structures only, while structures smaller than $\lambda_x^+ < 50$ and $\lambda_z^+ < 10$ were free to permeate, $dv^+/dy = 0$. However, this was not sufficient to impede the amplification of small length scales. A model that accounts for advective effects in the modelled

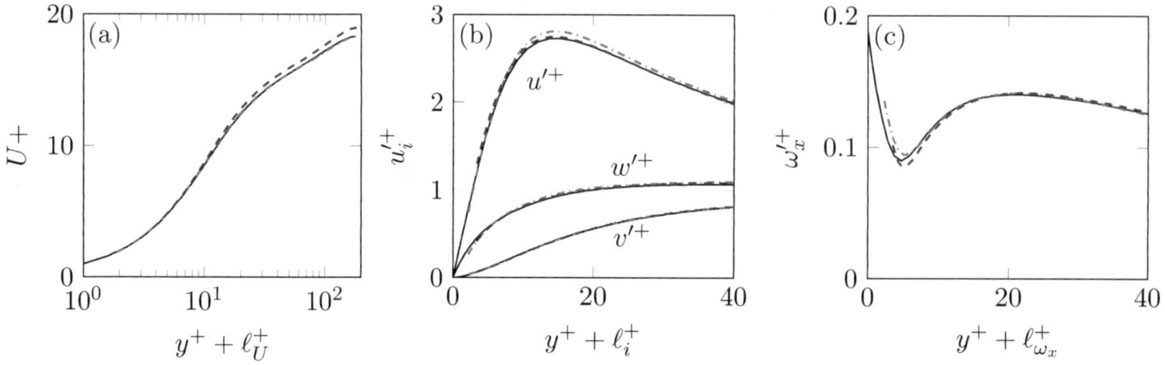


Figure 10: (a) Mean velocity profile; (b) rms velocity fluctuations; (c) streamwise vorticity fluctuations. Solid lines, smooth channel; dash-dotted, $DHV222$; dashed, $DHV444$. All results have been shifted by their corresponding virtual origins.

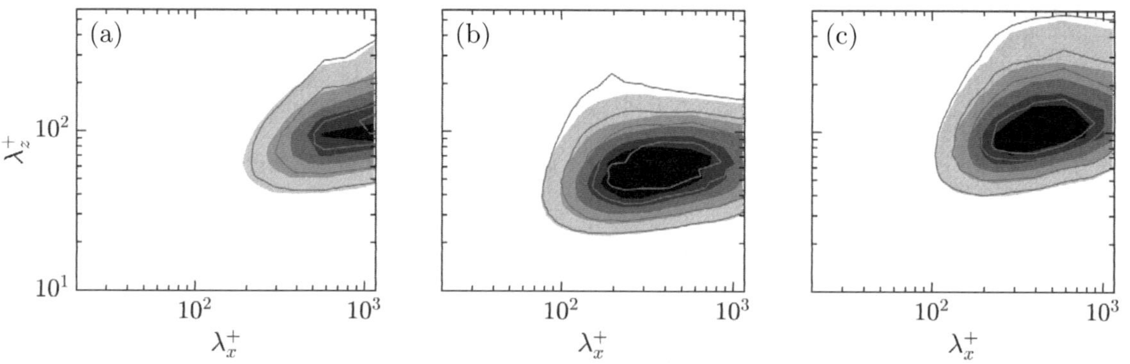


Figure 11: Premultiplied spectral energy densities of the velocity components for $DHV444$. Shaded contours represent the smooth channel at $y^+ \approx 7$ and the solid lines the present DNSs shifted by their corresponding virtual origin, at $y^+ \approx 7 - \ell_i^+$, $\ell_i^+$ being $\ell_u^+$, $\ell_v^+$ and $\ell_w^+$. (a) $k_x k_z E_{uu}$, (b) $k_x k_z E_{vv}$ and (c) $k_x k_z E_{ww}$.

layer may be more suitable for deep virtual origins, but this has been left for future work.

### *4.3. Robin conditions for the three velocities*

Since the model constructed from the Stokes equation breaks down for a virtual origin of 4 wall units, a different strategy has been pursued. By analogy with the slip-length model for the tangential velocities, we propose Robin conditions for the three velocities, so that the velocities $u^+$, $v^+$ and $w^+$ are related to their wall-normal gradients by three slip lengths, as in Eq. 3. However, in the wall-normal direction $\ell_y^+$ does not correspond to the virtual origin of $v$. The wall-normal velocity profile is not linear near the wall, unlike the tangential velocities, so the slip length $\ell_y^+$ should differ from the virtual origin $\ell_v^+$ which we intend to model. We set $\ell_y^+$ based on the offset with $\ell_v^+$ observed for smooth walls, as previously indicated in Figure 5b, that is, $\ell_y^+ = \ell_v^+ - \ell_{sm}^+$.

Comparing to the previous model, this technique produces a significant improvement on the results, as shown in Figures 10 and 11. This model reproduces with reasonable accuracy not only a virtual origin at $y^+ = -2$, but also that at $y^+ = -4$. For the latter, the displacements of the

$v'^+$ and $w'^+$ profiles are not exactly 4 wall units, but $\ell_v^+ \approx 3.5$ and $\ell_w^+ \approx 3.0$. The virtual origin for the quasi-streamwise vortices is also $\ell_{\omega_x}^+ \approx 3.0$, as observed in Figure 10c, which coincides with $\ell_w^+$. This agreement between the origin for the vortices and that for the spanwise velocity may be attributed to $\ell_w^+$ being the most restrictive origin between those of $v$ and $w$. Since in this case the virtual origin of $w$ is the shallowest of all the three velocities, i.e. $\ell_w^+ < \ell_v^+ < \ell_u^+$, the displacement of quasi-streamwise vortices would be limited by $w$, which would explain why $\ell_{\omega_x}^+ = \ell_w^+$. However, the virtual origins in this case are too close for the evidence to be conclusive. The relation between virtual origins will be analysed in the next section. The small difference between the virtual origin for the mean and that for the vortices translates into a slight shift upwards of the mean profile with respect to the smooth channel.

From these results, we can conclude that the Robin boundary conditions reproduce successfully, and in a simple manner, the effect of a virtual origin. Case $DHV444$ could be further improved by imposing a slightly larger $\ell_z^+$ than 4 wall units, since at 4 wall units the spanwise velocity profile is no longer linear, and $\ell_w^+$ differs from $\ell_z^+$. More accurate results would be obtained if the spanwise slip-length was defined similarly to how $\ell_y^+$ was calculated in this section, so that the origin of the spanwise velocity would lie at $y^+ = -4$.

### *4.4. Different virtual origins for the tangential velocities and for v*

In this section, we impose a different origin for $v$ than those for $u$ and $w$. Two cases will be presented, case $DHV220$ and $DHV440$ with $\ell_x^+ = \ell_z^+ = 2$ and 4, respectively, while keeping the impermeability condition at the simulation boundary, i.e. $\ell_y^+ = 0$, similar to [13].

For cases $DHV220$ and $DHV440$, the virtual origins for the mean flow are $\ell_U^+ \approx 2$ and 4, respectively, as shown in Figure 12a. However, the mean velocity profiles experience an upward shift of the logarithmic region relative to the smooth case, which directly relates to a reduction of drag. As imposed by $\ell_x^+$ and $\ell_z^+$, $u'^+$ and $w'^+$ profiles go to zero at $y^+ \approx -2$ and $y^+ \approx -4$ when linearly extrapolated, which correspond approximately to the virtual origins. The wall-normal velocity profiles are also shifted relative to the smooth profile, meaning that the virtual origin of $v$ is not zero. Although $v = 0$ at the wall, i.e. $\ell_y^+ = 0$, its slope $dv/dy$ is no longer zero due to continuity. Thus the vertical displacement of the $v'^+$ profile results in a non-zero virtual origin for $v$, $\ell_v^+ \approx 1.3$ for case $DHV220$ and $\ell_v^+ \approx 1.7$ for $DHV440$.

The resulting virtual origins for the quasi-streamwise vortices are $\ell_{\omega_x}^+ \approx 1.3$ for case $DHV220$ and $\ell_{\omega_x}^+ \approx 1.7$ for $DHV440$, which coincide with the corresponding virtual origins for $v$. The reason is that in these cases the downward displacement of the vortices is restricted by $v$ being zero at the simulation boundary, as discussed in Section 2. The vortices approach the wall, as they are allowed to slip in the spanwise direction, and their approach corresponds with the origin perceived by the wall-normal velocity, which is the shallowest origin between $\ell_v^+$ and $\ell_w^+$. Hence, the virtual origin for the vortices in these cases appears to be determined by the origin for the wall-normal velocity.

The fact that the origin for the mean flow is deeper than that for the vortices translates into a reduction of skin friction, as observed in the upward shift of the mean velocity profile in the log law region. Hence, a $\Delta U^+ > 0$ is obtained despite the slip lengths in the streamwise and spanwise directions being equal, due to the wall-normal velocity having a shallower virtual origin. Furthermore, as discussed in Section 2, in these cases the change $\Delta U^+$ should be well defined by $\Delta U^+ \propto \left(\ell_x^+ - \ell_{z,eff}^+\right)$. This is indeed the case. The values of $\ell_{\omega_x}^+$ agree reasonably well with the empirical function for the effective spanwise slip length in Eq. (2), which gives $\ell_{z,eff}^+ = 1.3$ and 2.0. Therefore, $\ell_{z,eff}^+$ can be interpreted as an expression to predict $\ell_v^+$, and equivalently $\ell_{\omega_x}^+$, when $\ell_y^+ = 0$. These results support the idea that the change in drag is due to the displacement of the mean flow with respect of the quasi-streamwise vortices, and clarify the saturation effect of the spanwise slip length observed for large $\ell_z^+$.

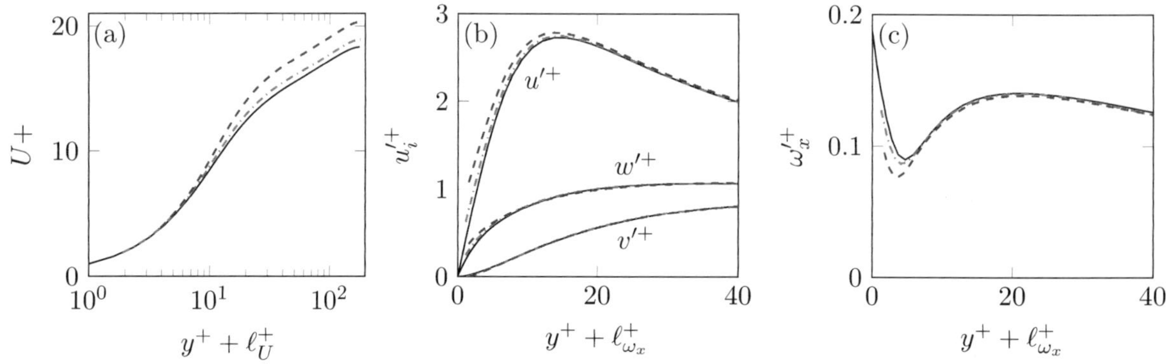


Figure 12: (a) Mean velocity profile shifted by the virtual origin of the mean flow, $\ell_U^+$; (b) rms velocity fluctuations shifted by the virtual origin of the vortices, $\ell_{\omega_x}^+$; (c) streamwise vorticity fluctuations shifted by $\ell_{\omega_x}^+$. Solid, smooth channel; dash-dotted, $DHV220$; dashed, $DHV440$.

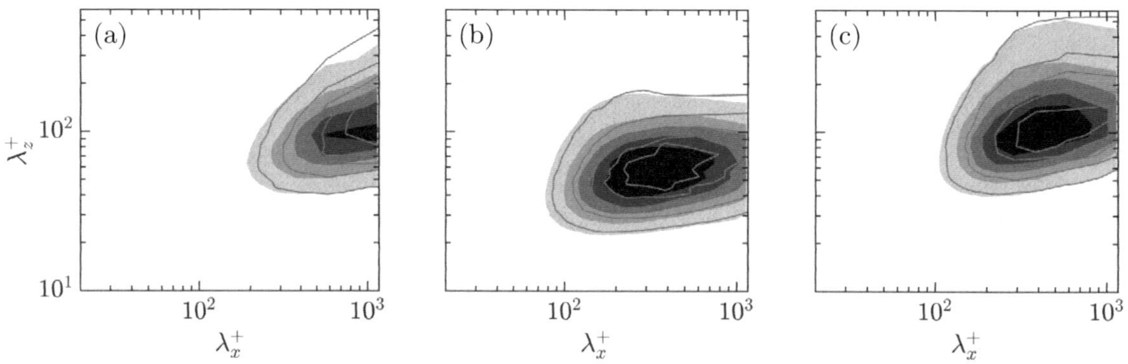


Figure 13: Premultiplied spectral energy densities of the velocity components for $DHV440$. Shaded contours represent the smooth channel at $y^+ \approx 7$ and the solid lines the present DNSs shifted by the virtual origin of the vortices, at $y^+ \approx 7 - \ell_{\omega_x}^+ = 5.3$. (a) $k_x k_z E_{uu}$, (b) $k_x k_z E_{vv}$ and (c) $k_x k_z E_{ww}$.

In Figure 12b, all the three velocity rms's have been shifted by the virtual origin of the vortices, achieving a better match with the smooth channel data than when shifting by their corresponding virtual origins. This implies that the origin of all three velocity rms's is represented well by the virtual origin of the vortices, even if the streamwise and spanwise profiles have a different slope at $y^+ = 0$, enforced by the boundary conditions implemented. The slip-lengths modify the origin of the near-wall turbulence and the near-wall cycle is shifted in block by $\ell_{\omega_x}^+$. This is confirmed by the good agreement of the spectral energy distributions observed in Figure 13 for $DHV440$, where the spectra of the present case are superimposed over the smooth channel data displaced by the virtual origin of the vortices, i.e. at $y^+ + \ell_{\omega_x}^+$.

## 5. Generalisation of the linear slip-length model

In this section we review the linear theory of slip lengths [6, 7]. From the present results, we confirm the existing understanding that the change in friction is caused by the difference between the virtual origins for the mean flow and that for quasi-streamwise vortices. However, the origin for the vortices is not completely defined by $\ell_z^+$ alone, as originally postulated in [5, 7]. We

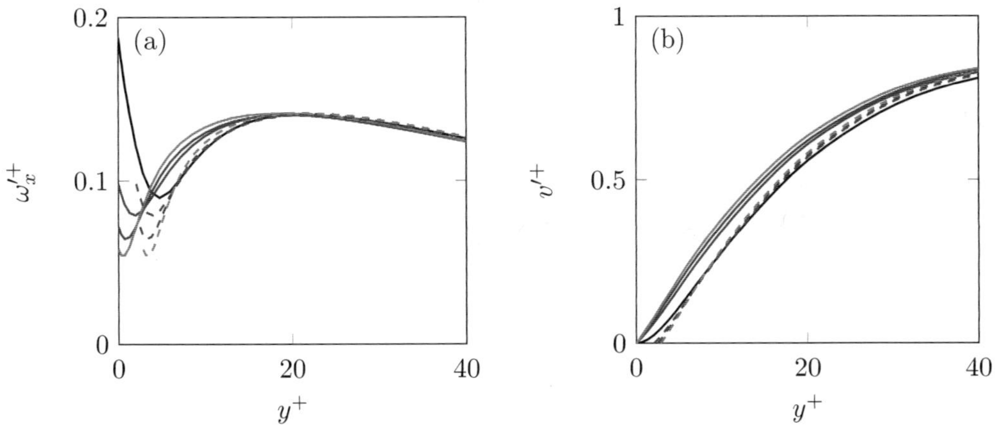


Figure 14: Rms fluctuations of (a) the streamwise vorticity and (b) the wall-normal velocity, for simulations with $\ell_x^+ = \ell_z^+$ and $\ell_y^+ = 0$. Black curves represent the smooth wall case for reference. Blue, $\ell_x^+ = \ell_z^+ = 4.2$; purple, $\ell_x^+ = \ell_z^+ = 6.7$; red, $\ell_x^+ = \ell_z^+ = 10$. The dashed lines correspond to the solid lines shifted by $\ell_{z,eff}^+$, as defined in Eq.2. Data from [24].

propose to generalise the existing linear theory to

$$\Delta U^+ \propto \left(\ell_U^+ - \ell_{\omega_x}^+\right). \tag{5}$$

We can expect $\ell_U^+ \approx \ell_x^+$, since in the viscous sublayer the mean velocity profile is essentially linear, and both $\ell_U^+$ and $\ell_x^+$ can be used interchangeably. However, contrary to the linear theory of slip lengths, the origin of the vortices is not given by the spanwise slip length alone, but by a combination of the spanwise and wall-normal slip lengths, $\ell_z^+$ and $\ell_y^+$. We suggest that $\ell_{\omega_x}^+$ is determined by the most restrictive virtual origin between that of $v$ and $w$, that is, by the shallowest between $\ell_w^+$ and $\ell_v^+$. Although $\ell_w^+$ and $\ell_v^+$ are only known a posteriori, we could predict $\ell_{\omega_x}^+$ from the slip-lengths $\ell_y^+$ and $\ell_z^+$. It has been empirically found that when the wall-normal velocity is set to zero at the simulation boundary (i.e. $\ell_y^+ = 0$), $\ell_{\omega_x}^+$ can be approximated by the expression in Eq. 2, as shown in Figure 2b.

To further support this idea, Figure 14 shows results for additional simulations with larger $\ell_x^+ = \ell_z^+$ and $\ell_y^+ = 0$ from [24]. The values of the slip lengths are chosen beyond the linear regime, where the saturation effect of the spanwise slip is clearly observed. As shown in Figure 14a, the vortices approach the wall as the slip lengths increase. Their displacement or virtual origin is captured here by $\ell_{z,eff}^+$ (Eq. 2), since when displaced by $\ell_{z,eff}^+$, the curves collapse with the smooth channel data. The profiles of $v'^+$ in Figure 14b also experience the same displacement, i.e. $\ell_v^+ \approx \ell_{\omega_x}^+ \approx \ell_{z,eff}^+$, as the displacement of the vortices is limited by $v$ being zero at the simulation boundary.

The expression in Eq. 2 could still be expanded for $\ell_y^+ \neq 0$, at least when $\ell_y^+ < \ell_z^+$. Assuming that only the difference between the origins for the different velocities can have an effect, the problem has only two degrees of freedom. A change of frame of reference to $y^+ = -\ell_y^+$ shows that the same result would be obtained for slip lengths 0 in $y$, $\ell_x^+ - \ell_y^+$ in $x$ and $\ell_z^+ - \ell_y^+$ in $z$. The resulting change in drag should then be given by

$$\Delta U^+ \propto \left(\ell_x^+ - \ell_y^+\right) - \frac{\left(\ell_z^+ - \ell_y^+\right)}{1 + \left(\ell_z^+ - \ell_y^+\right)/4}. \tag{6}$$

However, in order to generalise for any $\ell_z^+$ and $\ell_y^+$, cases with $\ell_v^+ > \ell_w^+$ should also be studied in the future. The distance of the quasi-streamwise vortices to the wall is essentially governed by

IOP Conf. Series: Journal of Physics: Conf. Series **1001** (2018) 012011 doi:10.1088/1742-6596/1001/1/012011

the ability of the induced cross-flow to approach the surface. We would therefore expect $\ell^+_{\omega_x}$ to depend only on $\ell^+_y$ and $\ell^+_z$. The *w-control* case from Choi *et al.* [18] provides some insight into this dependence when $\ell^+_v > \ell^+_w$. The drag reduction obtained with *w-control*, where $\ell^+_w \approx 0$ but $\ell^+_v$ is finite, was practically equal to that obtained with *v-w-control*, where both $\ell^+_v \approx \ell^+_w \approx 0$. This would suggest that with *w-control* the origin of the vortices is essentially the same as with *v-w-control*, that is, that in both cases $\ell^+_{\omega_x} \approx 0$. Therefore when $\ell^+_v > \ell^+_w$, it appears that the origin for the vortices is governed by that of the spanwise velocity, where $\ell^+_{\omega_x} \approx \ell^+_w \approx \ell^+_z$. This observation is however preliminary, and needs to be further investigated.

## 6. Conclusions

Complex surfaces can reduce drag if they induce a greater resistance to the induced, secondary cross-flow than to the mean, streamwise flow. Previous studies by Luchini *et al.* [6] and Jiménez [7] proposed that the drag reduction produced by these surfaces could be quantified by the difference between the virtual origins, measured from a reference 'wall' plane, of the mean velocity and of the spanwise flow induced by the quasi-streamwise vortices. This was expressed as the difference between the streamwise and spanwise slip lengths, $\ell^+_x - \ell^+_z$. The slip-length model, however, neglects the effect of non-zero wall-normal velocity at the reference plane, which can be significant for the drag. In terms of the virtual origin for the vortices, and for slip lengths greater than a few wall units, the wall-normal velocity induced is no longer negligible and the virtual origin of $v$ also needs to be considered. Otherwise, a saturation of the effect of the spanwise slip is observed [13]. In this paper, a distinction has been made between slip lengths and virtual origins. The slip lengths $\ell^+_x$, $\ell^+_z$ and $\ell^+_y$ are defined as the constants of proportionality between the respective velocity components and their wall-normal gradients, and the virtual origins $\ell^+_u$, $\ell^+_w$ and $\ell^+_v$ as the origins effectively perceived by the three velocity components.

In the present work we have explored two methods to represent virtual origins below the plane where boundary conditions are imposed, through a Stokes model for the flow beneath, and through Robin conditions for the three velocity components. We restrict ourselves to virtual origins of a few wall units, which modulate the near-wall turbulent cycle without completely destroying it. The models would therefore not be applicable, for instance, to fully rough surfaces. We first developed the models in the framework of equal virtual origins, that is, to model the presence of an underlying smooth wall. The first model gave satisfactory results for a virtual origin of up to 2 wall units, but failed for a virtual origin of 4 wall units. This was caused by the large effective permeability at small length scales, which excited short and wide, artificial structures in the flow. The second method, on the other hand, proved to hold at least up to a virtual origin 4 wall units below the simulation boundary, although further refinement is currently under investigation.

The latter model was then applied on simulations with different virtual origins for the different velocity components. We have conducted DNSs with equal slip lengths in the tangential directions, $\ell^+_x = \ell^+_z$, while keeping $v = 0$ at the simulation boundary, i.e. $\ell^+_y = 0$. The results are consistent with the change in drag being due to the difference between the virtual origins for the mean flow and for quasi-streamwise vortices. The origin for the mean flow is essentially determined by $\ell^+_x$, while the origin for the vortices is given by a combination of the spanwise and wall-normal slip lengths, i.e. $\ell^+_z$ and $\ell^+_y$. The results suggest that the origin of the vortices, $\ell^+_{\omega_x}$, is mainly determined by the smallest origin between $\ell^+_w$ and $\ell^+_v$, which is the one that restricts more intensely the cross flow induced by the vortices. Accounting for a non-zero $\ell^+_v$ and its effect on $\ell^+_{\omega_x}$ allows us to generalise the slip-length theory of Luchini *et al.* [6] and Jiménez [7]. When $\ell^+_y = 0$, we observe that a good approximation for $\ell^+_{\omega_x}$ is $\ell^+_{\omega_x} \approx \ell^+_v \approx \ell^+_z/(1 + \ell^+_z/4)$. The simulations of [18] suggest that $\ell^+_v < \ell^+_w$ results in $\ell^+_{\omega_x} \approx \ell^+_w \approx \ell^+_z$, but additional simulations need to be conducted to find a general expression for $\ell^+_{\omega_x}$. Further research also needs to be conducted to extend the models to surfaces which select certain modes at the interface, such as superhydrophobic posts [17] or deep transverse grooves [25].

**Acknowledgements**
This work was partially supported by the Coturb program of the European Research Council. GG was supported by an educational grant from Fundación Bancaria 'la Caixa'. CTF was supported by Engineering and Physical Sciences Research Council. The authors are grateful to C. Sanmiguel for insightful refereeing comments.

## References

[1] R. García-Mayoral and J. Jiménez. Hydrodynamic stability and breakdown of the viscous regime over riblets. *J. Fluid Mech.*, 678:317–347, 2011.
[2] J. P. Rothstein. Slip on superhydrophobic surfaces. *Annu. Rev. Fluid Mech.*, 42:89–109, 2010.
[3] S. Hahn, J. Je, and H. Choi. Direct numerical simulation of turbulent channel flow with permeable walls. *J. Fluid Mech.*, 450:259–285, 2002.
[4] N. Abderrahaman-Elena and R. García-Mayoral. Analysis of anisotropically permeable surfaces for turbulent drag reduction. *Phys. Rev. Fluids*, 2:114609, 2017.
[5] P. Luchini, F. Manzo, and A. Pozzi. Resistance of a grooved surface to parallel flow and cross-flow. *J. Fluid Mech.*, 228:87–109, 1991.
[6] P. Luchini. Reducing the turbulent skin friction. *Proc. 3rd ECCOMAS CFD Conf.*, pages 466–470, 1996.
[7] J. Jiménez. On the structure and control of near wall turbulence. *Phys. Fluids*, 6:944, 1994.
[8] F. H. Clauser. The turbulent boundary layer. *Adv. App. Mech.*, 4:1–51, 1956.
[9] D.W. Bechert, M. Bruse, W. Hage, J. G. T. Van der Hoeven, and G. Hoppe. Experiments on drag-reducing surfaces and their optimization with an adjustable geometry. *J. Fluid Mech.*, 338:59–87, 1997.
[10] J. Seo and A. Mani. On the scaling of the slip velocity in turbulent flows over superhydrophobic surfaces. *Phys. Fluids*, 28:025110, 2016.
[11] C. T. Fairhall and R. García-Mayoral. Inhomogeneity of the slip-lengths for simulations over superhydrophobic surfaces. *Flow Turbul. Combust.*, under revision.
[12] K. Fukagata, N. Kasagi, and P. Koumoutsakos. A theoretical prediction of friction drag reduction in turbulent flow by superhydrophobic surfaces. *Phys. Fluids*, 18(5):16–20, 2006.
[13] A. Busse and N. D. Sandham. Influence of an anisotropic slip-length boundary condition on turbulent channel flow. *Phys. Fluids*, 24:055111, 2012.
[14] T. Min and J. Kim. Effects of hydrophobic surface on skin-friction drag. *Phys. Fluids*, 16:L55, 2004.
[15] J. Jiménez, M. Uhlmann, A. Pinelli, and G. Kawahara. Turbulent shear flow over active and passive porous surfaces. *J. Fluid Mech.*, 442:89–117, 2001.
[16] W. P. Breugem and B. J. Boersma. Direct numerical simulations of turbulent flow over a permeable wall using a direct and a continuum approach. *Phys. Fluids*, 17:1–15, 2005.
[17] J Seo, R García-Mayoral, and A Mani. Turbulent flows over superhydrophobic surfaces: flow-induced capillary waves, and robustness of air-water interfaces. *J. Fluid Mech.*, 835:448–473, 2018.
[18] H. Choi, P. Moin, and J. Kim. Active turbulence control for drag reduction in wall-bounded flows. *J. Fluid Mech.*, 262:75–110, 1994.
[19] H. Le and P. Moin. An improvement of fractional step methods for the incompressible Navier-Stokes equations. *J. Comput. Phys.*, 92(2):369–379, 1991.
[20] Y. Mizuno and J. Jiménez. Wall turbulence without walls. *J. Fluid Mech.*, 723:429–455, 2013.
[21] M. P. Encinar, R. García-Mayoral, and J. Jiménez. Scaling of velocity fluctuations in off-wall boundary conditions for turbulent flows. *J. Phys. Conf. Ser.*, 506:012002, 2014.
[22] H. J. Bae, A. Lozano-Durán, S. T. Bose, and P. Moin. Turbulence intensities in large-eddy simulation of wall-bounded flows. *Phys. Rev. Fluids*, 3:014610, 2018.
[23] C. Sanmiguel Vila and O. Flores. Identifying vertical velocity eddies from wall-pressure. *J. Phys. Conf, Ser.*, this volume, 2018.
[24] C. T. Fairhall and R. García-Mayoral. The application of slip length models to larger textures in turbulent flows over superhydrophobic surfaces. *Bull. Am. Phys. Soc.*, 62(14):F13–9, 2017.
[25] M. MacDonald, A. Ooi, R. García-Mayoral, N. Hutchins, and D. Chung. Direct numerical simulation of high aspect ratio spanwise-aligned bars. *J. Fluid Mech.*, under revision.

IOP Conf. Series: Journal of Physics: Conf. Series **1001** (2018) 012012 doi:10.1088/1742-6596/1001/1/012012

# Turbulent flows over sparse canopies

**Akshath Sharma and Ricardo García-Mayoral**

Department of Engineering, University of Cambridge, Cambridge, CB2 1PZ, UK

E-mail: `r.gmayoral@eng.cam.ac.uk`

**Abstract.**
Turbulent flows over sparse and dense canopies exerting a similar drag force on the flow are investigated using Direct Numerical Simulations. The dense canopies are modelled using a homogeneous drag force, while for the sparse canopy, the geometry of the canopy elements is represented. It is found that on using the friction velocity based on the local shear at each height, the streamwise velocity fluctuations and the Reynolds stress within the sparse canopy are similar to those from a comparable smooth-wall case. In addition, when scaled with the local friction velocity, the intensity of the off-wall peak in the streamwise vorticity for sparse canopies also recovers a value similar to a smooth-wall. This indicates that the sparse canopy does not significantly disturb the near-wall turbulence cycle, but causes its rescaling to an intensity consistent with a lower friction velocity within the canopy. In comparison, the dense canopy is found to have a higher damping effect on the turbulent fluctuations. For the case of the sparse canopy, a peak in the spectral energy density of the wall-normal velocity, and Reynolds stress is observed, which may indicate the formation of Kelvin–Helmholtz-like instabilities. It is also found that a sparse canopy is better modelled by a homogeneous drag applied on the mean flow alone, and not the turbulent fluctuations.

## 1. Introduction

The study of flows over canopies has been a topic of interest for over five decades. Inoue [1] was one of the first to study the coherent bending of canopies on windy days and termed it *"honami"*. Some early experimental studies, notably Finnigan & Mulhearn [2], attributed this coherent bending to energetic gusts of air from the outer region of the atmospheric boundary layer. Raupach *et al.* [3] provided an alternate explanation for this phenomenon, theorising that a Kelvin–Helmholtz-like instability stemming from the inflection point in the mean velocity profile was responsible for the coherent motion of the crops. Since then, a large number of studies, summarised in [4, 5], have reported the formation of Kelvin–Helmholtz-like instabilities over canopies. The characteristic length scales of these instabilities is set by the shear length scale, $L_s = U(dU/dy)^{-1}$ at the canopy tips [4, 3]. Hence, by manipulating the shear length of a canopy it could be tuned to produce these instabilities at a particular scale, thereby providing a method to alter and possibly control these scales in a turbulent flow. This could potentially be used to design artificial canopies to harvest energy, enhance heat transfer, or alternatively, natural canopies to reduce crop loss [6].

Although Kelvin–Helmholtz-like instabilities are known to be spanwise coherent, in some cases they have been known to become distorted after their formation in turbulent flows. Finnigan *et al.* [7] and more recently, Bailey & Stoll [8] performed Large-Eddy Simulations of flow over dense canopies to determine the evolution of the Kelvin–Helmholtz rollers over the canopy.

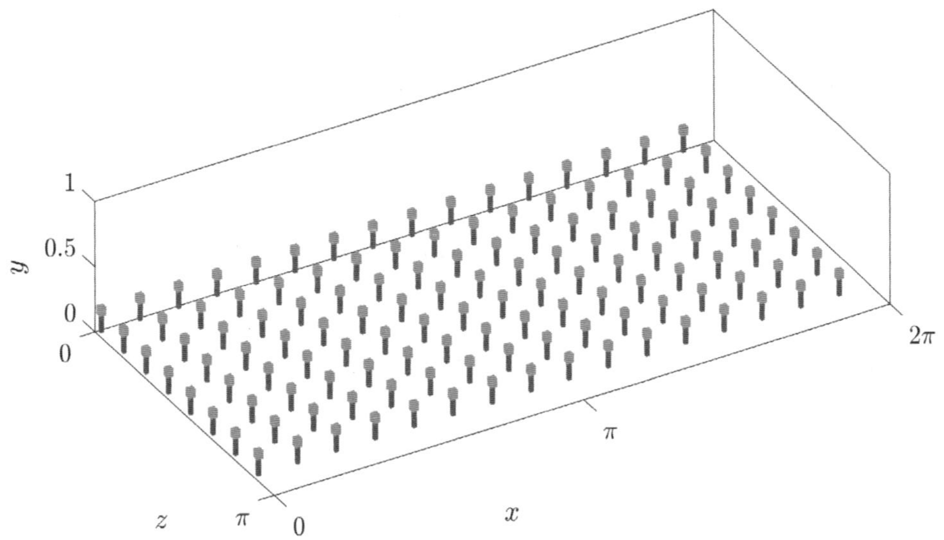


**Figure 1.** Schematic of numerical domain.

Finnigan *et al.* [7] proposed a model which involved the pairing of adjacent rollers to form large hairpin-like structures over the canopy. Bailey & Stoll [8] tried to establish how persistent the two-dimensional structures were after their formation, in the background turbulence. They argued that while hairpin structures of scales smaller than the Kelvin–Helmholtz rollers were present for their case, the overall two dimensional structure of the rollers was preserved in the developed flow. They also found that these rollers are responsible for the majority of the Reynolds stress near the canopy top.

All the above studies involved dense canopies, where the canopy could be modelled as a homogeneous drag force. As the spacing between canopy elements is increased, the homogeneous assumption eventually fails, leading up to the regime of sparse canopies. Such canopies are harder to characterise as there is no clear definition for the limit of canopy density where it can be called sparse. Based on empirical evidence, Nepf [5] defined a sparse canopy as one in which the ratio of frontal area of the canopy elements to the base area is $\lambda_f \leq 0.1$. Depending upon the level of sparsity, Kelvin–Helmholtz rollers have been reported over sparse canopies as well [9, 10], although there is no defined limit of sparsity till which the rollers can be sustained.

In the present study, we explore the use of a sparse canopy to manipulate large scales in a turbulent flow, while attempting not to affect the smaller scales. For this purpose, we simulate the flow over a canopy in which the spacing between the canopy elements is larger than the width of streaks. This should limit the effect of the canopy on the near-wall turbulence cycle, while also reducing its impact on smaller scales. In order to target the larger scales within the flow, the height of the canopy is kept relatively large so that the tips of the canopy elements protrude into the base of the logarithmic layer. The shear layer thickness at the canopy top, which controls the wavelength of the instabilities formed, is tuned by adjusting the profile of the drag force applied by the canopy in the vertical direction. This should theoretically allow us to control the scale of the instability being introduced into the flow. The canopy geometry and the distribution of drag force is discussed further in the next section. The drag force applied by the canopy also controls the strength of the inflection point in the mean velocity profile. Higher inflection generally results in a stronger instability, but, too large a drag can also damp the instability [11].

In addition to representing the individual canopy elements, we also explore whether a

**Table 1.** Simulation parameters. $u_\tau$ is the friction velocity based on the total drag in the channel, $\delta^+$ is the Reynolds number based on $u_\tau$, $h^+$ is the height of the canopy, $N_c$ and $\int D_y^+$ are the number of canopy elements and the drag force integrated over the canopy height. $\Delta x^+$, $\Delta z^+$ and $\Delta y^+$ are the resolutions in the streamwise, spanwise and wall-normal directions respectively.

| Case | $u_\tau$ | $\delta^+$ | $h^+$ | $N_c$ | $\int D_y^+$ | $\Delta x^+$ | $\Delta z^+$ | $\Delta y^+$ |
|---|---|---|---|---|---|---|---|---|
| S | 0.1363 | 504.3 | 105.9 | $16 \times 8$ | 0.809 | 8.25 | 4.12 | 0.20-2.0 |
| H | 0.1363 | 504.3 | 105.9 | – | 0.849 | 11.00 | 5.50 | 0.20-2.0 |
| H0 | 0.1405 | 520.5 | 109.3 | – | 0.823 | 11.36 | 5.68 | 0.21-2.1 |

homogeneous drag can be used to approximate the flow within sparse canopies. For this purpose, the flow in two more canopies is simulated, one exerting a homogeneous drag on the flow as in [8, 12, 13, 7], and the other only exerting drag on the mean flow and not on the fluctuations, that is, only on the zero mode in Fourier space. The drag of these canopies is set to exert roughly the same total drag on the flow as the sparse one.

The organisation of the paper is as follows. Section 2 briefly describes the numerical method used, and also elaborates on the canopy geometry. Section 3 discusses the results obtained from the simulations conducted before presenting the conclusions and summary in section 4.

## 2. Numerical simulations

The simulations are conducted in an open channel with the canopy elements protruding upwards from the wall. The streamwise, wall-normal and spanwise coordinates are $x$, $y$ and $z$ respectively, with the associated velocities being $u$, $v$ and $w$. The size of the simulation box is $2\pi \times 1 \times \pi$, with reference to the channel height, $\delta = 1$. The domain is periodic in the $x$ and $z$ directions. A schematic representation of the numerical domain for the present study is shown in figure 1. No-slip and impermeability conditions are applied at the bottom boundary along with free slip and impermeability conditions at the top. The flow is incompressible, with the density set to one for convenience. The numerical method used to solve the three dimensional Navier-Stokes equations is adapted from [14]. The streamwise and spanwise directions are decomposed using a Fourier spectral discretisation. The wall-normal direction is discretised using a second order centred difference scheme on a staggered grid. The grid in the wall normal direction is stretched to give a resolution ranging from $\Delta y^+_{min} \approx 0.2$ to $\Delta y^+_{max} \approx 2$. The time advancement is carried out using a three–step Runge–Kutta method [15], with continuity enforced using a fractional-step, pressure-correction method [16].

To simulate flow in sparse canopies the individual canopy elements need to be resolved. This is usually achieved by the means of an immersed boundary method, where a body force is applied opposing the flow within the roughness or canopy elements [17, 18]. In the present work, an 'immersed forcing' approach similar to that used by [9] is used to represent the individual canopy elements. It differs from the traditional immersed boundary method in the sense that the velocity within these boundaries is not strictly zero.

For the sparse canopy, case S, a forcing of the form $C_d\mathbf{u}|\mathbf{u}|$, where $C_d$ is the drag coefficient, is applied opposing the direction of flow at the points representing the canopy elements. This is similar to the method employed by [9] to represent sparse, but row-homogeneous canopies. In the present case, the canopy elements can be thought to be made from a porous material, which while allowing some fluid to pass though it, imparts a large drag force on the flow. For the first dense canopy, case H, a homogeneous forcing of the form $C_d\mathbf{u}|\mathbf{u}|$ is applied to all points below

IOP Conf. Series: Journal of Physics: Conf. Series **1001** (2018) 012012 doi:10.1088/1742-6596/1001/1/012012

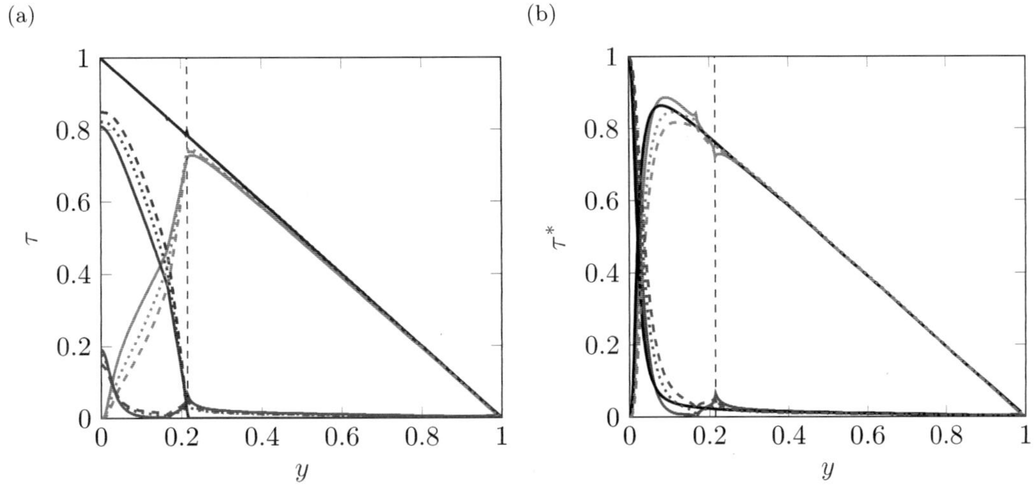


**Figure 2.** (a) The balance of total stress in the channel scaled with $u_\tau$ —— represents case S; - - - - case H; ······ case H0. The blue, orange, purple and green lines represent the total stress, Reynolds stress, cumulative canopy drag and viscous stress respectively. (b) The Reynolds stresses and viscous stresses rescaled by $u^*$. The solid black lines represent a smooth channel. The vertical dashed line depicts the height of the canopy.

the height of the canopy. In the last simulation, case H0, a forcing of form $C_d\langle u\rangle^2$ is applied to all points below the height of the canopy. Note that in case H0, the drag is only applied in the streamwise direction and has no fluctuating component. Case S was run first at a constant mass flow rate, with the viscosity adjusted to get the target friction Reynolds number. The subsequent cases were also run at the same mass flow rate, with the drag coefficients, $C_d$, adjusted to obtain similar mean pressure gradients as case S. The relevant parameters for each simulation are given in table 1. In order to tune the shear layer thickness, the head of the canopy, i.e. approximately the top 20 wall units, has a larger frontal area, and consequently exerts a larger drag force than the rest of the canopy. This causes a local increase in shear at the head of the canopy, as can be observed in figure 2, which in turn sets the shear layer thickness at the canopy top. Consequently, it should also set the wavelength of the Kelvin–Helmholtz-like instability formed over the canopy. For case S, the canopy elements are in a square arrangement, and the spacing between the elements is $\Delta x_f^+ \approx 200$. This results in a roughness density of $\lambda_f = 0.075$, which is within the sparse range empirically demarcated by [5]. In addition, the spanwise spacing between the canopy elements is roughly twice the width of streaks near the wall, implying that the canopy is sparse even from the point of turbulent fluctuations.

The data for the smooth wall was obtained from the simulation of a full channel at $Re_\tau = 550$ [19].

## 3. Results and discussion

The balance of stresses in the channel is shown in figure 2a, and can be written as

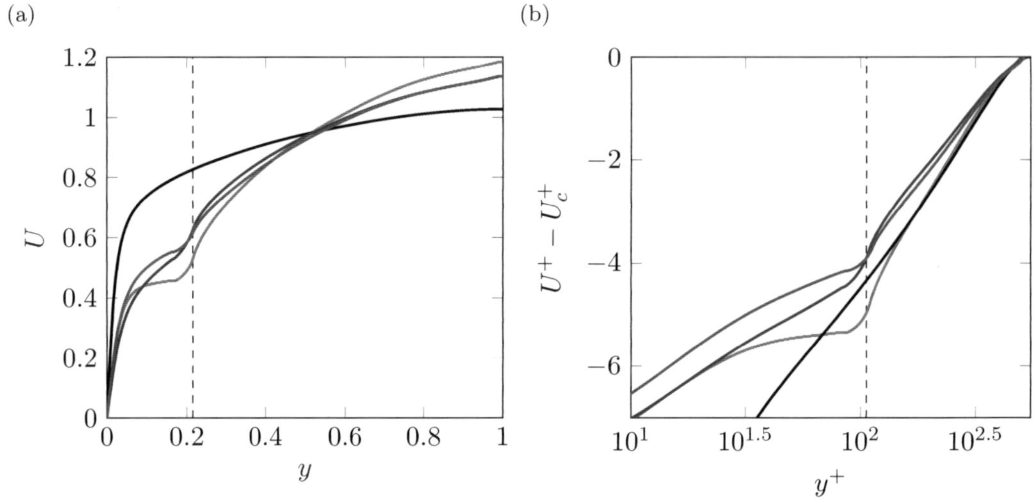


**Figure 3.** Mean velocity profiles, (a) scaled in outer units, (b) scaled in deficit law in inner units. The black, red, purple and blue lines represent smooth-wall, S, H0 and H respectively. The dashed line marks the canopy top.

$$\frac{dP}{dx}y + \tau_0 + \int_0^h D_y dy = -\overline{u'v'} + \frac{1}{Re}\frac{dU}{dy} + \int_y^h D_y dy,$$

$$D_y = \begin{cases} C_d(y)U|U|, & \text{for } y \leq h \\ 0, & \text{for } y > h \end{cases}$$

where $C_d(y)$ is the drag coefficient of the canopy, $\tau_0$ is the wall shear stress, $-\overline{u'v'}$ is the Reynolds stress, $h$ is the height of the canopy, and $U$ and $dP/dx$ are the mean streamwise velocity and mean pressure gradient respectively.

For a smooth channel in equilibrium, this balance essentially reduces to the mean pressure gradient countering the shear stress at the wall, and the friction velocity can be defined as

$$u_\tau^2 = -\delta\frac{dP}{dx}. \quad (1)$$

In the present case, we define a 'global' friction velocity using equation (1). This formulation takes into account the wall shear stress and the total drag exerted by the canopies. All quantities in figure 2a are normalised by this 'global' friction velocity, $u_\tau$. However, unlike the wall shear stress, the drag force has a vertical distribution, and the flow within the canopy at a given height feels the influence of the wall shear stress and a contribution of the drag up to that height. Tuerke & Jimenez [20] observed that, when the pressure gradient has to balance body forces additional to the wall shear stress in the channel, the turbulent fluctuations scale with the local shear at a given height instead of just the wall shear. They calculated the local friction velocity, $u^*$ corresponding to the local shear as

$$u^*(y) = \left(\frac{\delta\tau_l}{\delta - y}\right)^{1/2}, \quad (2)$$

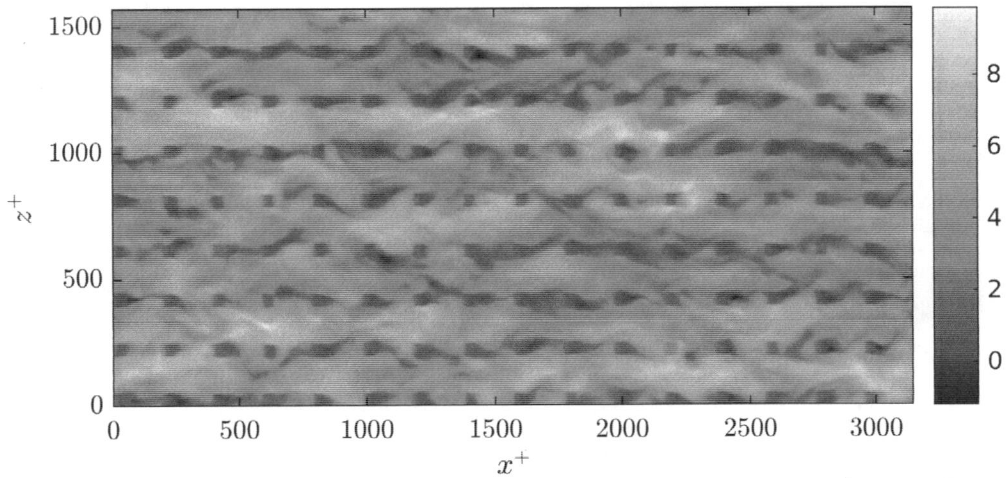


**Figure 4.** Contours of streamwise velocity normalised by $u_\tau$, at $y^+ = 105$ for case S.

where the local shear is the sum of the viscous and Reynolds stresses at a particular height, $\tau_l = -\overline{u'v'} + \frac{1}{Re}\frac{dU}{dy}$.

Figure 2b depicts the Reynolds and viscous stresses within the channel, renormalised by $u^*$. The resemblance of these renormalised stresses to the smooth wall data seems to suggest that the major affect of the addition of the canopy in the channel is the rescaling of the turbulent fluctuations. However, we will see later that this is not the case. It can also be observed that even after the rescaling, the Reynolds stresses in the dense canopies, cases H0 and H, are more damped than in case S. Although not unexpected, the drag damping in case H affects the turbulence structures within and outside it more than that for cases H0 and S.

The differences manifest themselves in the mean velocity profiles as well, as can be observed in figure 3. The maximum velocity within the dense canopies, H and H0, are larger than that in the sparse canopy, S. The likely factor behind this is the formation of wakes and regions of flow separation in case S, which are absent in the other cases. The wakes can be observed in the contours of streamwise velocity shown in figure 3. Other differences may be attributed to the minor variations in the drag and Reynolds stress profiles within the three canopies. It can also be observed in figure 3b that the expected Kármán constant is not recovered in cases H and H0. This can likely be attributed to not using the correct displacement height, which is generally used while fitting rough wall velocity profiles to the log-law [21]. Rather than fitting the log-law, Mizuno & Jiménez [22] proposed that the linearity of the mixing length scale, $L_m = u_\tau (dU/dy)^{-1}$, in the channel was a better indicator of the logarithmic region, especially in low Reynolds number flows such as the present cases. Although not shown here, a linear region in the mixing length, $L_m$, was obtained for all three cases.

An additional feature of the velocity profiles for all cases is the presence of an inflection point near the canopy top. Inflection points have been traditionally indicative of the presence of a Kelvin–Helmholtz-like instability in canopy flows [8, 4, 3]. As discussed in the previous section, the occurrence of such instabilities has been widely reported over dense canopies. Over sparse canopies however, this would depend on the level of sparsity. It is also not clear whether, if formed, these instabilities would be able to maintain their spanwise coherent structure.

### 3.1. Turbulence statistics

Figures 5 (a), (c), (g) and (e) show turbulent fluctuations scaled with the 'global' $u_\tau$. Outside the canopy, the canopy sublayer, which we take here as the height at which the fluctuations recover to approximately the smooth wall values, extends to about twice the canopy height, which is in the limit observations made by [23, 24, 25] for roughness sublayers of canopies and roughness. Within the canopy, the fluctuations are highly damped for all cases, as compared to the smooth-

IOP Conf. Series: Journal of Physics: Conf. Series **1001** (2018) 012012 doi:10.1088/1742-6596/1001/1/012012

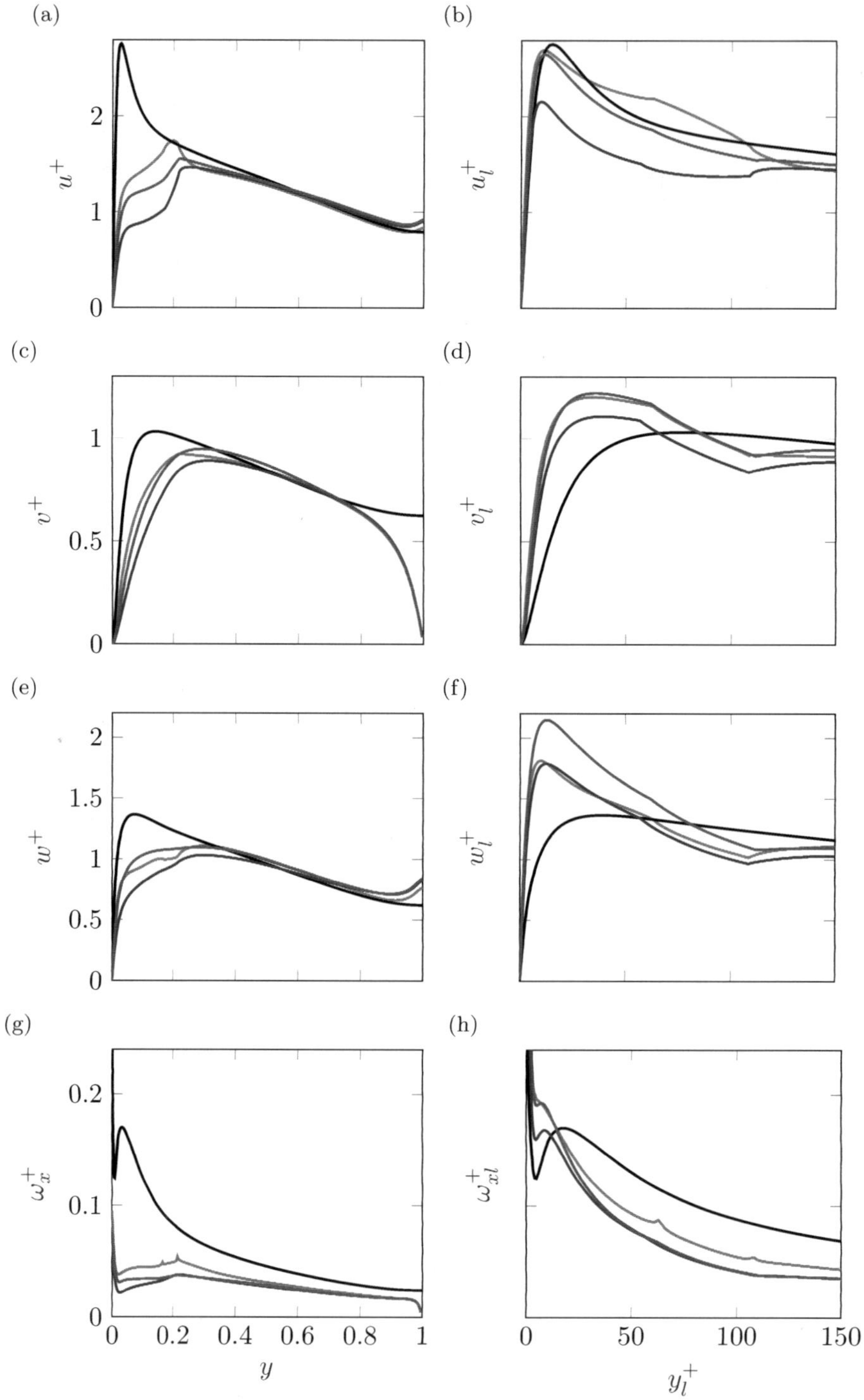


**Figure 5.** Root mean squares of turbulent fluctuations for flow for smooth channel (black), case S (red), case H0 (purple) and case H (blue). Plots in the left (right) column are scaled with $u_\tau$ ($u^*$). The smooth-wall data is of a channel at $Re_\tau = 550$.

wall. However, as discussed previously, $u_\tau$ may not be the appropriate scale for comparing the fluctuations. Figures 5 (b), (d), (f) and (h) show the fluctuations rescaled with $u^*(y)$ against the local height in wall units, defined as $y_l^+ = yu^*/\nu$. Note that $u^*$ is the same as $u_\tau$ above the height of the canopies, where no drag force being applied. In this local scaling we observe that the streamwise velocity fluctuations for cases S and H0 are similar to those for the smooth-wall. The fluctuations for case S are larger near the head of the canopy, which may be a result of the flow separation around it. However, even though the magnitude of drag force applied in the three cases are within 5 percent of each other, the fluctuations for case H are significantly damped. This can be attributed to the difference in the way the three cases disrupt the near-wall turbulence cycle. As noted by [26], the near wall peak in the streamwise fluctuations and the streamwise vorticity can be considered as indicators of the near wall streaks and quasi-streamwise vortices respectively. In case H0, there is no drag on the streamwise perturbations, and consequently the streaks formed are consistent with the local shear. Similarly, in case S, the streaks can rest between the canopy elements and do not experience much damping due to drag. For case H however, the drag force damps the perturbations through the entire canopy, which results in weaker streaks as compared to the other two cases. We observe a similar trend in the streamwise vorticity, where the off-wall peaks for cases S and H0 coincide, while the peak for case H is more damped. However, the peak in the streamwise vorticity for cases S and H0 in local scaling exceeds the smooth-wall value. The off-wall peaks of spanwise and wall-normal velocity fluctuations, which are mainly generated by vortices, also exceed the corresponding smooth-wall values. In addition, the location of the peaks is also not concurrent with that of the smooth-wall case. This could be due to the local scale for the shear having a wider support in $y$ for the cross velocity components, and not being set merely by the shear at the particular height. Nevertheless, more profound modifications of the dynamics cannot be ruled out with the present evidence.

As case H0 only has a drag in the streamwise direction, as opposed to case S which applies a drag in all three directions, a comparison between them can be used compare the relative importance of the drag in different directions. A comparison between cases S and H0 reveals that the drag force in the streamwise direction alone is sufficient to modulate the level of fluctuations in all directions, and that the spanwise and wall-normal drag forces do not show a significant effect on the flow. This phenomenon can partly be explained by continuity. In addition, as the drag force applied is proportional to the square of the velocity, and as the streamwise velocity is much larger than the other two components, so is the streamwise drag. The streamwise drag modulates the value of the local shear, and consequently sets the scale of near-wall streaks. The near wall-streaks affect the formation of quasi-streamwise vortices, which in turn modulates the spanwise and wall-normal fluctuations [27].

### 3.2. *Spectral analysis*

In this section, we analyse the energetically relevant scales within the flow to gauge whether the turbulence structure in the canopy has been modified by the Kelvin–Helmholtz-like instability. This instability is spanwise coherent, and a clear signature of this instability has been known to be observed even in turbulent conditions [8, 14, 28]. In the spectral energy density, $k_x k_y E_{vv}$, these instabilities cause the appearance of a peak, at fixed streamwise wavelength and elongated in the spanwise direction, which is most notable in the spectral energy density of the wall-normal velocity [14, 19]. The streamwise wavelength of the peaks corresponds to the wavelength of the instability.

In figure 6 we compare the spectral energy densities of all cases with that of a smooth channel at a height of $y^+ = 105$, which roughly corresponds to the tips of the canopy elements, where we would expect the centres of the Kelvin–Helmholtz-like instabilities to lie. For case S, the canopy wavelength, $\lambda_x^+ = \lambda_z^+ \approx 200$, and its harmonics are visible at the bottom left of the plots. At

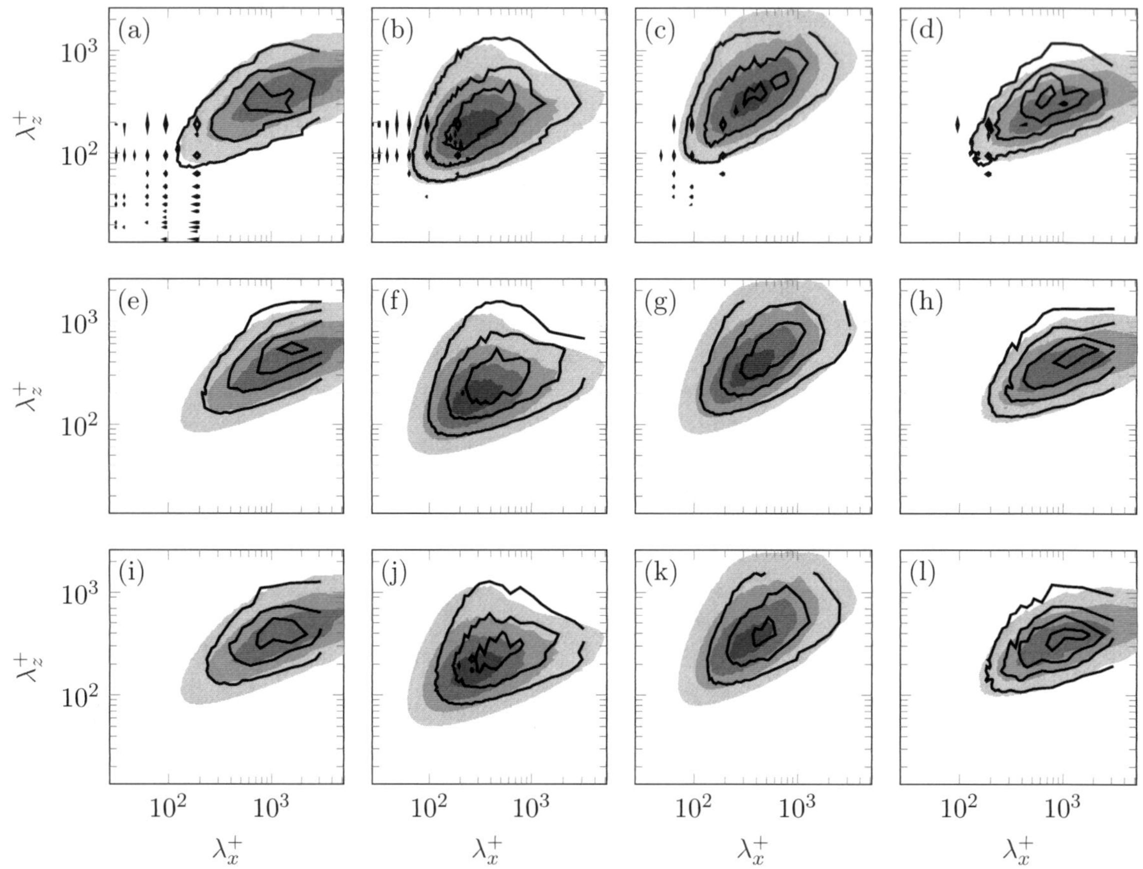


**Figure 6.** Spectral energy densities, $k_x k_z E_{uu}$ (a,e,i), $k_x k_z E_{vv}$ (b,f,j), $k_x k_z E_{ww}$ (c,g,k), and (d) $k_x k_z E_{uv}$ (d,h,l), scaled in friction units, at $y_l^+ \approx 105$. The filled contours correspond to smooth-wall data, and the superimposed lines represent from (a) to (d), case S; (e) to (h), case H0; and (i) to (l) for case H. The contour increments, in wall units, are 0.113, 0.039, 0.059 and 0.040 respectively.

this height, the canopy wavelength does not appear to change the energy distribution, with the contours of energy density of the smooth-wall case coinciding with that of case S, for smaller scales. In figure 6a, the damping of larger scales can be observed. As noted by [26], this may imply the shortening of streaks at this height. The damping of large scales in $k_x k_z E_{uu}$ is also reflected in in the co-spectra of the Reynolds stress shown in figure 6d. In $k_x k_z E_{vv}$, shown in figure 6b, we observe the opposite, i.e. a redistribution of energy to larger scales, with a peak in the spectra at $\lambda_x^+ \approx 600$, with a corresponding peak in the $uv$ co-spectra. Although it is not certain, this can be a possible reflection of the Kelvin–Helmholtz-like instability.

For cases H0 and H, the turbulent fluctuations appear to have an 'origin' which is different to that of the smooth-wall case, and therefore, the smaller scales in the spectra do not coincide with those of the smooth-wall. Even so, some qualitative comparisons between the cases can still be made. The core region of $k_x k_z E_{vv}$ and $k_x k_z E_{ww}$ in case H is smaller than that for the other cases. The reason for this stems from nature of the drag term for this case. As the drag is

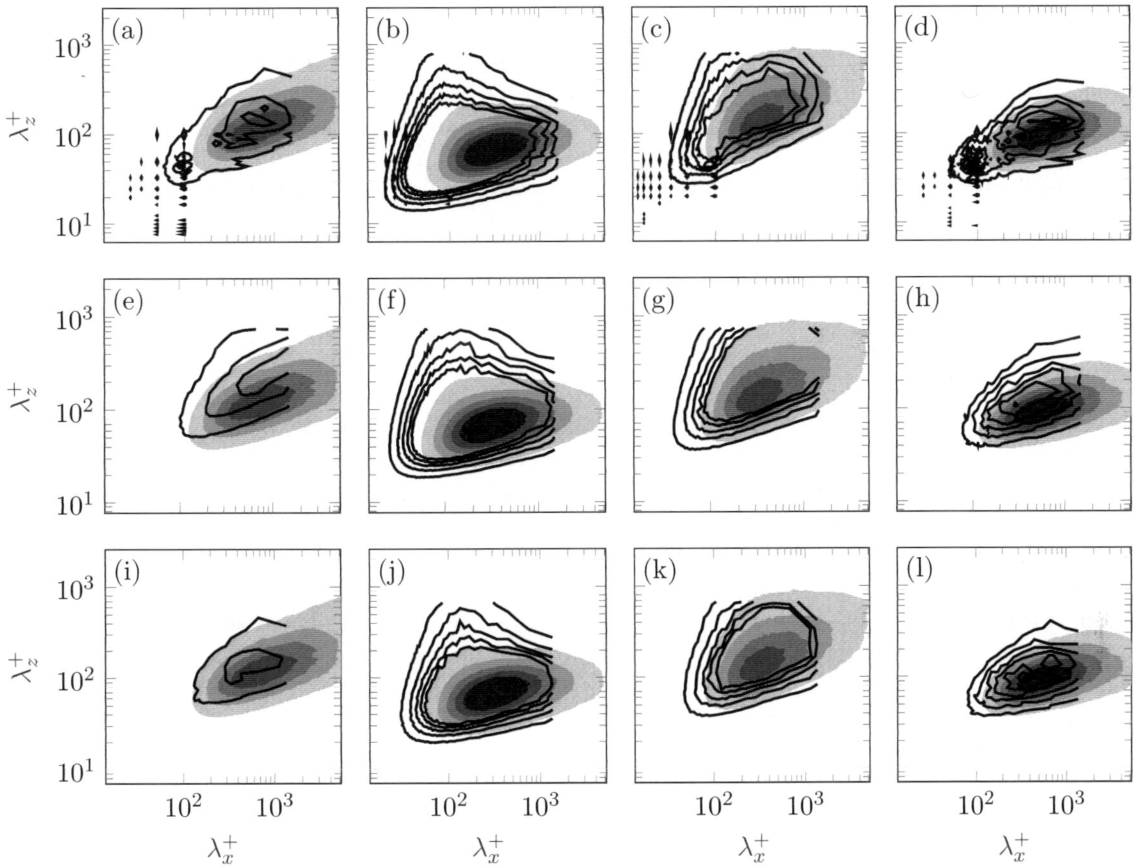


**Figure 7.** Same as figure 6 except, scaled with the local friction velocity, at height $y_l^+ = 15$, with the contour increments, in wall units, being 0.282, 0.014, 0.059 and 0.036 respectively.

proportional the square of the total velocity, the core region of the spectra, which has the highest velocity, experiences the highest damping for case H. As discussed previously, the wall-normal and spanwise fluctuations for the other cases are not much affected by the drag force.

Interestingly, a clear peak in the spectrum of the wall normal velocity, which would be indicative of the Kelvin–Helmholtz-like instability, is not observed in cases H and H0. However, this does not imply that no instabilities have formed in these case. The more probable reason for this is that the instabilities in cases H and H0 are weaker than in case S, owing to weaker inflection in their mean velocity profiles.

Figure 7 shows the spectral energy densities of all cases within the canopy at an effective height $y_l^+ = 15$. The height is chosen as it approximately corresponds to the near wall peaks in the averaged turbulent fluctuations. Focusing only on the spectra of $u^2$ and co-spectra of $uv$, we observe for case S, that the fluctuations in the vicinity of the canopy wavelength are distorted. It is also worth noting that, owing to scaling with the local friction velocity, $u^*$, the canopy wavelength has now decreased to $\lambda_x^+ = \lambda_z^+ \approx 115$. Similar to the effect seen at the canopy tips, cases S and H exhibit shortening of the spectrum of $k_x k_z E_{uu}$, as opposed to case H0 where the larger scales do not appear to be affected. Although, they do not exactly collapse, the spectrum of the streamwise velocity for all cases appears somewhat similar to that of the smooth-wall. However, in agreement with the observations in the r.m.s velocity fluctuations of $u$ and $v$, the spectra of these velocities are also more intense compared to smooth-wall values at this height. Again, these changes may be attributed to the fact that the local friction velocity is not the appropriate scale for these fluctuations, or, due to the forcing intrinsically changing the near–wall turbulence.

## 4. Conclusions

The results obtained from preliminary DNSs of flow over canopies with height $h^+ \approx 100$ are presented. Three cases are simulated, the first where the geometry of the canopies is resolved (case S), the second with the canopies modelled as a homogeneous drag force (case H), and the third with the drag force only applied on the mean streamwise flow (case H0). The magnitude and distribution of the average canopy drag force for all cases was maintained to be similar. We found that the streamwise drag had the most impact on the flow within the canopy, and large structures in $u$ were damped for all cases run. In case S, we also observe a redistribution of energy to larger scales, manifested in the formation of peaks at $\lambda_x \approx 600$ in $k_x k_z E_{vv}$ and $k_x k_z E_{uv}$, probably as result of a Kelvin–Helmholtz-like instability. The RMSs of streamwise velocity and Reynolds stress in the sparse canopy were found to be similar to a smooth-wall, when scaled with the local friction velocity at each height within the canopy. However, this local scaling was observed to break down for the wall-normal and spanwise fluctuations. The averaged turbulent fluctuations were found to be similar for cases S and H0, implying that applying a drag on the mean flow alone, and not on the fluctuations, is a better representation of a sparse canopy than applying a homogeneous drag. The drag in case H was observed to have a much larger effect on the streamwise perturbations and, by extension on the near-wall turbulence cycle [27], due to excessive damping of the turbulent fluctuations.

## Acknowledgments

This work was supported, in part, by the Coturb program of the European Research Council. AS was supported by a scholarship from the Cambridge Commonwealth, European and International Trust. The authors are grateful to S. Ghebali for insightful refereeing comments.

## References

[1] E Inoue. Studies of the phenomena of waving plants (honami) caused by wind. *J. Agric. Meteorol.*, 11:87–90, 1955.

[2] J J Finnigan and P J Mulhearn. Modelling waving crops in a wind tunnel. *Bound.-Lay. Meteorol.*, 14:253–277, 1978.

[3] M R Raupach, J J Finnigan, and Y Brunet. Coherent eddies and turbulence in vegetation canopies: the mixing-layer analogy. In J. R. Garratt and P. A. Taylor, editors, *Bound.-Lay. Meteorol. 25th Anniversary Vol.*, pages 351–382. Springer, 1996.

[4] J J Finnigan. Turbulence in plant canopies. *Ann. Rev. Fluid Mech.*, 32:519–571, 2000.

[5] H M Nepf. Flow and transport in regions with aquatic vegetation. *Ann. Rev. Fluid Mech.*, 44:123–142, 2012.

[6] E De Langre. Effects of wind on plants. *Ann. Rev. Fluid Mech.*, 40:141–168, 2008.

[7] J J Finnigan, R H Shaw, and E G Patton. Turbulence structure above a vegetation canopy. *J. Fluid Mech.*, 637:387–424, 2009.

[8] B N Bailey and R Stoll. The creation and evolution of coherent structures in plant canopy flows and their role in turbulent transport. *J. Fluid Mech.*, 789:425–460, 2016.

[9] B N Bailey and R Stoll. Turbulence in sparse, organized vegetative canopies: a large-eddy simulation study. *Bound.-Lay. Meteorol.*, 147:369–400, 2013.

[10] M Böhm, J J Finnigan, M R Raupach, and D Hughes. Turbulence structure within and above a canopy of bluff elements. *Bound.-Lay. Meteorol.*, 146:393–419, 2013.

[11] A Sharma, G Gómez-de Segura, and R García-Mayoral. Linear stability analysis of turbulent flows over dense filament canopies. In *Proc. TSFP10*, 2017.

IOP Conf. Series: Journal of Physics: Conf. Series **1001** (2018) 012012 doi:10.1088/1742-6596/1001/1/012012

[12] A Busse and N D Sandham. Parametric forcing approach to rough-wall turbulent channel flow. *J. Fluid Mech.*, 712:169–202, 2012.

[13] S Dupont, F Gosselin, C Py, E De Langre, P Hemon, and Y Brunet. Modelling waving crops using large-eddy simulation: comparison with experiments and a linear stability analysis. *J. Fluid Mech.*, 652:5–44, 2010.

[14] R García-Mayoral and J Jiménez. Hydrodynamic stability and breakdown of the viscous regime over riblets. *J. Fluid Mech.*, 678:317–347, 2011.

[15] H Le and P Moin. An improvement of fractional step methods for the incompressible Navier–Stokes equations. *J. Comput. Phys.*, 92:369–379, 1991.

[16] J Kim and P Moin. Application of a fractional-step method to incompressible Navier–Stokes equations. *J. Comput. Phys.*, 59:308–323, 1985.

[17] E A Fadlun, R Verzicco, P Orlandi, and J Mohd-Yusof. Combined immersed-boundary finite-difference methods for three-dimensional complex flow simulations. *J. Comput. Phys.*, 161:35–60, 2000.

[18] J Mohd-Yusof. Development of immersed boundary methods for complex geometries. *CTR Ann. Res. Briefs*, pages 325–336, 1998.

[19] R García-Mayoral and J Jiménez. Scaling of turbulent structures in riblet channels up to $\mathrm{Re}_\tau \approx 550$. *Phys. Fluids*, 24:105101, 2012.

[20] F Tuerke and J Jiménez. Simulations of turbulent channels with prescribed velocity profiles. *J. Fluid Mech.*, 723:587–603, 2013.

[21] P S Jackson. On the displacement height in the logarithmic velocity profile. *J. Fluid Mech.*, 111:15–25, 1981.

[22] Y Mizuno and J Jiménez. Mean velocity and length-scales in the overlap region of wall-bounded turbulent flows. *Phys. Fluids*, 23:085112, 2011.

[23] K A Flack, M P Schultz, and J S Connelly. Examination of a critical roughness height for outer layer similarity. *Phys. Fluids*, 19:095104, 2007.

[24] I N Harman and J J Finnigan. A simple unified theory for flow in the canopy and roughness sublayer. *Bound.-Lay. Meteorol.*, 123:339–363, 2007.

[25] M R Raupach, R A Antonia, and S Rajagopalan. Rough-wall turbulent boundary layers. *Appl. Mech. Rev.*, 44:1–25, 1991.

[26] O Flores and J Jiménez. Effect of wall-boundary disturbances on turbulent channel flows. *J. Fluid Mech.*, 566:357–376, 2006.

[27] J Jiménez and A Pinelli. The autonomous cycle of near-wall turbulence. *J. Fluid Mech.*, 389:335–359, 1999.

[28] J Jiménez, M Uhlmann, A Pinelli, and G Kawahara. Turbulent shear flow over active and passive porous surfaces. *J. Fluid Mech.*, 442:89–117, 2001.

IOP Conf. Series: Journal of Physics: Conf. Series **1001** (2018) 012013 doi:10.1088/1742-6596/1001/1/012013

# Causal analysis of self-sustaining processes in the logarithmic layer of wall-bounded turbulence

**H. J. Bae[1], M. P. Encinar[2], A. Lozano-Durán[1]**
[1]Center for Turbulence Research, Stanford University, CA 94305, USA
[2]School of Aeronautics, Universidad Politécnica de Madrid, 28040 Madrid, Spain
E-mail: adrianld@stanford.edu

**Abstract.**
Despite the large amount of information provided by direct numerical simulations of turbulent flows, their underlying dynamics remain elusive even in the most simple and canonical configurations. Most common approaches to investigate the turbulence phenomena do not provide a clear causal inference between events, which is essential to determine the dynamics of self-sustaining processes. In the present work, we examine the causal interactions between streaks, rolls and mean shear in the logarithmic layer of a minimal turbulent channel flow. Causality between structures is assessed in a non-intrusive manner by transfer entropy, i.e., how much the uncertainty of one structure is reduced by knowing the past states of the others. We choose to represent streaks by the first Fourier modes of the streamwise velocity, while rolls are defined by the wall-normal and spanwise velocity modes. The results show that the process is mainly unidirectional rather than cyclic, and that the log-layer motions are sustained by extracting energy from the mean shear which controls the dynamics and time-scales. The well-known lift-up effect is also identified, but shown to be of secondary importance in the causal network between shear, streaks and rolls.

## 1. Introduction

Turbulence is an important nonlinear dynamical phenomenon and one of the most challenging problems in classical physics. However, it has not been systematically investigated from the point of view of causality and, to date, the only attempts are the works by [1, 2, 3]. Indeed, most standard methods in turbulence research (one-point statistics, correlations, Fourier analysis, mode decomposition, exact coherent structures, energy budgets, data interrogation, etc.) do not provide a clear causal inference between events. Understanding the causal relations in turbulent flows is a necessary requirement (although maybe not sufficient) in order to comprehend the physical behavior of the system in an intuitive manner. In this study, we investigate the causal interaction between different coherent structures in the logarithmic layer of fully developed wall-bounded turbulence.

Streamwise rolls and streaks are ubiquitous in wall-shear flows. They are considered to be the most representative coherent structures that are able to explain the dynamics of the flow as a whole. Since the experiment by [4] and the discovery of sublayer streaks and ejections by [5], among others, the roll-streak structure has attracted enormous interest within the fluid mechanics community. Their spatially and temporally varying structure is generally believed to play an important role in maintaining and carrying shear-driven turbulence [6, 7, 8, 9, 10, 11].

Most works have focused on the buffer layer where the streaks and rolls (or vortices) are essentially one-scale objects, which simplifies their analysis. Farther from the wall, streaks and rolls persist but their internal Reynolds numbers are higher, making their characterization and understanding a more challenging task.

Although it is widely agreed that in both the buffer and logarithmic layers the streaks and rolls are involved in a regeneration cycle, alternative mechanisms have been derived from different simplified scenarios. In the buffer layer, it has been hypothesized that streamwise vortices near the wall may collect the fluid from the inner region, where the flow is very slow, and organize it into streaks [12, 11]. Other alternatives suggest that the generation of streaks are due to the structure-forming properties of the linearized Navier–Stokes operator, independent of any organized vortices [13]. Conversely, the streaks are hypothesized to trigger the formation of vortices by losing their stability [10, 14]. For a review of self-sustained processes in the buffer layer see [15].

A similar but more disorganized scenario is believed to take place in the logarithmic layer [16, 17, 18]. The existence of a self-similar streak/roll structure consistent with the Townsend attached-eddy hypothesis [19] has been documented by [20, 21, 22], and [23] from direct numerical simulation data at moderate Reynolds numbers. However, the most interesting results are not the kinematic description of the structures in individual flow realizations, but rather the elucidation of how they relate to each other and how they evolve in time. A growing body of evidence indicates that the generation of these log-layer streaks has its origins in the linear lift-up effect [20, 24, 25, 26] in conjunction with the Orr's mechanism [27, 11]. On the other hand, rolls are speculated to be the consequence of a sinuous secondary instability of the streaks that collapse through a rapid meander until breakdown [28, 29, 26, 30]. Other scenarios advocate for the formation of rolls via a stochastic structural instability [31].

The different mechanisms, each capable of leading to the observed turbulence structure, are rooted in theoretical or conceptual arguments. Whether the flow follows any or a combination of these mechanisms is in fact unclear. Most of the theories stem from linear stability theory, which has proved very successful in providing a theoretical framework to explain the lengths and time scales observed in the flow. However, an appropriate base flow for the linearization must be selected *a priori* depending on the flow state of interest, introducing some degree of arbitrariness. Moreover, quantitative results are known to be sensitive to the details of the base state [32]. Another criticism is that turbulence is a highly nonlinear phenomenon, and a full self-sustained cycle is not expect to be unraveled from a single set of linearized equations. For example, in turbulent channel flows, the classic linearization around the mean velocity profile does not account for the redistribution of energy from the streamwise velocity component to the cross-flow, which is the prevailing energy transfer on average [33]. In order to capture different energy transfer mechanisms, the base state for linearization should be selected accordingly. Moreover, optimal solutions should not be taken as representative of the actual flow and, if so, the time and length-scales for which linearization remains meaningful becomes a relevant issue, which is barely discussed in the literature. The causal analysis presented here aims to clarify what accounts for driving the fully nonlinear self-sustaining processes in the log-layer in a non-intrusive manner.

Causal inference is an important subject in many scientific disciplines. Given that what is usually known about events in question is in the form of time series, causal analysis of temporal signals is of particular importance. Frequently, causal inference is understood in terms of time-correlation between pairs of signals. However, it is well known that correlation lacks the directionality and asymmetry required to guarantee causation [34]. Another common method for causality assessment is the so-called Granger test [35], which is a statistical measure of the usefulness of one time series in forecasting another. However, this provides only a yes-or-no judgment, without the quantitative information that may be needed in many circumstances.

IOP Conf. Series: Journal of Physics: Conf. Series **1001** (2018) 012013 doi:10.1088/1742-6596/1001/1/012013

In an attempt to remedy this deficiency, recent works have centered their attention to an information-theoretic measure, namely, transfer entropy [36]. This measure is notoriously challenging to evaluate, requiring long time series and a high associated computation cost [37]. However, recent advancements in entropy estimation from insufficient datasets [38, 39] have made transfer entropy a viable method to quantify causality.

The paper is organized as follows. The first section contains the methods subdivided in three parts (Section 2). First, we define transfer entropy as a tool to quantify causality. In the other two, we introduce the numerical experiment and the flow decomposition into mean flow, streaks, and rolls. The results are presented and discussed in Section 3 in terms of the characteristic time-scales for causality and the causal network between coherent structures. Finally, conclusions are offered in Section 4.

## 2. Methods

### *2.1. Quantitative causality analysis: transfer entropy*

We use the framework provided by information theory to quantify the causality between a set of temporal signals by their associated transfer entropy [36]. Given a set of time-dependent random variables $x_k(t)$, $k = 1, \cdots, n$, the transfer entropy from $x_j$ to $x_i$ is defined as

$$T_{j\to i}(\Delta t) = H(x_i(t)|x_{1,\cdots,j-1,j+1,\cdots,n}(t-\Delta t)) - H(x_i(t)|x_{1,\cdots,n}(t-\Delta t)), \quad (1)$$

where $x_{1,\cdots,n} = (x_1, \cdots, x_n)$, and $\Delta t$ is the time-lag in the signal considered to evaluate causality at the current state. $H(x|y)$ is the conditional Shannon entropy [40], that is, the uncertainty of a variable $x$ given $y$, and is defined as

$$H(x|y) = E[\log(f(x,y))] - E[\log(f(y))], \quad (2)$$

where $f(\cdot)$ is the probability density function, $f(\cdot|\cdot)$ denotes the conditional probability density function, and $E[\cdot]$ signifies the expected value. In this form, the transfer entropy (or causality) from $x_j$ to $x_i$ can be interpreted as the decrease in uncertainty in $x_i$ by knowing the past state of $x_j$. We define the self-induced transfer entropy as $T_{i\to i}$, and the cross-induced transfer entropy as $T_{j\to i}$ for $j \neq i$.

To guarantee statistical convergence, transfer entropy calculation requires data from long time series, especially when the dimension of the data set, $n$, is large. To alleviate the computational cost associated with computing $H$, the Shannon entropy is approximated using the Kozachenko-Leonenko estimator [38], which is known to provide reasonable improvements for high dimensional datasets from a relatively short time series [39].

Another key aspect for causality quantification is the normalization of the measurement. Time correlation, for example, guarantees a value between 0 and 1 when properly normalized, making the correlation coefficient relevant in terms of determining relative importance. In the same manner, causality must be normalized to allow for meaningful comparisons. We define the normalized transfer entropy [41] as

$$NT_{j\to i}(\Delta t) = \frac{T_{j\to i}(\Delta t) - T^{\text{Shuffled}}_{j\to i}(\Delta t)}{T_{(1,\cdots,n)\to i}(\Delta t)}. \quad (3)$$

The term $T^{\text{Shuffled}}_{j\to i}$ aims to remove spurious contributions due to statistical errors, and it is the transfer entropy computed from the random variables $x_1, \cdots, x_{j-1}, x^{\text{Shuffled}}_j, x_{j+1}, \cdots, x_n$, where $x^{\text{Shuffled}}_j$ is $x_j$ randomly permuted in time in order to retain first order statistics but break any time-delayed causal links. Eq. (3) is used as the measure of causality for the remainder of the paper.

The causality measure from Eq. (3) is advantageous compared to the classic time-correlations. One desirable property is the asymmetry of the measurement, i.e., if a variable $x_1$ is causal to $x_2$, it does not imply that $x_2$ is causal to $x_1$. Moreover, as $NT_{2\to1}(\Delta t)$ is based on the probability density function of $x_k$, it is invariant under shifting and rescaling of the signals. In addition, transfer entropy accounts only for direct causality, that is, if $x_2$ is only caused by $x_1$ and $x_3$ is only caused by $x_2$, there is no causality from $x_1$ to $x_3$ [41].

### 2.2. *Numerical experiment*

The data are obtained from direct numerical simulation of a plane turbulent channel flow with two periodic directions and no-slip condition at the wall [42]. In the following, the streamwise, wall-normal and spanwise directions are denoted by $x$, $y$, and $z$, respectively, and the corresponding velocity components by $u$, $v$, and $w$. The friction Reynolds number of the simulation is $\mathrm{Re}_\tau = u_\tau \delta/\nu = 934$, where $u_\tau$ is the friction velocity, $\delta$ is the channel half-height, and $\nu$ is the kinematic viscosity.

The incompressible flow is integrated in the form of evolution equations for the wall-normal vorticity and the Laplacian of the wall-normal velocity [43]. The spatial discretization is dealiased Fourier in the two wall-parallel directions and Chebychev polynomials in $y$. Time stepping is third-order semi-implicit Runge-Kutta [44]. The number of grid points in the streamwise and spanwise directions are $N_x = 128$ and $N_z = 128$ such that the resolutions are $\Delta x^+ = 11.5$ and $\Delta z^+ = 5.8$, respectively, where the superscript + denotes wall units defined in terms of the friction velocity and the kinematic viscosity. The maximum wall-normal resolution is $\Delta y^+_{max} = 7.7$. The simulation was run with $CFL = 0.5$ for $140\delta/u_\tau$ (after transients), and the velocity fields were stored every $\sim 25$ wall units to generate a time-resolved dataset.

The length, height and width of the computational domain are $L_x = \pi\delta/2$, $L_y = 2\delta$ and $L_z = \pi\delta/4$, respectively. These dimensions correspond to a minimal box simulation for the log-layer and are considered to be sufficient for isolating the relevant dynamical structures involved in the bursting process [16, 11]. Minimal simulation boxes have demonstrated their ability to reproduce statistics of full-size turbulence computed in much larger domains. [16] showed that turbulence remains "healthy" roughly below $y \approx 0.3L_z$, corresponding in our case to $y \approx 0.25\delta$, that is the height chosen in the present study to analyze the causality transfer between rolls, streaks and mean shear.

Figure 1 shows instantaneous snapshots of the three velocity components. The streaky elongated nature of $u$ is clearly observed along the streamwise direction, as opposed to the shorter structure of $v$ and $w$. Figure 1 also highlights the fact that only a few large scale eddies are contained and isolated in this computational domain, enabling the causality analysis between individual objects.

### 2.3. *Representation of streaks and rolls*

We define $\hat{u}_{nm}(y,t)$ (respectively, $\hat{v}_{nm}$ and $\hat{w}_{nm}$) as the two-dimensional streamwise and spanwise Fourier transform of $u(x,y,z,t)$ (respectively, $v$ and $w$) where $k_{x_n} = 2\pi n/L_x$ and $k_{z_m} = 2\pi m/L_z$ are the corresponding wavenumbers, with $n = -N_x/2,\ldots,N_x/2-1$ and $m = -N_z/2,\ldots,N_z/2-1$. The causal analysis is applied to snapshots of $\hat{u}_{nm}(y,t)$, $\hat{v}_{nm}(y,t)$ and $\hat{w}_{nm}(y,t)$ averaged in the range of wall-normal distances $y \in [0.2\delta, 0.3\delta]$, consistent with the discussion in Section 2.2. Quantities averaged in the wall-normal direction are denoted by $\langle\cdot\rangle$.

We focus the analysis on the shear-dominated motions carrying most of the turbulent kinetic energy. The mean time-varying velocity profile is defined as $\langle\hat{u}_{00}(y,t)\rangle$ and along the text we will occasionally refer to this component as *mean shear.* Streaks are characterized by the first modes of the streamwise velocity fluctuation, and are classified into two categories as shown in Figure 1. The straight streak is defined as the $L_2$-norm of $\langle\hat{u}_{0(\pm1)}(y,t)\rangle$. The meandering streak is represented by $\langle\hat{u}_{(\pm1)(\pm1)}(y,t)\rangle$, which are indicators of the streak breakdown. An example

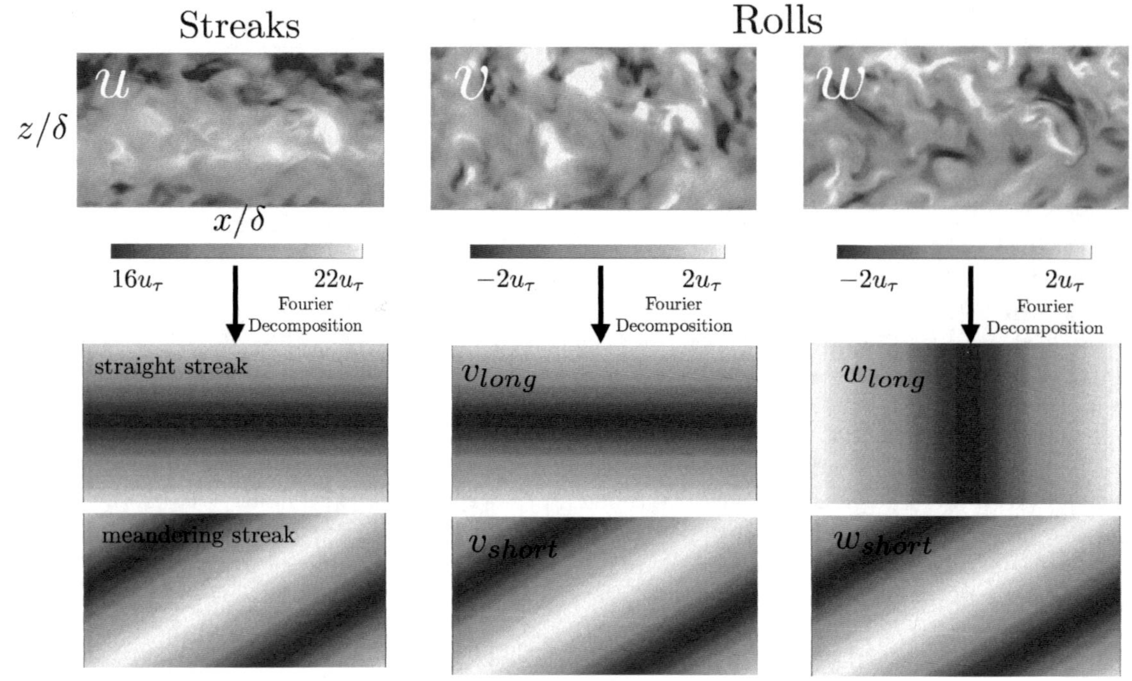


**Figure 1.** Snapshots of the instantaneous velocities $u, v$, and $w$ (top), and the representative Fourier modes of the streaks and rolls used for the causal analysis (bottom).

illustrating the two configurations of the streaks is shown in Figure 2. Rolls are defined as the cross-flow component of the velocity. Analogously to the streaks, $\hat{v}$ and $\hat{w}$ are divided into long and short motions (as shown in Figure 1). Long and short modes for $\hat{v}$ are $(n, m) = (0, \pm 1)$ and $(n, m) = (\pm 1, \pm 1)$, respectively. Whereas long and short modes for $\hat{w}$ are $(n, m) = (\pm 1, 0)$ and $(n, m) = (\pm 1, \pm 1)$. The final set of signals $x_k$ for $k = 1, \cdots, 7$, is composed of the mean streamwise velocity, the straight streak, the meandering streak, the long $v$, the short $v$, the long $w$, and the short $w$, respectively.

## 3. Results

### 3.1. Times-scales for causal interactions

The first goal is to establish the time horizon for causal influence between variables. This is achieved by computing the transfer entropy as a function of the time-lag, $\Delta t$. The behavior of $NT_{j\to i}(\Delta t)$ may differ for each $(i, j)$ pair, but for the sake of simplicity, we define two global measures: an average over all cross-induced transfer entropies $\overline{NT}_{j\to i}(\Delta t)$ $(j \neq i)$ and self-induced transfer entropies $\overline{NT}_{i\to i}(\Delta t)$.

The results are shown in Figure 3(a), and times are normalized by the average shear, $S = \langle \partial u/\partial y \rangle$. As expected, the maximum self-induced entropy peaks at $\Delta t = 0S^{-1}$, implying that the maximum reduction of uncertainty on a variable takes place by knowing the current state of the variable itself. More interestingly, the causality of a variable on itself has a characteristic time-span of $\sim 5S^{-1}$ measured by the time $\overline{NT}_{i\to i}(\Delta t)$ has decayed by 50%. The average cross transfer entropy exhibits a different behavior, with a peak at $\Delta t = 10S^{-1}$ and a characteristic time-span of $\sim 20S^{-1}$. The time scales discussed above are comparable to those in [45], who reported that the bursting time scale for shear flows is approximately universal and equal to

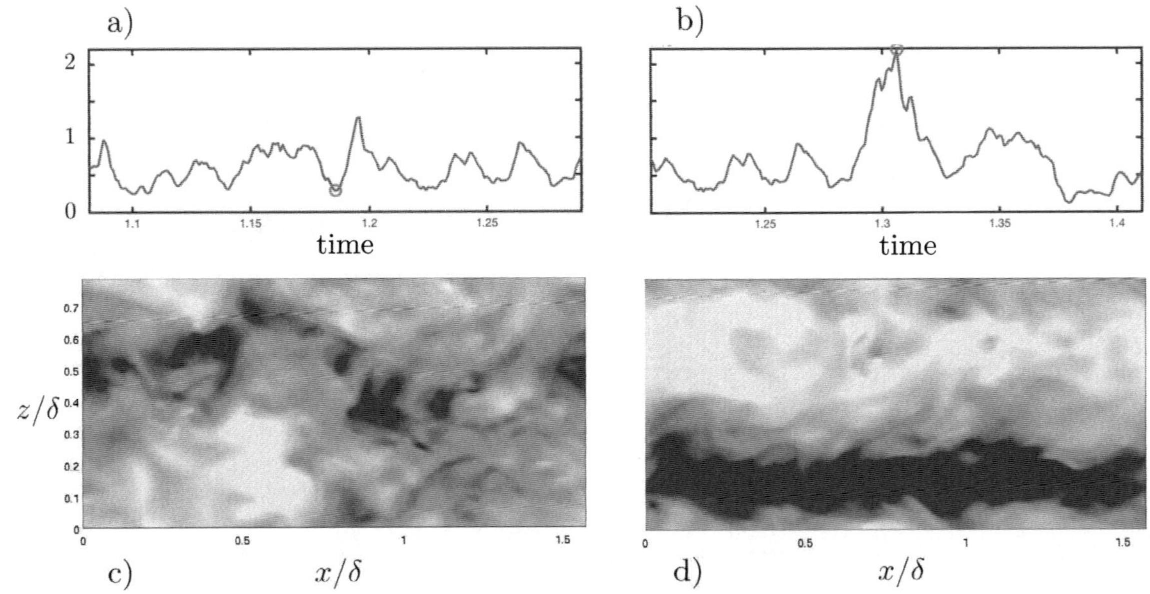


**Figure 2.** Two instances (red dot) in the time-series of the straight streak signal (a) and (b), and the corresponding streamwise velocity snapshots (c) and (d). Panels (a) and (c) feature a time of low intensity of the signal, whereas (b) and (d) correspond to the formation of a strong streamwise straight streak.

$20S^{-1}$.

The previous analysis is analogous to the classic time-correlation of two variables that, for comparison purposes, is shown in Figure 3(b). Only cross correlations with a maximum value larger than 0.4 are included. The characteristic time scales of the time-correlations are of the order of $10S^{-1}$ and consistent with those from the transfer entropy analysis, meaning that both methods are valid to extract the most representative time scales of the flow. However, transfer entropy is specifically designed for building bi-directional causality maps among variables, whereas the time-correlation approach is limited by its symmetry for this purpose. The reader is referred to [46] for a more detailed analysis of the time-correlations in minimal channel flows

*3.2. Causal network of the self-sustained streak-roll interaction*

We examine the causal network ($NT_{j\to i}, \forall j, i$) between variables at three different time scales, $\Delta t = 2S^{-1}, 5S^{-1}$ and $10S^{-1}$. This choice aims to capture the transition in the causality network from a self-induced dominated to a cross-induced dominated scenario. Longer time-lags could be considered; however, the statistical errors become significant to draw meaningful conclusions and the extension to larger time scales will be relegated to future work. The full causality maps are shown in Figure 4 for the three time scales. As it could be anticipated given the complexity of turbulent flows, many causal relations are significant and there is no obvious dominant pattern. In order to attain a simplified physical interpretation of the results from Figure 4, the dominant cross-induced causal connectivities have been compiled in three causality diagrams (Figure 5). Dominant causalities are defined as those with $NT > 0.7$. In addition to this constraint and to avoid spurious solutions due to statistical errors, those causalities which did not remain above 0.7 when considering half of the time history were excluded.

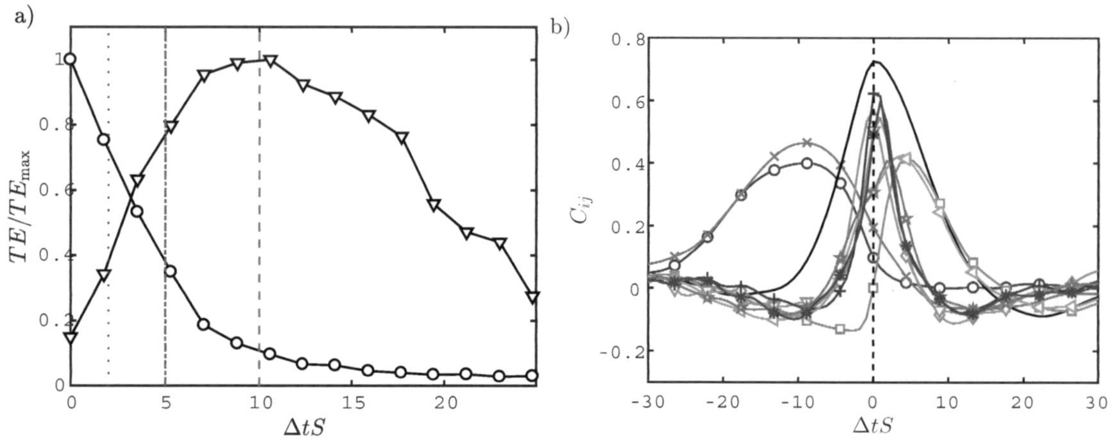


**Figure 3.** (a) Normalized self- (○) and cross-induced (▽) transfer entropy as a function of the time-lag. The blue dotted line, green dot-dash line, and red dash line are $\Delta T = 2S^{-1}, 5S^{-1}$ and $10S^{-1}$, respectively. (b) Time cross-correlations. Colors and symbols are: mean to straight streak (blue ○); mean to $v$ long (red ×); straight streak to $v$ long (black); straight streak to $w$ long (magenta □); meandering streak to $v$ short (green ◇); meandering streak to $w$ long (cyan ▽); meandering streak to $w$ short (blue +); $v$ long to $w$ long (gold ◁); $v$ short to $w$ long (red ⋆); $v$ short to $w$ short (green ∗).

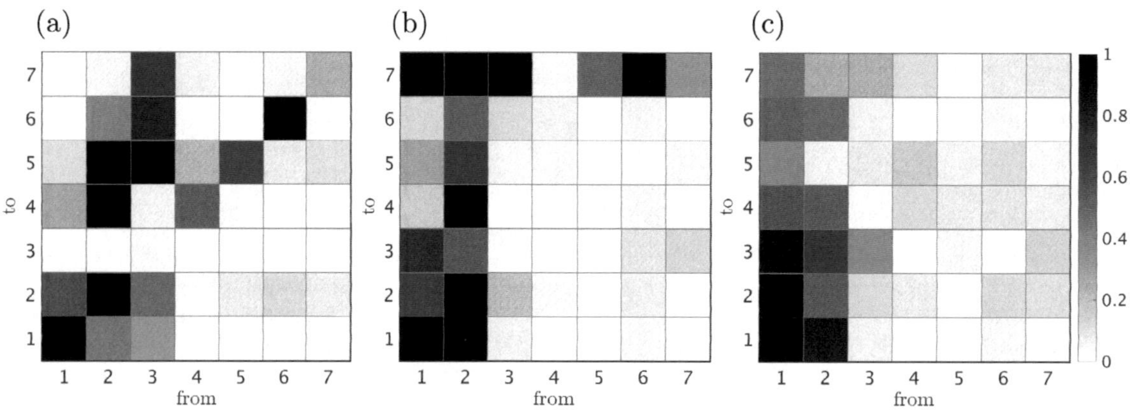


**Figure 4.** Causality maps for (a) $\Delta t = 2S^{-1}$, (b) $5S^{-1}$, and (c) $10S^{-1}$. Signal 1: time-varying mean velocity; 2: streamwise straight streak; 3: meandering streak; 4: long $v$; 5: short $v$; 6: long $w$; and 7: short $w$.

The causality diagram for $\Delta t = 2S^{-1}$ is shown in Figure 5 (a). The diagram reveals that most of the information flows from the streamwise streaks to the cross flow components whereas the mean flow remains inactive. At $\Delta t = 5S^{-1}$ (Figure 5 b), the straight streaks are still the dominant element with a strong causal effect on the mean flow, which begins to influence the meandering streaks. Finally, at $\Delta t = 10S^{-1}$, the causality diagram is reverted and the causal effect flows from the mean flow to the streamwise streaks, while the effect of the streamwise streaks are lessened. For all three time scales considered, there is no causal influence from the cross flow to the other components. Many weaker causal influences ($NT < 0.7$) are omitted

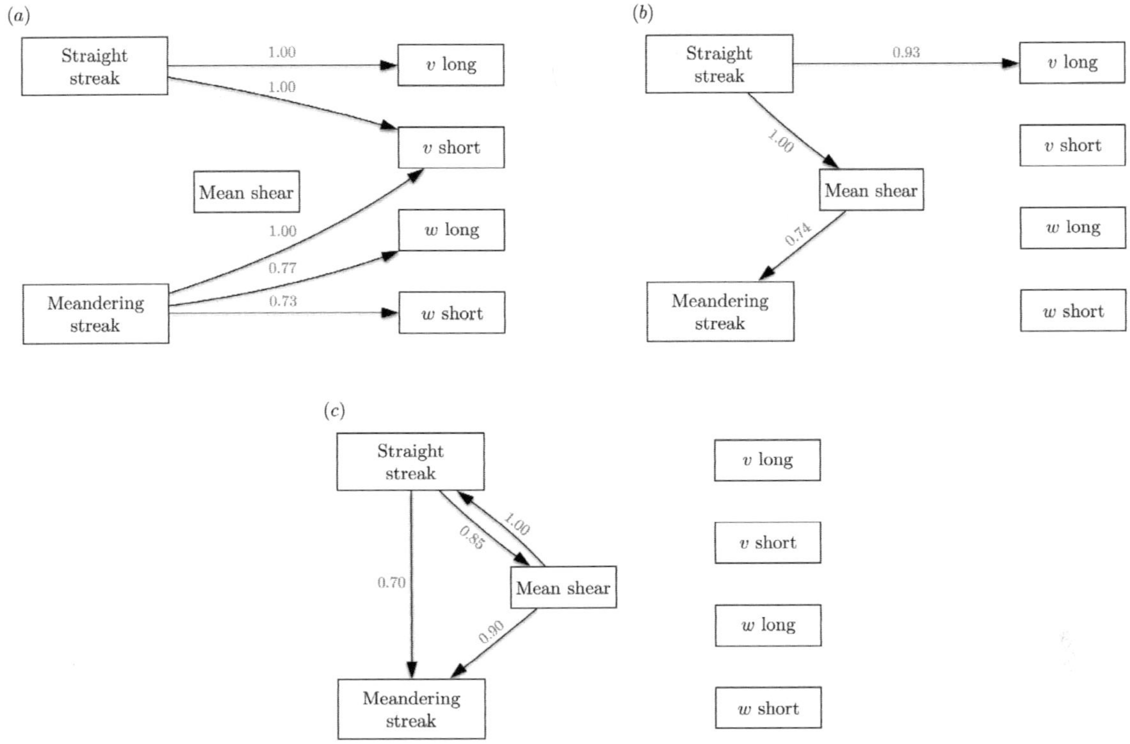


**Figure 5.** Causality diagrams for (a) $\Delta t = 2S^{-1}$, (b) $5S^{-1}$, and (c) $10S^{-1}$. Only connections with $NT > 0.7$ are included.

in the description above, the most significant one being the lift-up mechanism. This can be identified, for example, as the causal influence of the long $v$ to the straight streaks, that is also present and of the order $\sim 0.3$.

Finally, the self-sustaining process of the logarithmic layer is summarized in Figure 6 and interpreted in the context of different known mechanisms. The process is mostly unidirectional, emanating from the time-varying mean velocity profile, which generates straight streaks in a characteristic time scale of $10S^{-1}$. We refer to this process as a parametric instability. The effect of the straight streaks on the flow is twofold. First, it generates long velocity rolls which interact weakly ($NT < 0.7$) with the straight streaks through the lift-up mechanism. Secondly, straight streaks meander, marking the onset of its instability. Both processes take place at a time scale of $5S^{-1}$. The instability results in the breakdown of the streamwise streaks into short rolls, in a relatively fast process that spans along the time period of $2S^{-1}$.

It is important to remark that the causality discussed above is a measure based on the reduction of the statistical uncertainty of the variables, and it is not directly linked to energy transfer. Furthermore, the current method detects direct causality only if all the intermediate causal variables are accounted for. Otherwise, the causality may flow indirectly through other

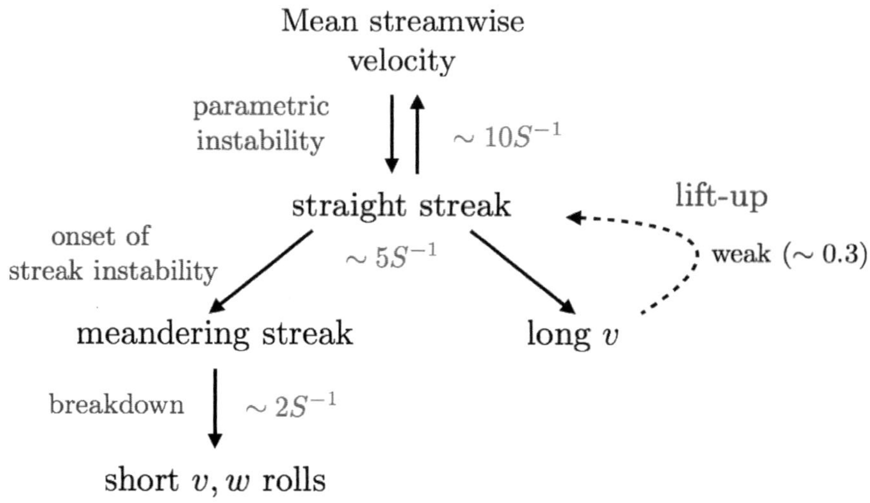


**Figure 6.** Summary of the most relevant causal relations for shear-dominated scales in the logarithmic layer. See text for details.

signals not taken into consideration. For example, the dynamic equation for $\hat{v}_{01}$ is

$$\frac{\partial \hat{v}_{01}}{\partial t} = -\sum_{n,m} ik_{x_n}\hat{v}_{nm}\hat{u}_{(-n)(1-m)} - \sum_{n,m}\frac{\partial \hat{v}_{nm}}{\partial y}\hat{v}_{(-n)(1-m)} \tag{4}$$

$$-\sum_{n,m} ik_{z_m}\hat{v}_{nm}\hat{w}_{(-n)(1-m)} + \nu\left(\frac{\partial^2 \hat{v}_{01}}{\partial y^2} - k_{z_1}^2\hat{v}_{01}\right), \tag{5}$$

which does not directly depend on $\hat{u}_{0(\pm 1)}$ (similarly for $\hat{v}_{0(-1)}$). Hence, the causal connection from the straight streak to $v$ long shown in Figure 5 follows necessarily an indirect path.

## 4. Conclusions

Despite the extensive information provided by direct numerical simulation of turbulent flows, the causal relation between different flow structures has been overlooked in turbulence research. In the present work, we frame the causal analysis between turbulent signals from a transfer entropy perspective.

We have examined the causal structure between the streamwise rolls, streaks and mean shear in the logarithmic layer of a turbulent channel flow. The velocity components of the minimal channel of the log layer was decomposed in wall-parallel Fourier modes, resulting in a time series of the mean flow, two different configurations of streaks (straight and meandering), and rolls (short and long). Transfer entropy analysis was then conducted between the series, providing the causality flow between flow structures.

The detailed analysis of the causal network shows that the self-sustaining process is mainly unidirectional rather than cyclic, meaning that the logarithmic layer motions are sustained by the temporal change of the mean shear which controls the dynamics and time-scales. The causality flows then to the straight streaks which become unstable and break into smaller scale rolls. The well-known lift-up effect is present but shown to be weak from a causal point of view. The temporal causality horizon spans $20S^{-1}$ since the initialization of the process from the mean shear until the breakdown of the flow into smaller scale rolls.

**Acknowledgments**
This work was funded in part by the Coturb program of the European Research Council. H.J.B. and A.L.D. acknowledge the support of the Center for Turbulence Research and the NASA Transformative Aeronautics Concepts Program (grant no. NNX15AU93A). We also thank Prof. P. J. Ioannou for his helpful comments on the manuscript.

## References

[1] R T Cerbus and W I Goldburg. Information content of turbulence. *Phys. Rev. E*, 88:053012, 2013.
[2] G Tissot, A Lozano-Durán, L Cordier, J Jiménez, and B R Noack. Granger causality in wall-bounded turbulence. *J. Phys.: Conf. Ser.*, 506:012006, 2014.
[3] X San Liang and A Lozano-Durán. A preliminary study of the causal structure in fully developed near-wall turbulence. In *CTR Proc. of the Summer Prog.*, page 233, 2016.
[4] P S Klebanoff, K D Tidstrom, and L M Sargent. The three-dimensional nature of boundary-layer instability. *J. Fluid Mech.*, 12:1–34, 1962.
[5] S J Kline, W C Reynolds, F A Schraub, and P W Runstadler. The structure of turbulent boundary layers. *J. Fluid Mech.*, 30:741–773, 1967.
[6] H T Kim, S J Kline, and W C Reynolds. The production of turbulence near a smooth wall in a turbulent boundary layer. *J. Fluid Mech.*, 50:133–160, 1971.
[7] J Jiménez and P Moin. The minimal flow unit in near-wall turbulence. *J. Fluid Mech.*, 225:213–240, 1991.
[8] J M Hamilton, J Kim, and F Waleffe. Regeneration mechanisms of near-wall turbulence structures. *J. Fluid Mech.*, 287:317–348, 1995.
[9] F Waleffe. On a self-sustaining process in shear flows. *Phys. Fluids*, 9:883–900, 1997.
[10] W Schoppa and F Hussain. Coherent structure generation in near-wall turbulence. *J Fluid Mech.*, 453:57–108, 2002.
[11] J Jiménez. Cascades in wall-bounded turbulence. *Annu. Rev. Fluid Mech.*, 44, 2012.
[12] K M Butler and B F Farrell. Optimal perturbations and streak spacing in wall-bounded turbulent shear flow. *Phys. Fluids A*, 5:774–777, 1993.
[13] S I Chernyshenko and M F Baig. The mechanism of streak formation in near-wall turbulence. *J. Fluid Mech.*, 544:99–131, 2005.
[14] B F Farrell and P J Ioannou. Dynamics of streamwise rolls and streaks in turbulent wall-bounded shear flow. *J. Fluid Mech.*, 708:149–196, 2012.
[15] R L Panton. Overview of the self-sustaining mechanisms of wall turbulence. *Prog. Aerosp. Sci*, 37:341–383, 2001.
[16] O Flores and J Jiménez. Hierarchy of minimal flow units in the logarithmic layer. *Phys. Fluids*, 22:071704, 2010.
[17] Y Hwang and C Cossu. Self-sustained processes in the logarithmic layer of turbulent channel flows. *Phys. Fluids*, 23:061702, 2011.
[18] C Cossu and Y Hwang. Self-sustaining processes at all scales in wall-bounded turbulent shear flows. *Phil. Trans. R. Soc. A*, 375(2089):20160088, 2017.
[19] A A Townsend. Equilibrium layers and wall turbulence. *J. Fluid Mech.*, 11:97–120, 1961.
[20] J C Del Álamo, J Jimenez, P Zandonade, and R D Moser. Self-similar vortex clusters in the turbulent logarithmic region. *J. Fluid Mech.*, 561:329–358, 2006.
[21] A Lozano-Durán, O Flores, and J Jiménez. The three-dimensional structure of momentum transfer in turbulent channels. *J. Fluid Mech.*, 694:100–130, 2012.
[22] A Lozano-Durán and J Jiménez. Time-resolved evolution of coherent structures in turbulent channels: characterization of eddies and cascades. *J. Fluid Mech.*, 759:432–471, 2014.
[23] A Lozano-Durán and G Borrell. Algorithm 964: An efficient algorithm to compute the genus of discrete surfaces and applications to turbulent flows. *ACM Trans. Math. Softw.*, 42:34:1–34:19, 2016.
[24] Y Hwang and C Cossu. Self-sustained process at large scales in turbulent channel flow. *Phys. Rev. Lett.*, 105:044505, 2010.
[25] R Moarref, A S Sharma, J A Tropp, and B J McKeon. A foundation for analytical developments in the logarithmic region of turbulent channels. *arXiv preprint arXiv:1409.6047*, 2014.
[26] F Alizard. Linear stability of optimal streaks in the log-layer of turbulent channel flows. *Phys. Fluids*, 27:105103, 2015.
[27] W M F Orr. The stability or instability of the steady motions of a perfect liquid and of a viscous liquid. part I: A perfect liquid. *Proc. R. Ir. Acad. A*, 27:9–68, 1907.
[28] P Andersson, L Brandt, A Bottaro, and D S Henningson. On the breakdown of boundary layer streaks. *J. Fluid Mech.*, 428:29–60, 2001.

[29] J Park, Y Hwang, and C Cossu. On the stability of large-scale streaks in turbulent couette and poiseulle flows. *C. R. Mécanique*, 339:1–5, 2011.

[30] A Cassinelli, M de Giovanetti, and Y Hwang. Streak instability in near-wall turbulence revisited. *J. Turb*, 18:443–464, 2017.

[31] B F Farrell, P J Ioannou, J Jiménez, N C Constantinou, A Lozano-Durán, and M-A Nikolaidis. A statistical state dynamics-based study of the structure and mechanism of large-scale motions in plane poiseuille flow. *J. Fluid Mech.*, 809:290–315, 2016.

[32] N J Vaughan and T A Zaki. Stability of zero-pressure-gradient boundary layer distorted by unsteady klebanoff streaks. *J. Fluid Mech.*, 681:116–153, 2011.

[33] N N Mansour, J Kim, and P Moin. Reynolds-stress and dissipation-rate budgets in a turbulent channel flow. *J. Fluid Mech.*, 194:15–44, 1988.

[34] H Beebee, C Hitchcock, and P Menzies. *The Oxford handbook of causation.* Oxford University Press, 2009.

[35] C W J Granger. Investigating causal relations by econometric models and cross-spectral methods. *Econometrica*, 424–438, 1969.

[36] T Schreiber. Measuring information transfer. *Phys. Rev. Lett.*, 85:461, 2000.

[37] K Hlaváčková-Schindler, M Paluš, M Vejmelka, and J Bhattacharya. Causality detection based on information-theoretic approaches in time series analysis. *Phys. Rep.*, 441:1–46, 2007.

[38] L F Kozachenko and N N Leonenko. Sample estimate of the entropy of a random vector. *Probl. Peredachi Inf.*, 23:9–16, 1987.

[39] A Kraskov, H Stögbauer, and P Grassberger. Estimating mutual information. *Phys. Rev. E*, 69:066138, 2004.

[40] C E Shannon. A mathematical theory of communication. *ACM Sigmobile MC2R*, 5:3–55, 2001.

[41] P Duan, F Yang, T Chen, and S L Shah. Direct causality detection via the transfer entropy approach. *IEEE Trans. Control Syst. Technol.*, 21:2052–2066, 2013.

[42] J Jiménez. Direct detection of linearized bursts in turbulence. *Phys. Fluids*, 27:065102, 2015.

[43] J Kim, P Moin, and R Moser. Turbulence statistics in fully developed channel flow at low reynolds number. *J. Fluid Mech.*, 177:133–166, 1987.

[44] P R Spalart, R D Moser, and M M Rogers. Spectral methods for the navier-stokes equations with one infinite and two periodic directions. *J. Comp. Phys.*, 96:297–324, 1991.

[45] A Sekimoto, S Dong, and J Jiménez. Direct numerical simulation of statistically stationary and homogeneous shear turbulence and its relation to other shear flows. *Phys. Fluids*, 28:035101, 2016.

[46] J Jiménez. How linear is wall-bounded turbulence? *Phys. Fluids*, 25:110814, 2013.

# The mechanism by which nonlinearity sustains turbulence in plane Couette flow

**M-A. Nikolaidis**[1], **B. F. Farrell**[2], **P. J. Ioannou**[1]

[1]National and Kapodistrian University of Athens, Department of Physics, Panepistimiopolis, Zografos, 15784 Athens, Greece
[2]Department of Earth and Planetary Sciences, Harvard University, Cambridge, MA 02138, USA

E-mail: pjioannou@phys.uoa.gr

**Abstract.** Turbulence in wall-bounded shear flow results from a synergistic interaction between linear non-normality and nonlinearity in which non-normal growth of a subset of perturbations configured to transfer energy from the externally forced component of the turbulent state to the perturbation component maintains the perturbation energy, while the subset of energy-transferring perturbations is replenished by nonlinearity. Although it is accepted that both linear non-normality mediated energy transfer from the forced component of the mean flow and nonlinear interactions among perturbations are required to maintain the turbulent state, the detailed physical mechanism by which these processes interact in maintaining turbulence has not been determined. In this work a statistical state dynamics based analysis is performed on turbulent Couette flow at $R = 600$ and a comparison to DNS is used to demonstrate that the perturbation component in Couette flow turbulence is replenished by a non-normality mediated parametric growth process in which the fluctuating streamwise mean flow has been adjusted to marginal Lyapunov stability. It is further shown that the alternative mechanism in which the subspace of non-normally growing perturbations is maintained directly by perturbation-perturbation nonlinearity does not contribute to maintaining the turbulent state. This work identifies parametric interaction between the fluctuating streamwise mean flow and the streamwise varying perturbations to be the mechanism of the nonlinear interaction maintaining the perturbation component of the turbulent state, and identifies the associated Lyapunov vectors with positive energetics as the structures of the perturbation subspace supporting the turbulence.

## 1. Introduction

Turbulence is widely regarded as the primary exemplar of an essentially nonlinear phenomenon. However, the mechanism by which energy is transferred in shear flows from the externally forced component of the flow to the broad spectrum of spatially and temporally varying perturbations is through linear non-normal interaction between these flow components [1, 2]. Nevertheless, nonlinearity participates in an essential way in this cooperative interaction by which turbulence self-sustains. Our goal in this study is to provide a more comprehensive understanding of the role of nonlinearity and its interaction with linear non-normality in the maintenance of turbulence.

In order to study the mechanism by which nonlinearity between streamwise varying components participates in the maintenance of turbulence in shear flow we begin by partitioning the velocity field of plane parallel Couette flow into streamwise mean and perturbation components, or equivalently into the $k_x = 0$ and the $k_x \neq 0$ components of the Fourier decomposition of the flow field, where $k_x$ is the wavenumber in the streamwise, $x$, direction. In this decomposition the flow field is partitioned as:

$$\mathbf{u} = \mathbf{U}(y, z, t) + \mathbf{u}'(x, y, z, t) \ , \tag{1}$$

with cross-stream direction $y$ and spanwise direction $z$. Note that in this decomposition the mean flow retains temporal variation in its spanwise structure as would occur e.g. in the presence of time-dependent streaks.

The non-dimensional Navier-Stokes equations expressed using this mean and perturbation partition are:

$$\partial_t \mathbf{U} + \underbrace{\mathbf{U} \cdot \nabla \mathbf{U}}_{N_1} + \nabla P - \Delta \mathbf{U}/R = \underbrace{-\langle \mathbf{u}' \cdot \nabla \mathbf{u}' \rangle_x}_{N_2} \ , \tag{2a}$$

$$\partial_t \mathbf{u}' + \underbrace{\mathbf{U} \cdot \nabla \mathbf{u}' + \mathbf{u}' \cdot \nabla \mathbf{U}}_{N_3} + \nabla p' - \Delta \mathbf{u}'/R = \underbrace{-\left(\mathbf{u}' \cdot \nabla \mathbf{u}' - \langle \mathbf{u}' \cdot \nabla \mathbf{u}' \rangle_x\right)}_{N_4} \ , \tag{2b}$$

$$\nabla \cdot \mathbf{U} = 0 \ , \quad \nabla \cdot \mathbf{u}' = 0 \ , \tag{2c}$$

where $R = U_w h/\nu$ is the Reynolds number and $\pm U_w$ the wall velocity at $y = \pm h$. The flow satisfies no-slip boundary conditions in the cross-stream direction: $\mathbf{U}(x, \pm h, z, t) = (\pm U_w, 0, 0)$, $\mathbf{u}'(x, \pm h, z, t) = (0, 0, 0)$ and periodic boundary conditions in the $z$ and $x$ directions. From now on, lengths will be normalised with $h$, velocities with $U_w$, and time with $h/U_w$. Averaging is denoted with a bracket $\langle \cdot \rangle$ with the bracket subscript indicating the averaging variable, so that e.g. the streamwise mean velocity is $\mathbf{U} \equiv \langle \mathbf{u} \rangle_x = L_x^{-1} \int_0^{L_x} \mathbf{u} \; dx$, where $L_x$ is the streamwise length of the channel. The Navier-Stokes equations with this decomposition will be referred to as the NL system and direct numerical simulations of this system as NL. We have indicated with an underbrace the terms in the NL system of primary relevance to our study. The advective term in the streamwise mean flow equation (2a) comprising nonlinear interactions among $k_x = 0$ flow components will be referred to as $N_1$, the Reynolds stress divergence term in (2a), produced by nonlinear interaction between the $k_x$ and $-k_x$ flow components with $k_x \neq 0$, will be referred to as $N_2$. In the perturbation equation (2b) $N_3$ gives the linear interaction between the mean flow and the $k_x \neq 0$ flow components, which is the source of the non-normality giving rise to the transfer of energy from the mean flow to the perturbations, and $N_4$ is the nonlinear interactions between $k_{x_1} \neq 0$ and $k_{x_2} \neq 0$, with $k_{x_1} \neq -k_{x_2}$.

Transition to and maintenance of a self-sustained turbulent state results even when only nonlinearity $N_2$ is retained [3]. By retaining nonlinearities $N_1$ and $N_2$ we obtain the restricted non-linear system (RNL):

$$\partial_t \mathbf{U} + \mathbf{U} \cdot \nabla \mathbf{U} + \nabla P - \Delta \mathbf{U}/R = -\langle \mathbf{u}' \cdot \nabla \mathbf{u}' \rangle_x \ , \tag{3a}$$

$$\partial_t \mathbf{u}' + \mathbf{U} \cdot \nabla \mathbf{u}' + \mathbf{u}' \cdot \nabla \mathbf{U} + \nabla p' - \Delta \mathbf{u}'/R = 0 \ , \tag{3b}$$

$$\nabla \cdot \mathbf{U} = 0 \ , \quad \nabla \cdot \mathbf{u}' = 0 \ . \tag{3c}$$

It has been confirmed that this RNL system supports a realistic self-sustaining process (SSP) which maintains a turbulent state in minimal channels [3, 4], in channels of moderate sizes at both low and high Reynolds numbers (at least for $R_\tau \leq 1000$) [5–7], and also in very long channels [8].

Consider in isolation the time varying mean flow $\mathbf{U}$ obtained from a state of turbulence either of the NL or of the RNL system. Sufficiently small perturbations, $\mathbf{u}'$, on this mean flow evolve according to

$$\partial_t \mathbf{u}' + \mathbf{U} \cdot \nabla \mathbf{u}' + \mathbf{u}' \cdot \nabla \mathbf{U} + \nabla p' - \Delta \mathbf{u}'/R \ = 0 \ , \quad \nabla \cdot \mathbf{u}' = 0 \ , \tag{4}$$

which is the perturbation equation (3b) of the RNL system, while in the NL system the finite perturations $\mathbf{u}'$ obey the different equation (2b) with the $N_4$ term included. In the self-sustained RNL turbulent state the perturbations, $\mathbf{u}'$, that evolve under the linear dynamics

(3b) or equivalently under (4) remain finite and bounded. Therefore the mean-flow, $\mathbf{U}$, of the RNL turbulent state is stable in the sense that perturbations, i.e. the streamwise varying flow components, $\mathbf{u}'$, that evolve under (4), have zero asymptotic growth rate and the mean flow can be considered to be in the critical state of neutrality, poised between stability and instability. A question that will be addressed in this paper is whether the mean flow, $\mathbf{U}$, that is obtained from a simulation of NL turbulence shares this property of being adjusted similarly to neutrality in the sense that perturbations, $\mathbf{u}'$, that evolve under (4), remain bounded and therefore have vanishing asymptotic growth rate and the mean flow of the NL system can therefore be considered to be similarly in a critical state of neutrality when proper account is taken of dissipation in the NL system. If the turbulent mean flow $\mathbf{U}$ of NL can be shown to be neutral, in this sense of parametric neutrality, then the mechanism of turbulence identified analytically in the RNL system, in which the perturbations arise from parametric instability of the mean flow with the mean flow being regulated to neutrality through quasi-linear interaction with the perturbation field, will have been extended to NL and the turbulent dynamics of NL identified to be essentially the same as that of the analytically characterized RNL dynamics. For this program to succeed it is required to show that the dynamically substantive difference between the RNL system and the NL system, which is the appearance in the NL system of the perturbation-perturbation nonlinearity $N_4$, does not fundamentally change the dynamics of turbulence operating in the RNL system.

The mechanism determining the statistical state of turbulence in the RNL system and its extension to the NL system we are studying can be related to an influential conjecture of Malkus: "First, that the mean flow will be statistically stable if an Orr-Sommerfeld type equation is satisfied by fluctuations of the mean; second, that the smallest scale of motion that can be present in the spectrum of the momentum transport is the scale of the marginally stable fluctuations of the mean" [9]. In this quote, Malkus conjectures that the turbulent mean state is adjusted to a state of neutrality, as he defines neutrality, and the turbulent perturbations responsible for the Reynolds stresses are in the subspace of the neutral modes of the Orr-Sommerfeld operator about the mean flow, as he defines it. We have in common with Malkus the concept of adjustment by quasi-linear interaction between the mean flow and perturbations as the general mechanism determining the statistical state of turbulence in shear flow and we have succeeded to show that RNL turbulence operates with this program. Extension of the program to the NL system is the goal of the present work. We differ with Malkus in the particulars of this general program, including considering the parametric instability of the time dependent streamwise mean state as that being adjusted to neutrality rather than the time, spanwise and streamwise mean flow. While e.g. turbulent convection [10, 11] and the baroclinic turbulence in the midlatitude atmosphere [12] display a usefully close approximate adherence to this conjecture when both the spatial and temporal means are taken to define the mean flow, the turbulent mean state of wall-bounded turbulence, defined as the streamwise, spanwise and temporal mean, $\langle \mathbf{U} \rangle_{z,t}$, is hydrodynamically stable and far from neutrality in apparently strong violation of the Malkus conjecture [13]. However, RNL turbulence suggests that the program of Malkus [9] was essentially correct and can be extended to wall-turbulence requiring only the additional recognition that the instability to be equilibrated is the instability of the time-dependent operator associated with linearization about the temporally varying streamwise mean flow, $\mathbf{U}$.

The maximum growth rate of perturbations to the streamwise mean $\mathbf{U}$ that are governed by the linear dynamics of (4) is given by the top Lyapunov exponent of $\mathbf{u}'$ defined as:

$$\lambda_{Lyap} = \lim_{t\to\infty} \frac{\log|\mathbf{u}'|}{t} \ . \tag{5}$$

RNL turbulence with the definition of mean (1) satisfies the Malkus conjecture precisely under

IOP Conf. Series: Journal of Physics: Conf. Series **1001** (2018) 012014 doi:10.1088/1742-6596/1001/1/012014

our interpretation because for RNL

$$\lambda_{Lyap} = 0 \ . \tag{6}$$

An issue we wish to examine in this work is whether NL turbulence (with the $N_4$ term included) is similarly neutral in the Lyapunov sense with its perturbations being similarly supported by parametric growth on its fluctuating mean flow with perturbation structure being associated with the predicted Lyapunov vector structure. Specifically, we mean whether fluctuations, $\mathbf{u}'$, evolving under (4) on the time dependent mean flow, $\mathbf{U}$, that has been obtained from a turbulent simulation of the NL equations, have $\lambda_{Lyap}$, as defined in (5), zero when proper account is taken of dissipative processes and the predicted Lyapunov structure can be verified to be maintaining the perturbations. We caution the reader that the Lyapunov exponents we are calculating are not the familiar Lyapunov exponents of small perturbations from a turbulent trajectory, which is associated with the growth of perturbations $\delta\mathbf{U}$, $\delta\mathbf{u}'$ to the tangent linear dynamics of the full NL system linearized about a turbulent trajectory $\mathbf{U}$, $\mathbf{u}'$. The turbulent trajectory $\delta\mathbf{U}$, $\delta\mathbf{u}'$ is chaotic and will have typically many positive Lyapunov exponents. We instead calculate only the Lyapunov exponents and structures of perturbations, $\mathbf{u}'$, evolving under the linear dynamics (4) with time dependent mean flow $\mathbf{U}$. It is important to recognize that the parametric perturbation evolution equation (4) governing the perturbation Lyapunov vector dynamics of RNL is not limited to small perturbation amplitude and the nonlinearity required to regulate the perturbations to their finite statistical equilibrium state, although not present in (4), is contained in the Reynolds stress feedback term $N_2$ appearing in the mean equation.

Choice of the mean used in the cumulant expansion is fundamental to formulating an SSD to gain an understanding of the mechanism of turbulence in shear flow. SSD analysis reveals that transition to and maintenance of turbulence occurs in association with the breaking of statistical symmetries of the laminar state through a sequence of bifurcations. The laminar state in Couette flow has spanwise, streamwise and temporal statistical homogeneity. When the SSD of Couette flow is closed at second order it has been shown that at a first critical Reynolds number the spanwise symmetry is broken, and at a second higher Reynolds number the temporal statistical homogeneity is broken, coincident with transition to the turbulent state [3, 4]. The centrality of spanwise variation of the mean flow, which is associated with the streak component, to the maintenance of turbulence has been demonstrated by numerical experiments that show turbulence is not sustained when the streaks are sufficiently damped or removed [14]. Given that in the SSD turbulent state the spanwise and temporal homogeneity are broken, it is dynamically inconsistent to use either the spanwise or the temporal mean to separate the flow field into mean and perturbation components. It is necessary to allow both spanwise and temporal variations in the mean structure which requires representation (1).

It remains an open question whether in the turbulent state the streamwise statistical homogeneity is also broken. The evidence that is usually presented to support breaking of the streamwise statistical homogeneity is that no arbitrarily long streamwise structures are observed in simulations, but this evidence is based on individual realizations which may not reflect the symmetries of the statistical state. The evidence supporting the non-breaking of the streamwise symmetry is that RNL has been shown to support a realistic turbulent state under the assumption of a streamwise mean ($k_x = 0$) in long turbulent channels, demonstrating that there exists a statistical turbulent state with statistical streamwise homogeneity [8].

Comparing RNL and NL dynamics provides a new perspective on the role of the perturbation-perturbation nonlinearity $N_4$ in the maintenance and regulation of turbulence. The $N_4$ term in (2b) does not contribute directly to maintaining the perturbation energy because the perturbation-perturbation interactions redistribute energy internally among the streamwise $k_x \neq 0$ components of the flow and the term $\langle \mathbf{u}' \cdot N_4 \rangle_{x,y,z}$ is zero[1]. Consequently, from (2b) we obtain that the

[1] In our simulations time discretization produces a $\langle \mathbf{u}' \cdot N_4 \rangle_{x,y,z,t}$ of the order of $-0.0005 U_w^3/h$ which provides an

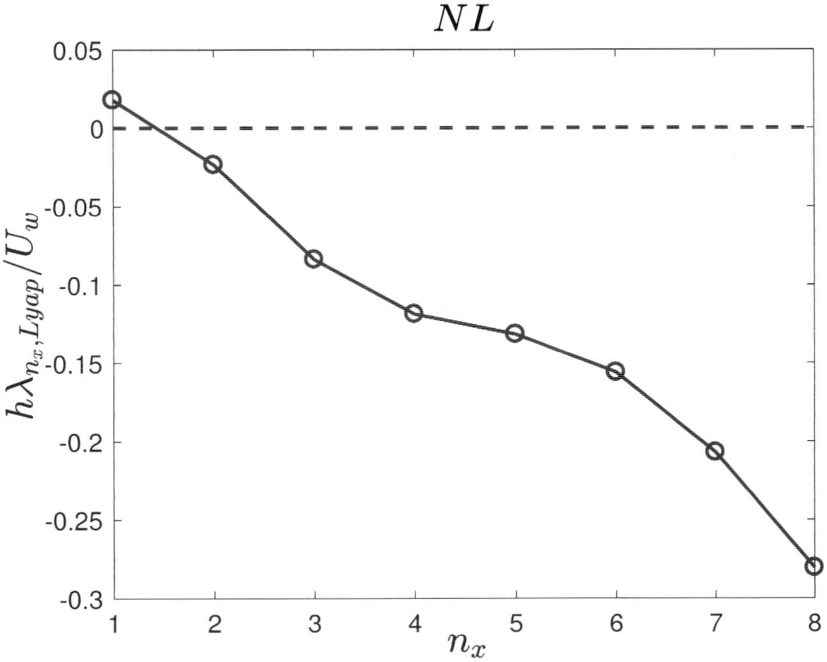


**Figure 1.** The top Lyapunov exponent of perturbations with channel wavenumbers $n_x = 1, \ldots, 8$ evolving under the time-dependent turbulent mean flow in NL, $\mathbf{U}$. The Lyapunov exponent of all $n_x \geq 2$ components of $\mathbf{u}'$ is negative. For comparison, the least stable mode of the streamwise-spanwise-temporal mean of $\mathbf{U}$ has decay rate $\sigma = -0.12\ U_w/h$ at $n_x = 1$ and $n_z = 3$. The Lyapunov exponent was nondimensionalized using advective time units, $h/U_w$. A plane Couette channel at $R = 600$ was used.

perturbation energy density, $E_p = \langle |\mathbf{u}'|^2/2 \rangle_{x,y,z}$, evolves according to:

$$\frac{dE_p}{dt} = \underbrace{\langle \mathbf{u}' \cdot \left( -\mathbf{U} \cdot \nabla \mathbf{u}' - \mathbf{u}' \cdot \nabla \mathbf{U} + \Delta \mathbf{u}'/R \right) \rangle_{x,y,z}}_{\dot{E}_{linear}} . \tag{7}$$

just as in RNL turbulence. The term $\dot{E}_{linear}$ gives the energy transfer to the streamwise-varying perturbations by interaction with the mean $\mathbf{U}$. The Lyapunov exponent of the $\mathbf{u}'$ associated with the mean flow, $\mathbf{U}$, taken from an NL simulation, and defined in (5) is also given by the time-average of the instantaneous energy growth rates:

$$\lambda_{Lyap} = \left\langle \frac{1}{2E_p} \frac{dE_p}{dt} \right\rangle_t . \tag{8}$$

Equation (8) gives the top Lyapunov exponent of the NL perturbation dynamics, but a full spectrum of exponents can also be obtained using orthogonalization techniques [15].

This top Lyapunov exponent should be contrasted with the exponent obtained by inserting into (7) and (8) the $\mathbf{u}'$ taken from a simulation of the NL turbulent state. This $\mathbf{u}'$ is bounded because it is the perturbation state vector and therefore this exponent is exactly $\lambda_{state} = 0$. While only the top Lyapunov vector is excited in RNL, in NL a spectrum of Lyapunov vectors

error estimate for the accuracy of our results.

comprise the $\mathbf{u}'$ of the NL state and because these are orthogonal in Fourier representation it is useful to consider the energetics of each of the streamwise Fourier components of $\mathbf{u}'$ separately. If we decompose the perturbation field into its streamwise components:

$$\mathbf{u}' = \sum_{n_x=1}^{N} \mathbf{u}'_{n_x}(y,z,t)\mathrm{e}^{\mathrm{i}k_x x} , \tag{9}$$

with $k_x = 2\pi n_x/L_x$, the effective growth rate of each of the streamwise components $\mathbf{u}'_{n_x}$ will also be individually zero. This requires that

$$\lambda_{n_x,state} = \left\langle \frac{\mathrm{Re}\langle \mathbf{u}'_{n_x} \cdot (-\mathbf{U}\cdot\nabla\mathbf{u}'_{n_x} - \mathbf{u}'_{n_x}\cdot\nabla\mathbf{U} + \Delta\mathbf{u}'_{n_x}/R + N_{4,n_x})\rangle_{y,z}}{2E_{n_x}} \right\rangle_t = 0. \tag{10}$$

where $E_{n_x} = \langle|\mathbf{u}'_{n_x}|^2/2\rangle_{y,z}$ is the kinetic energy of the $n_x$ streamwise component and $N_{4,n_x}$ is the $n_x$ streamwise component of the $N_4$ term in (2b) (Re denotes the real part of a quantity). In the energetics of NL in addition to the rate of energy transfer to the perturbations from the mean flow:

$$\dot{E}_{def,n_x} = \mathrm{Re}\left\langle \mathbf{u}'_{n_x} \cdot \left(-\mathbf{U}\cdot\nabla\mathbf{u}'_{n_x} - \mathbf{u}'_{n_x}\cdot\nabla\mathbf{U}\right)\right\rangle_{y,z} , \tag{11}$$

and perturbation energy dissipation rate:

$$\dot{E}_{dissip,n_x} = \frac{1}{R}\mathrm{Re}\left\langle \mathbf{u}'_{n_x}\cdot\Delta\mathbf{u}'_{n_x}\right\rangle_{y,z} , \tag{12}$$

which are the only terms in (7) that are involved in the determination of the Lyapunov exponent, there is the additional term

$$\dot{E}_{nonlin,n_x} = \mathrm{Re}\left\langle \mathbf{u}'_{n_x}\cdot N_{4,n_x}\right\rangle_{y,z} , \tag{13}$$

that is only present in the NL equations and specifies the net energy transfer rate to the other nonzero streamwise components.

The dynamical significance of $N_4$ in sustaining the turbulent state is revealed by comparing the perturbation energetics under the influence of the NL mean flow $\mathbf{U}$ with and without the term $N_4$. This can be achieved by calculating the Lyapunov exponents $\lambda_{Lyap}$ of the $\mathbf{U}$ obtained from an NL simulation and the associated Lyapunov vectors together with the contributions of each to the growth rate from $\dot{E}_{def,n_x}$, $\dot{E}_{dissip,n_x}$, $\dot{E}_{nonlin,n_x}$ and comparing these rates to those obtained when the term $N_4$ is included. Although the $N_4$ term is energetically neutral it may have a profound impact on the energetics by modifying the perturbations to extract more or less energy from the mean flow. If the term $N_4$ is not fundamental to sustaining the turbulence and regulating the mean flow but instead the mechanism of RNL is fundamentally responsible for maintaining NL turbulence the following three conditions should be satisfied: ($i$) the top Lyapunov exponent, $\lambda_{Lyap,n_x}$, associated with streamwise components $n_x$ of the turbulent field should be neutral after accounting for the transfer of energy to the other streamwise perturbation components, which would indicate that the turbulent state has been regulated to neutralize the top Lyapunov vector growth rate (maximum Lyapunov exponent) coincident with the (necessary) neutralization of the state vector, ($ii$) the transfer of energy from the mean flow by the top Lyapunov vector $\dot{E}_{def,n_x}$ should exceed that by the state vector indicating that $N_4$ has disrupted the Lyapunov vector making it less effective at transferring energy from the mean flow, and ($iii$) the Lyapunov vectors span the energy and the energetics of the NL perturbation field in a convincingly efficient manner, most tellingly if they span it in the order of their growth rate. Satisfying these conditions would strongly support the conclusion that the turbulence

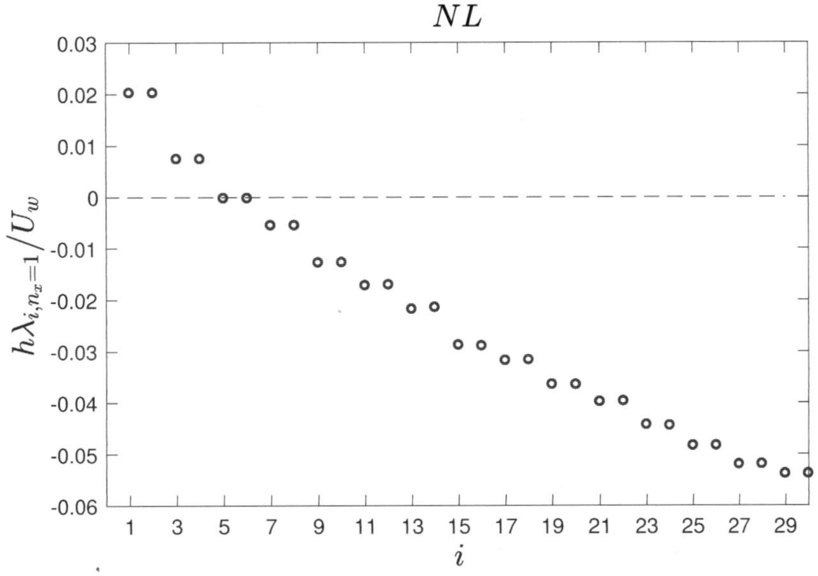


**Figure 2.** The first 30 Lyapunov exponents of $n_x = 1$ perturbations to the NL turbulent flow. Note that each exponent corresponds to two states that allow for streamwise translation of the Lyapunov vector.

**Table 1.** The channel is periodic in the streamwise, $x$, and spanwise, $z$, direction and at the channel walls $y = \pm h$ the velocity is $\mathbf{u} = (\pm U_w, 0, 0)$. The channel length is $L_x$ and $L_z$ in the streamwise and spanwise directions respectively. The number of streamwise and spanwise Fourier components is $N_x$ and $N_z$ after dealiasing in the streamwise and spanwise direction by the 2/3 rule, and we use $N_y$ grid points in the wall-normal direction. $R = U_w h/\nu$ is the Reynolds number of the simulation, with $\nu$ the kinematic viscosity.

| Parameter | $[L_x, L_z]/h$ | $N_x \times N_z \times N_y$ | $R$ |
|---|---|---|---|
| NS600 | $[1.75\pi\ ,\ 1.2\pi]$ | $17 \times 17 \times 35$ | 600 |

is being maintained primarily through the parametric perturbation growth process associated with the temporal variation of $\mathbf{U}$ that supports RNL, without substantial contribution from the $N_4$ nonlinearity. The alternative is that the $N_4$ term has a first-order effect on the energetics which would imply centrality in the dynamics of turbulence of the alternative role for $N_4$ which is to replenish the subset of perturbations lying in the directions of growth. This distinction in mechanism can be clarified by observing that, if the mean flow is chosen to be the time-independent streamwise-spanwise-temporal mean, which in a boundary layer flow is the stable Reynolds-Tiederman profile [13], the $N_4$ nonlinearity must assume this role if turbulence is to be sustained as the parametric mechanism is not available. Turbulence could be in principle sustained by this mechanism if $N_4$ were sufficiently effective in scattering perturbations back into the directions of non-normal growth as is commonly hypothesized in toy models [16–19]. However, it is known that this is not the case [14].

## 2. The Lyapunov exponent of the mean flow in Couette turbulence at $R = 600$

Consider a Couette turbulence simulation at $R = 600$ in a periodic channel with parameters

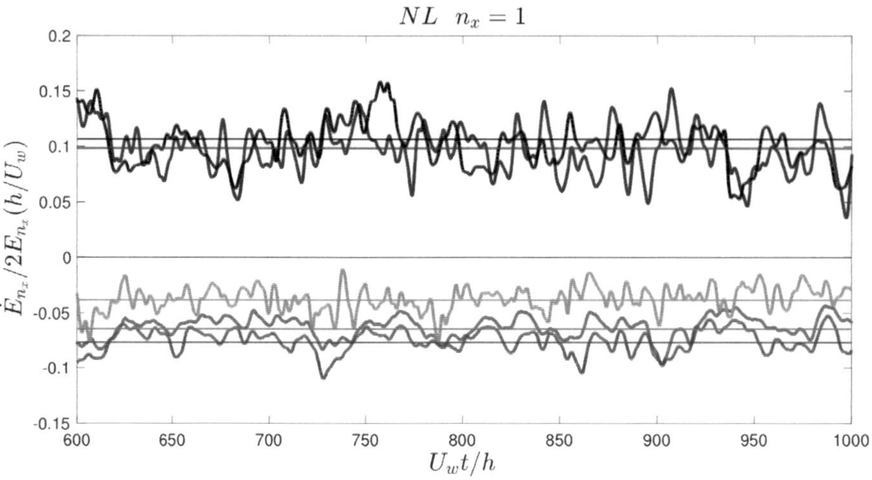


**Figure 3.** Contribution to the instantaneous energy growth rate of the $n_x = 1$ perturbation component of the NL simulation: extraction from the fluctuating $n_x = 0$ mean component $\dot{E}_{def,n_x}/(2E_{n_x})$ (blue, solid); loss to dissipation $\dot{E}_{dissip,n_x}/(2E_{n_x})$ (red, solid); transfer to the other $n_x > 1$ streamwise components $\dot{E}_{nonlin,n_x}/(2E_{n_x})$ (green). The mean values of these rates are indicated with the dashed lines with the corresponding color. These average rates sum to $\lambda_{state} = 0$. The corresponding rates for the first Lyapunov vector are shown in black and purple dash-dotted lines (there is no energy transfer to the other components as $N_4$ is absent in this calculation). These rates sum to the Lyapunov exponent $\lambda_{Lyap} = 0.02U_w/h$, which is comparable to $\dot{E}_{nonlin,n_x}/(2E_{n_x})$.

given in Table 1. This is a larger channel compared to the minimal Couette flow that was studied by Hamilton, Kim & Waleffe [20] at $R = 400$. RNL turbulence with these parameters at $R = 600$ was systematically examined recently [21].

We first calculate the Lyapunov exponent $\lambda_{Lyap}$ of the NL mean flow by estimating (8) from a long integration of (4) with the mean flow $\mathbf{U}$ obtained from a turbulent NL simulation. The initial state $\mathbf{u}'$ is inconsequential because, with measure zero exception, any random initial condition converges in this system with exponential accuracy to the structure associated with the largest Lyapunov exponent. The full spectrum of Lyapunov exponents can be obtained by an orthogonalization procedure. For a discussion of the calculation and properties of Lyapunov exponents and the structures associated with them refer to Refs. [15, 21–24]. Because of the streamwise independence of $\mathbf{U}$, the different streamwise Fourier components of $\mathbf{u}'$ in this Lyapunov exponent calculation, in which the $N_4$ term is absent, evolve independently and the structure (the Lyapunov vector) associated with a given Lyapunov exponent has streamwise structure confined to a single streamwise wavenumber $k_x = 2\pi n_x h/L_x$, corresponding to the $n_x$ streamwise wavenumber.

The top Lyapunov exponent at each $n_x$ is shown in Fig. 1. This plot reveals that the time dependent streamwise mean flow $\mathbf{U}$ is asymptotically stable to all perturbations with $n_x > 1$ with only the $n_x = 1$ streamwise component supporting a positive Lyapunov exponent of $\lambda_{Lyap} \approx 0.02U_w/h$. Recall that in RNL the top Lyapunov exponent also has wavenumber $n_x = 1$ and is exactly zero consistent with mean $\mathbf{U}$ being adjusted by the feedback through the Reynolds stress term $N_2$ to exact neutrality. The top Lyapunov exponent obtained using the $\mathbf{U}$ of NL must be positive only to the degree required to account for the energy exported to other perturbations. The degree of positivity of $\lambda_{Lyap}$ is necessarily adjusted by feedback through $N_2$ to account for this loss. With this consideration the NL mean flow $\mathbf{U}$ can be verified to be neutral.

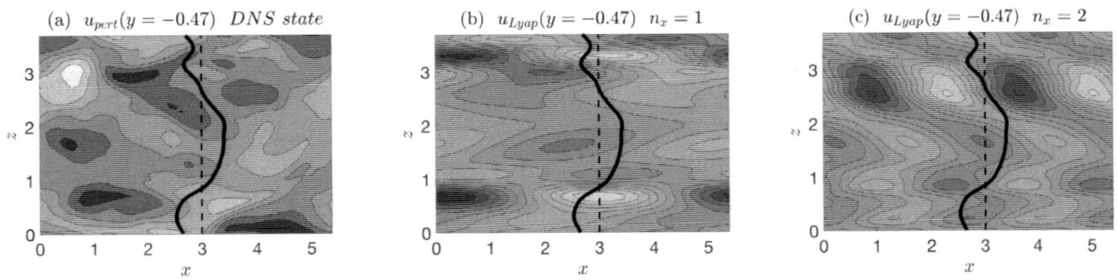


**Figure 4.** Snapshots at the same time showing the streamwise perturbation velocity $u'$ of the NL state (panel (a)), of the top Lyapunov vector which has streamwise channel wavenumber $n_x = 1$ (panel (b)) and the decaying top Lyapunov vector associated with $n_x = 2$ (panel (c)). Also shown is the streak component of the streamwise mean flow $U - \langle U \rangle_z$ (black line).

Strictly speaking, the $n_x = 1$ component is associated with a pair of Lyapunov exponents corresponding to the real and imaginary part of a single structure. This degeneracy of the Lyapunov exponents results from the streamwise homogeneity of the linear operator which implies that the streamwise phase of the Lyapunov vectors is arbitrary. The growth rates of the first 30 Lyapunov exponents associated with $n_x = 1$ are shown in Fig. 2. Specifically, the perturbation subspace is spanned by a set of Lyapunov vectors comprised of pairs of structures sharing a single spanwise/cross-stream structure but having respectively $\sin(k_x x)$ and $\cos(k_x x)$ streamwise dependence. From Fig. 2 it is apparent that at this Reynolds number and for this minimal channel the NL turbulent mean flow supports two positive Lyapunov exponent pairs.

Contributions to the Lyapunov exponent from mean flow energy transfer and from dissipation are plotted as a function of time in Fig. 3. The growth rate associated with energy transfer from the fluctuating streamwise mean is on average $0.1066U_w/h$, while the dissipation rate is on average $0.0862U_w/h$ resulting in the positive Lyapunov exponent $\lambda_{Lyap} = 0.0204U_w/h$. These transfers occur when perturbations evolve under the dynamics of the fluctuating streamwise mean flow $\mathbf{U}$ associated with NL turbulence but in the absence of two effects: ($i$) disturbance to the perturbation structure by the perturbation-perturbation nonlinearity $N_4$ and ($ii$) transfer of energy to other perturbations by $N_4$. This result demonstrates that the parametric growth mechanism is able to maintain the perturbation turbulence component against dissipation with sufficient additional energy extraction to account for transfer to the other scales. The last can be roughly estimated by the energy transfer rate from $n_x = 1$ to $n_x \geq 2$ in the full NL simulation, which is on average $-0.0386\ U_w/h$ or $0.8u_\tau/h$ in terms of friction velocity, as shown in Fig. 3.

We now contrast the energetics of the Lyapunov vectors on the turbulent NL mean flow just shown with the corresponding energetics of the $n_x = 1$ Fourier component of the state vector obtained from the turbulent NL simulation itself in order to determine whether the $N_4$ term has the effect of influencing the perturbations to be in a more or less favorable configuration for extracting energy from the mean flow. These results are also shown in Fig. 3 from which it can be seen that the transfer from the mean $\mathbf{U}$ with the influence of the $N_4$ term included is slightly less than that achieved by the first Lyapunov vector in the absence of the influence of $N_4$: the transfer to the NL state produces growth rate $0.0985U_w/h$ compared to $0.1066U_w/h$ for the case of the unperturbed Lyapunov vector on the NL mean flow. This demonstrates that the nonlinear term does not configure the flow states so as to extract more energy from the mean. However, despite the fact that the energy extracted from the mean flow by the NL perturbation state and the first Lyapunov vector are nearly equal when averaged over time, the correlation coefficient of their time series, shown in Fig. 3, is low (0.26) indicating that the $N_4$ term has disrupted the first Lyapunov vector and spread its energy to other Lyapunov vectors. The fact that this

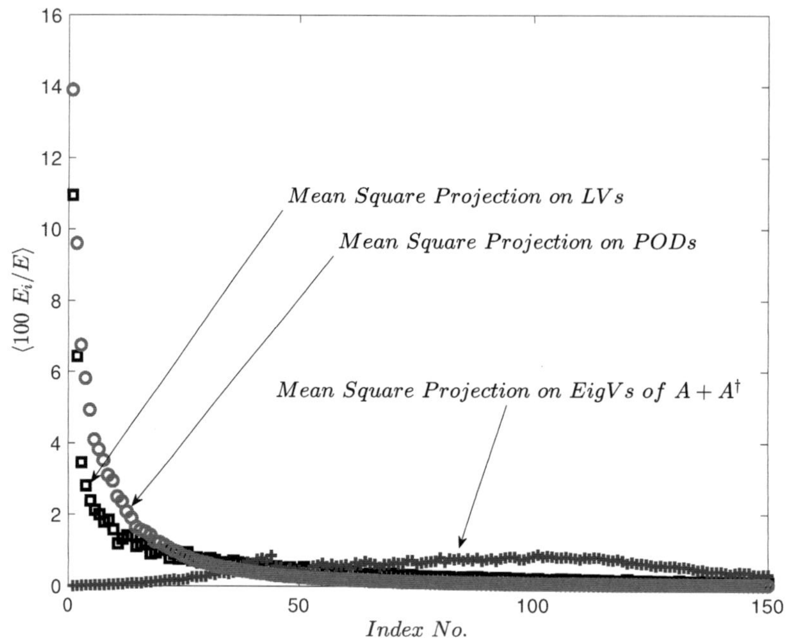


**Figure 5.** Average energy percentage of the $n_x = 1$ flow accounted for by each pair of Lyapunov vectors (black squares). Average energy percentage of the $n_x = 1$ flow accounted for by the eigenvectors of $A + A^\dagger$ ordered in descending order of their eigenvalue (blue crosses). $A$ is the operator in (4) governing the linear evolution of the perturbations about $\mathbf{U}$. The eigenvectors of $A + A^\dagger$ are the orthogonal directions of stationary instantaneous perturbation energy growth rate, with this growth rate given by the corresponding eigenvalue. The perturbation component of the turbulent flow is adjusted so as to have a small projection on the first eigenvectors of $A + A^\dagger$ associated with large instantaneous energy growth rates. Also shown is the energy percentage accounted for by the PODs of the $n_x = 1$ component of the NL perturbation state (green circles).

disruption does not substantially alter the time-mean energy extraction from the streamwise mean flow suggests that the time mean energetics resulting from projection on the Lyapunov vectors of $\mathbf{U}$ is not substantially altered by $N_4$ while the projection at an instant in time is (recall that in RNL $N_4$ is absent and this projection is onto the single top Lyapunov vector). To investigate this possibility we need to project the energetics on the Lyapunov vectors rather than on the growth directions associated with the eigenvectors of $A + A^\dagger$, with $A$ the operator in (4) governing the linear evolution of the perturbations about $\mathbf{U}$. Note also that the decay rate of the NL flow state due to dissipation is $0.0646U_w/h$ which is less then the corresponding decay rate of the first Lyapunov vector. This difference in decay rate is explained by reference to Fig. 4, from which it is apparent that the $n_x = 1$ perturbation component of the NL state is of larger scale than the $n_x = 1$ Lyapunov vector. Also, consider that, in the energetics of the $n_x = 1$ perturbation component in NL there is a term not present in the corresponding Lyapunov vector: the energy interchanged with the remaining $n_x \neq 0$ components, which is also shown in Fig. 3. The $n_x = 1$ perturbation component of the NL state exports energy to the other streamwise components of the flow and this nonlinear transfer contributes $0.0386U_w/h$ at this wavenumber to the decay rate. This additional decay is just sufficient to reduce the mean growth rate of the NL state to the expected value $\lambda_{state} = 0$.

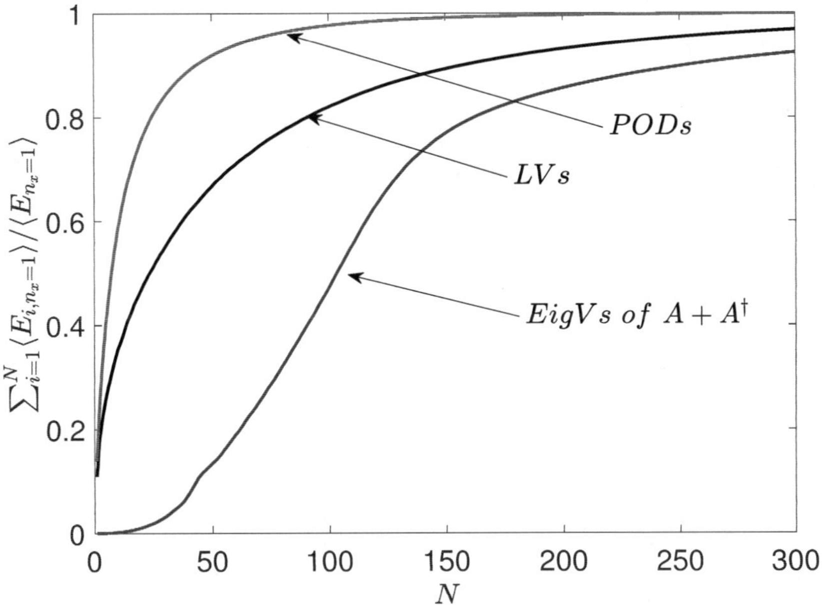


**Figure 6.** Average fraction of the energy of the $n_x = 1$ flow accounted for by the first $N$ pairs of Lyapunov vectors (black), by the first $N$ pairs of eigenvectors of of $A + A^\dagger$ (blue) ordered in descending order of their eigenvalue, and by the first $N$ pairs of POD's of the $n_x = 1$ component of the NL state (green).

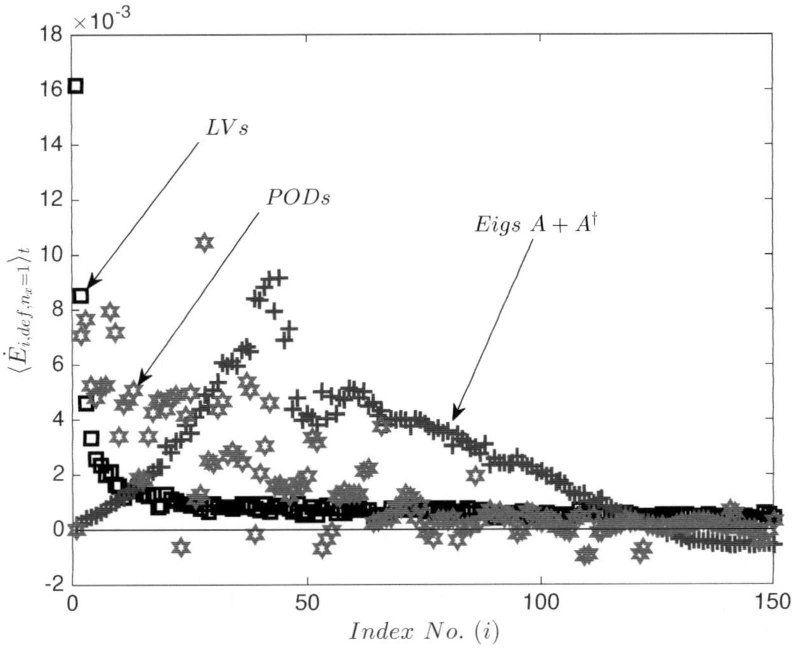


**Figure 7.** Contribution of the Lyapunov vectors (black squares), the PODs (green stars) and the eigenvectors of $A + A^\dagger$ (blue crosses) to the energy extraction rate, $\dot{E}_{def}$, from the mean flow.

We conclude that the Lyapunov exponent of the fluctuating streamwise mean flow **U** in NL turbulence has been adjusted to near neutrality which is consistent with the parametric growth mechanism fully accounting for the maintenance of the perturbation component of the turbulent state. The perturbation-perturbation nonlinearity, $N_4$, does not configure the perturbations to extract more energy from the mean flow than they would in the absence of this term implying that $N_4$ acts as a disruption to the parametric growth process supporting the Lyapunov vector rather than augmenting the perturbation maintenance by the frequently hypothesized mechanism in which perturbation-perturbation nonlinearity replenishes the subspace of perturbations configured to transfer energy from the mean flow to the perturbations.The fact that the mean NL flow has been adjusted to near neutrality indicates that the first Lyapunov vector should be a dominant component of the NL perturbation state. This will be examined in the next section.

## 3. Analysis of perturbation energetics by projection onto the Lyapunov vector basis

Despite the persuasive correspondence between the mean energetics of the NL perturbation state and the mean energetics of the top Lyapunov vector calculated using the associated fluctuating streamwise mean flow, **U**, time series of perturbation growth rate for these shown in Fig. 3 and snapshots of the perturbation state and the Lyapunov vectors shown in Fig. 4 reveal considerable differences. This suggests further analysis to clarify the relation between the perturbation state and the Lyapunov vectors. The orthogonality property imposed on the Lyapunov vectors makes them an attractive basis for decomposing the $n_x = 1$ component of the NL perturbation state for the purpose of analyzing the relation between perturbation structure and energetics. Expanding the Fourier amplitude of the $n_x = 1$ NL perturbation state $\hat{\mathbf{u}}'$ in the basis of the orthonormal in energy $n_x = 1$ Lyapunov vectors:

$$\hat{\mathbf{u}}'(t) = \sum_i a_i(t)\mathbf{u}_{LV_i}(t), \tag{14}$$

with projection coefficient:

$$a_i(t) = \left\langle \mathbf{u}_{LV_i}(t) \cdot \hat{\mathbf{u}}'(t) \right\rangle_{x,y,z}, \tag{15}$$

we obtain that the contribution to the energy of the perturbations accounted for by projection of the perturbation state on Lyapunov vector $u_{LV_i}$ is $E_i = a_i^2(t)/2$. The projection of the energy of the $n_x = 1$ component of the perturbation state on the first 150 Lyapunov vectors is shown in Fig. 5. The percentage of energy accounted for by projection on the most unstable Lyapunov vector is 11%, significantly larger than the energy in each of the remaining Lyapunov vectors. Adding the second unstable Lyapunov vector raises this value to 17.4% and the first 100 $n_x = 1$ Lyapunov vectors account for 82% of the energy of the $n_x = 1$ component of the perturbation state as seen in Fig. 6. In order to understand the significance of the Lyapunov vectors as a basis for representing the NL perturbation state we have determined the orthonormal structures of the proper orthogonal decomposition (PODs) of the $n_x = 1$ component of the NL flow with the methods discussed in [25]. Fig. 5 and Fig. 6 demonstrates that the Lyapunov vectors provide a good representation of the NL perturbation state. We conclude that the energy of the perturbation state is partitioned into the Lyapunov vectors in the order of their Lyapunov exponent and that the Lyapunov vectors are a good basis to represent the perturbation turbulent state. While the energy accounted by the POD basis is by construction expected to decrease monotonically (but not necessarily strictly monotonically) with the order of the POD this is not required of the Lyapunov vectors. Also note that the Lyapunov vectors are not constrained by the optimality of the POD basis to be an inferior basis for spanning the energy, because the Lyapunov vectors are time dependent and could theoretically span all the perturbation energy, as indeed is the case in RNL for which the entire energetics is explained by the first Lyapunov vector.

In Fig. 5 we also show the average projection of the NL state on the eigenvectors of the time-mean operator $A + A^\dagger$ ordered in descending order of its eigenvalue. $A$ is the linear operator in (4) governing the linear evolution of the perturbations about $\mathbf{U}$. The eigenvectors of $A + A^\dagger$ are the orthogonal directions associated with instantaneous energy growth rate of perturbations on the flow, $\mathbf{U}$. The perturbation component has small projection on the first eigenvectors of $A + A^\dagger$ which are the structures producing greatest instantaneous energy growth rates. The turbulent mean flow $\mathbf{U}$ is such that perturbations that lead to large instantaneous growth rate have large spanwise wavenumber and are located at the very high shear regions next to the channel boundaries. Consequently, the state has a small projection on the structures producing the most rapid instantaneous growth and the turbulent perturbation component is concentrated on the directions with small positive instantaneous growth rate. Small projection of the perturbation state on the directions of maximum instantaneous growth rate was also seen in RNL simulations at $R = 600$ [21]. What is remarkable and indicative of the fundamental role of the Lyapunov vectors in the dynamics of NL is the monotonic ordering of the perturbation energy in the Lyapunov basis; comparable ordering does not occur in the case of e.g. the dynamically important basis of the $A + A^\dagger$.

In RNL simulations at $R = 600$ the perturbation turbulent state is entirely supported by the top Lyapunov vector and the energetics of the perturbation state consequently are the energetics of this single Lyapunov vector. The $N_4$ nonlinearity distributes the perturbation energy over a subspace spanned primarily by the first few Lyapunov vectors, as shown in Fig. 5. We can determine the distribution of the first $N$ Lyapunov vectors ordered in contribution to the perturbation state energy growth rate, $\dot{E}_{i,def}$ by calculating

$$\sum_{i=1}^{N} \dot{E}_{i,def} \equiv \left\langle \mathbf{u}'_N \cdot \left( -\mathbf{U} \cdot \nabla \mathbf{u}'_N - \mathbf{u}'_N \cdot \nabla \mathbf{U} \right) \right\rangle_{x,y,z,t} , \tag{16}$$

where

$$\mathbf{u}'_N(t) = \mathrm{Re} \left( \sum_{\alpha=1}^{N} a_\alpha(t) \mathbf{u}_{LV_\alpha}(t) e^{2\pi i x / L_x} \right) , \tag{17}$$

is the truncation of the representation of the $n_x = 1$ perturbation state, given in (14), to the first $N$ Lyapunov vectors. From that calculation, we can then obtain the incremental contribution, $E_{i,def}$, of each Lyapunov vector. We can similarly determine the contribution of each of the eigenvectors of $A + A^\dagger$ and of the PODs to the energetics of the perturbation state. The results, shown in Fig. 7, reveal that the Lyapunov vectors provide the primary support for the perturbation energetics and their energetic contribution follows the Lyapunov vector growth rate ordering. If the $N_4$ term were dominant in determining the structures supporting the perturbation state the energetics of the NL state would not be expected to so closely reflect the asymptotic structures of the Lyapunov vectors.

## 4. Conclusions

Statistical state dynamics (SSD) based analysis of the transition to turbulence in Couette flow has revealed the sequence of symmetry breaking bifurcations of the statistical state of the flow as the Reynolds number increases [3, 4]. First the spanwise statistical symmetry of the laminar state is broken and later the temporal statistical symmetry is broken coincident with transition to the turbulent state. Analyses made using SSD systems closed at second order have also demonstrated that a realistic self-sustained turbulent state is maintained by the parametric growth mechanism arising from interaction between the temporally and spanwise spatially varying streamwise-mean flow and the associated perturbation component of the flow [5, 6, 8]. In these second order SSD systems the streamwise-mean flow is necessarily adjusted exactly to neutral stability, with the understanding that the time dependent streamwise mean flow is considered neutral when the first Lyapunov exponent is zero. This result allows reinterpretation of the Malkus conjecture that the statistical state of inhomogeneous turbulence has mean flow adjusted to neutral hydrodynamic stability.

In this work the identification of the parametric mechanism supporting the perturbation component of turbulence obtained using statistical state dynamics in the RNL system has been extended to NL. While support of both the energy and energetics is on a single feedback neutralized Lyapunov vector in the case of RNL, in the case of NL the energy and energetics are not confined to a single Lyapunov vector but rather are spread by nonlinearity over the Lyapunov vectors. Importantly, this support of the perturbation structure and energetics is ordered in the Lyapunov vectors descending in their associated exponents. The neutrality of the top Lyapunov vector in both RNL and NL, when account is taken of the transfer of energy to other scales, is interpreted as implying that the Malkus conjecture is valid if neutrality of the mean state is interpreted as neutrality of the top Lyapunov vector(s). Consistent with the parametric mechanism sustaining the turbulence, the perturbation structure is concentrated on the top Lyapunov vectors of the time varying streamwise-mean flow and ordered in their Lyapunov exponents. Identification of the dynamical support of RNL and NL turbulence by the marginally stable Lyapunov vectors with associated parametric growth mechanism vindicates the conjecture that the dynamically relevant mean flow in turbulent shear flow is not the streamwise-spanwise-temporal mean flow but rather the time and spanwise varying streamwise-mean flow which incorporates the statistical symmetries obtained from study of second order closures of the statistical state dynamics of wall-bounded turbulence. The perturbation-perturbation nonlinearity, does not configure the perturbations to extract more energy from the mean flow than they would in the absence of this term implying that the nonlinearity acts as a disruption to the parametric growth process supporting the perturbation field rather than augmenting the perturbation maintenance by the frequently hypothesized mechanism in which perturbation-perturbation nonlinearity replenishes the subspace of perturbations configured to transfer energy from the mean flow. The fact that the mean NL flow has been adjusted to Lyapunov neutrality and that the Lyapunov vectors support both the energy and the energetics of the perturbation component of the turbulent state indicates that the parametric growth mechanism on the fluctuating streamwise mean flow and its regulation by Reynolds stress feedback which has been identified to support RNL turbulence is also the mechanism underlying the support of NL turbulence.

## Acknowledgments

This work was funded in part by the Coturb program of the European Research Council. We thank Javier Jiménez for his reviewing comments. Marios-Andreas Nikolaidis acknowledges the support of the Hellenic Foundation for Research and Innovation (HFRI) and the General Secretariat for Research and Technology (GSRT). Brian F. Farrell was partially supported by the U.S. National Science Foundation under Grant Nos. NSF AGS-1246929 and NSF AGS-1640989.

## References

[1] Henningson D S and Reddy S C 1994 On the role of linear mechanisms in transition to turbulence *Phys. Fluids* **6** 1396–1398

[2] Kim J and Lim J 2000 A linear process in wall bounded turbulent shear flows *Phys. Fluids* **12** 1885–1888

[3] Farrell B F and Ioannou P J 2012 Dynamics of streamwise rolls and streaks in turbulent wall-bounded shear flow *J. Fluid Mech.* **708** 149–196

[4] Farrell B F, Ioannou P J and Nikolaidis M A 2017 Instability of the roll–streak structure induced by background turbulence in pretransitional Couette flow *Phys. Rev. Fluids* **2** 034607

[5] Thomas V, Lieu B K, Jovanović M R, Farrell B F, Ioannou P J and Gayme D F 2014 Self-sustaining turbulence in a restricted nonlinear model of plane Couette flow *Phys. Fluids* **26** 105112

[6] Farrell B F, Ioannou P J, Jiménez J, Constantinou N C, Lozano-Durán A and Nikolaidis M A 2016 A statistical state dynamics-based study of the structure and mechanism of large-scale motions in plane Poiseuille flow *J. Fluid Mech.* **809** 290–315

[7] Farrell B F, Gayme D F and Ioannou P J 2017 A statistical state dynamics approach to wall-turbulence *Phil. Trans. R. Soc. A* **375** 20160081

[8] Thomas V, Farrell B F, Ioannou P J and Gayme D F 2015 A minimal model of self-sustaining turbulence *Phys. Fluids* **27** 105104

[9] Malkus W V R 1956 Outline of a theory of turbulent shear flow *J. Fluid Mech.* **1** 521–539

[10] Malkus W V R 1954 The heat transport and spectrum of thermal turbulence *Proc. Roy. Soc. Lond. A.* **225** 196–212

[11] Malkus W V R and Veronis G 1958 Finite amplitude cellular convection *J. Fluid Mech.* **4** 225–260

[12] Stone P H 1978 Baroclinic adjustment *J. Atmos. Sci.* **35** 561–571

[13] Reynolds W C and Tiederman W G 1967 Stability of turbulent channel flow, with application to Malkus's theory *J. Fluid Mech.* **27** 253–272

[14] Jiménez J and Pinelli A 1999 The autonomous cycle of near-wall turbulence *J. Fluid Mech.* **389** 335–359

[15] Farrell B F and Ioannou P J 1996 Generalized stability. Part II: Non-autonomous operators *J. Atmos. Sci.* **53** 2041–2053

[16] Trefethen L N, Trefethen A E, Reddy S C and Driscoll T A 1993 Hydrodynamic stability without eigenvalues *Science* **261** 578–584

[17] Gebhardt T and Grossmann S 1994 Chaos transition despite linear stability *Phys. Rev. E* **50** 3705–3711

[18] Baggett J S and Trefethen L N 1997 Low-dimensional models of subcritical transition to turbulence *Phys. Fluids* **9** 1043–1053

[19] Grossmann S 2000 The onset of shear flow turbulence *Rev. Mod. Phys.* **72** 3705–3711

[20] Hamilton K, Kim J and Waleffe F 1995 Regeneration mechanisms of near-wall turbulence structures *J. Fluid Mech.* **287** 317–348

[21] Farrell B F and Ioannou P J 2017 Statistical state dynamics-based analysis of the physical mechanisms sustaining and regulating turbulence in Couette flow *Phys. Rev. Fluids* **2** 084608

[22] Farrell B F and Ioannou P J 1999 Perturbation growth and structure in time dependent flows *J. Atmos. Sci.* **56** 3622–3639

[23] Wolfe C L and Samelson R M 2007 An efficient method for recovering Lyapunov vectors from singular vectors *Tellus A* **59** 355–366

[24] Ding X, Chaté H, Cvitanović P, Siminos E and Takeuchi K A 2016 Estimating the dimension of an inertial manifold from unstable periodic orbits *Phys. Rev. Lett.* **117**(2) 024101

[25] Nikolaidis M A, Farrell B F, Ioannou P J, Gayme D F, Lozano-Durán A and Jiménez J 2016 A POD-based analysis of turbulence in the reduced nonlinear dynamics system. *J. Phys.: Conf. Ser.* **708** 012002

IOP Conf. Series: Journal of Physics: Conf. Series **1001** (2018) 012015 doi:10.1088/1742-6596/1001/1/012015

# Description and detection of burst events in turbulent flows

**P J Schmid[1], A García-Gutierrez[2] and J Jiménez[2]**
[1]Department of Mathematics, Imperial College London, London SW7 2AZ, UK
[2]School of Aeronautics, U. Politécnica Madrid, 28040 Madrid, Spain
E-mail: pjschmid@imperial.ac.uk

**Abstract.** A mathematical and computational framework is developed for the detection and identification of coherent structures in turbulent wall-bounded shear flows. In a first step, this data-based technique will use an embedding methodology to formulate the fluid motion as a phase-space trajectory, from which state-transition probabilities can be computed. Within this formalism, a second step then applies repeated clustering and graph-community techniques to determine a hierarchy of coherent structures ranked by their persistencies. This latter information will be used to detect highly transitory states that act as precursors to violent and intermittent events in turbulent fluid motion (e.g., bursts). Used as an analysis tool, this technique allows the objective identification of intermittent (but important) events in turbulent fluid motion; however, it also lays the foundation for advanced control strategies for their manipulation. The techniques are applied to low-dimensional model equations for turbulent transport, such as the self-sustaining process (SSP), for varying levels of complexity.

## 1. Introduction and motivation

Turbulent fluid motion is characterized by a wide range of motion, involving broad spatial as well as temporal scales (e.g. [1]): while large, coherent structures are known to play a critical and persistent role in the sustainment of turbulence, intermittent events linked to their breakdown and subsequent reconstitution are also observed. This co-existence of different dynamic events is perhaps best encapsulated by minimal flow-units [2] or self-sustaining model processes [3, 4, 5]. In the former case, spatial scales are progressively removed until turbulent fluid motion can no longer be sustained, thus restricting the essential dynamics to its minimal ingredients. In the latter case, hydrodynamic stability results are harnessed together with a feedback loop that describes the re-triggering (and eventual maintenance) of the instabilities. In either case, the final result shows the interplay of *slowly* moving coherent structures that intermittently – and often violently – collapse via a *fast* nonlinear process, before rising again and starting another cycle [6, 7].

Describing this combined, multiscale process is challenging and shall be attempted in this study. To this end, it seems advantageous to describe the dynamic evolution of the flow fields as a path in phase space. In this description, the coherent part of the quasi-cyclic process can be described by motion in certain regions of phase space, while intermittent events are represented by fast and far excursions from these regions. Characteristic turbulence features and quantities, such as production, dissipation and intermittency, can be associated with these events, and the frequency of their occurrence in simulations and experiments can be gauged. Identifying these building blocks of persistent and intermittent fluid structures is thus instrumental in a complete description and categorization of turbulent fluid motion; at the same time, the same structures and their catchment basins in phase space can be used to design efficient control strategies to avoid or mitigate violent events.

The distinction between persistence and intermittency is made within a probabilistic framework, together with the above phase-space description. A phase-space trajectory that remains (with high probability) in the vicinity of a given region describes a coherent structure. On the other hand, a trajectory that shows a high probability of diverging from its current neighbourhood in phase space represents a transitory state. The transition probabilities of phase space trajectories are thus closely related to the persistence of the structures they represent.

The same probabilities, together with the geometric properties of the phase-space trajectories, can be used to combine structures that behave similarly. To this end, we interpret the fluid motion as a weighted graph in phase space – connecting one region in phase space (a node in the graph) with a neighbouring region (a connected node in the graph), with a given transition probability between them (the weight of the edge between the two nodes). This point of view allows the application of advanced clustering algorithms to reorganize the graph (and thus the fluid motion) into communities (distinct and well-defined structures). In the end, we aim at the identification of the building blocks of turbulent flows and a minimal description of motion involving these blocks.

## 2. Mathematical background

The above formulation of fluid motion, given in the form of snapshot sequences, proceeds in three steps. First, the data need to be embedded in phase space, commonly via projection onto an appropriate lower-dimensional basis. Second, in this phase space the transition probabilities are computed by following the phase-space trajectory. In a third step, the raw transition probability matrix is deflated (lumped) by detecting and combining states of similar dynamic behavior. This last step is performed repeatedly, until a desired level of coarse-graining is reached.

### *2.1. Embedding in phase space*

Numerical simulations or experimental measurements of turbulent flows require the resolution of a wide range of scales; this yields data sequences of many degrees of freedom. However, not all degrees of freedom are independent and dynamically relevant. This fact justifies the projection of the full motion onto an appropriate basis and the approximation of the fluid motion by tracking the coefficients of the basis vectors. Various choices for the basis vectors exist and have been explored over past decades. Proper orthogonal decomposition (POD) [8, 9, 10] is a popular option; POD-modes preserve orthogonality of the spatial structure and can be shown to optimally preserve the energy in the retained modes. They are computed directly from the data matrix, i.e., the matrix whose columns represent the snapshots of the flow field, by a singular value decomposition or, alternatively, by forming correlation matrices and computing their eigenvalues. While POD-modes focus on spatial orthogonality, they yield multi-frequential time traces. If single-frequency structures are desired (for example, to describe the shedding structures behind a bluff body in terms of single Strouhal numbers), a dynamic mode decomposition (DMD) can be taken [11, 12, 13]. In this case, spatial orthogonality is sacrificed to gain a single-frequency decomposition of the data. Other alternatives are resolvent modes [14], that is, the principal singular vectors of a harmonically driven linear system.

In either case, we can describe a temporally evolving system with many spatial degrees of freedom by a set of temporally evolving coefficients. These coefficients form a phase space, and the original time-evolving flow field can be represented by a trajectory in this phase space. By truncating the number of basis vectors, we have control over the dimensionality of this phase space. As a representation of a general phase space embedding, we decompose the data matrix

$\mathsf{D}$ into a product of three matrices: the spatial basis $\mathsf{U}$, the amplitude matrix $\Sigma$, and the time-dynamics $\mathsf{V}$.

$$\mathsf{D} = \mathsf{U}\Sigma\mathsf{V}^H = \mathsf{U}\mathsf{W}^H \tag{1}$$

We assume that the matrix $\Sigma$ is diagonal, as it allows the one-to-one association of a spatial structure (column in $\mathsf{U}$) with a specific dynamic signature (column in $\mathsf{V}$). Without any additional constraints on $\mathsf{U}$ and $\mathsf{V}$, the above decomposition is not unique. Nonetheless, we will not further specify constraints and accept the decomposition in its generality. The columns of $\mathsf{W}$ describe the time traces in phase space and will form the foundation for further analysis.

### *2.2. Transition probabilities*

The phase-space trajectory is then used to determine state transition probabilities. To this end, we seek a gridding of the high-dimensional phase space which ascribes a finite hyper-cube volume in phase space to each instance in the data sequence. A naïve gridding approach would be prohibitively expensive, as the phase space would grow in dimension and the number of hyper-cube volumes would increase exponentially. A different approach has to be adopted.

Rather than gridding the entire phase space, we define hyper-cubes around time instances of the phase-space trajectory and introduce new hyper-cubes as the trajectory leaves any of the current hyper-cubes. In this process, we also keep a linked list that stores information about the temporal evolution along the trajectory as to the propagation direction from hyper-cube to hyper-cube. In this manner, the curse of dimensionality associated with the naïve approach can be avoided, and the final set of hyper-cubes is always less than the number of snapshots in our data matrix. More specifically, the tessellation of the phase-space trajectory is accomplished by a subdivision algorithm that, starting with a coarse assignment to large hyper-cubes, iteratively refines the hyper-cube volumes (by factors of two in subsequent coordinate directions) until a sufficient degree of fidelity of the phase-space motion is assured. As a result, we have access to information about the location and the motion of the trajectory in phase space, discretized by our set of hyper-cubes.

With this information, we next compute a transition probability matrix. Each row of the matrix represents a box containing a trajectory at a given time, while each column represents a box into which the trajectory moves within a single time-step. The entry in the matrix at a given row $i$ and column $j$ then measures the probability of having moved from box $j$ to box $i$ over one time step. This probability can be approximated by measuring the number of instances where motion between $i$ and $j$ is detected (within one time step) normalized by the total number of instances in $i$. Mathematically, we can express this as

$$\mathsf{P}_{ij} = \frac{m(\mathsf{B}_i \cap \mathcal{F}^{-1}(\mathsf{B}_j))}{m(\mathsf{B}_i)} \tag{2}$$

where $m(B)$ denotes a suitable density measure in phase space (in our case, the number of instances contained in the box $\mathsf{B}$), and $\mathcal{F}^{-1}$ is the temporal backstep operator, i.e., we evaluate its argument at the previous time-step (see, e.g., [15]). The above formula states that we count the number of occurrences, where instances move from box $j$ to box $i$ over one time step and divide it by the number of total instances in box $i$. This assigns a probability measure to the event that instances along the trajectory that end up in box $i$ have come from box $j$ over one time step. This procedure provides an approximation of the (Markov) transition probability matrix between all boxes in our tessellation of phase space and is referred to as the Ulam–Galerkin method [16]. It is interesting to note that the transition probability matrix $\mathsf{P}_{ij}$ is a finite-dimensional approximation of the Frobenius–Perron operator [17] (i.e., the adjoint of the Koopman operator).

IOP Conf. Series: Journal of Physics: Conf. Series **1001** (2018) 012015 doi:10.1088/1742-6596/1001/1/012015

This Markov transition matrix $\mathsf{P}$ allows the description of phase-space dynamics in probabilistic terms. Starting in one particular hypercube $\mathsf{B}_i$, we can compute the probabilities of remaining within the same hypercube or transitioning to a neighbouring hypercube. By repeated multiplication by $\mathsf{P}$ we arrive at a probability distribution of possible outcomes. Furthermore, the eigenvectors of the $\mathsf{P}$-matrix associated with near-unity eigenvalues constitute quasi-invariant flow states that represent persistent and highly observable coherent structures.

### *2.3. Community clustering and deflation*

With the tessellation of the occupied phase space established and the transition probability matrix computed, we are now in a position to re-organize the various identified flow states to account for the temporal dynamics between them. This is best accomplished by interpreting the transition probability matrix as an adjacency matrix of a directed, weighted graph [18]: the flow states (high-dimensional boxes) are connected by non-zero transition probabilities. At this stage, the dynamics or persistence of flow states has not been taken into account. This latter property is brought out by regrouping similar flow states into a compound flow state; in the language of graphs, this regrouping is known as community clustering. It consists of a repartition of the network into groups of nodes such that intra-connectivity within the same group is maximized, while inter-connectivity from group to group is minimized [19, 20]. Using the transition probabilities above, the community clustering is equivalent to lumping dynamically similar structures and to generating a hierarchy of structures ranging from quasi-invariant structures (with low transition probabilities) to highly transitory structures (with high transition probabilities).

Computationally, we apply a simple yet efficient algorithm to optimize the community structure of our phase-space network, proposed in [21]. This algorithm targets a measure of inter-connectedness in networks based on the number and weight of links between communities. Mathematically, we define as

$$Q = \frac{1}{2m} \sum_{i,j} \left[ \mathsf{P}_{ij} - \frac{k_i^{(in)} k_j^{(out)}}{2m} \right] \delta_{c_i, c_j} \tag{3}$$

the modularity of the graph. In the above expression, $\mathsf{P}_{ij}$ stands for the transition probability between node $i$ and $j$ and $k_i^{(in,out)} = \sum_j \mathsf{P}_{ij}$ is the sum of all probabilities associated with node $i$, going into or out of node $i$, respectively. The variable $c_i$ denotes the $i$-th community and $\delta$ is the standard Kronecker symbol. Finally, $m$ is defined as $m = \frac{1}{2}\sum_{ij} \mathsf{P}_{ij}$, the sum of the weights of all links in an undirected network [22, 23].

The algorithm proceeds in two steps. First, a greedy optimization step maximizes the modularity $Q$ by placing the neighbors of node $i$ into neighbouring communities, if it increases $Q$. Once, no more improvement can be achieved by swapping nodes into adjacent communities, a second step commences which replaces the new communities by single nodes and updates the transition probabilities. After this step, we return to the maximization of $Q$ and continue to iterate until no further improvement can be made. The two-step process performs very efficiently, in particular, for sparsely connected graphs, as is the case for us.

Alternative clustering algorithms can be and have been used [15], such as the classical k-means method. In this case, however, the number of communities is prescribed and the resulting community association is rather uniform. These properties defy the purpose of an objective clustering where strongly non-uniform communities are expected.

We apply the community clustering algorithm to our transition probability matrix $\mathsf{P} \in \mathbb{R}^{N \times N}$, which results into a regrouping of our $N$ initial boxes into $k$ communities, by maximizing the modularity of the underlying graph. The result of this process is a rectangular matrix

$\mathsf{Z}_{(1)} \in \mathbb{R}^{N\times k}$ which uniquely assigns a given box $\mathsf{B}_i$ to one of the identified communities; mathematically, we have

$$\mathsf{Z}_{(1),ij} = \begin{cases} 1 & \text{if} \quad \mathsf{B}_i \in c_j, \\ 0 & \text{otherwise.} \end{cases} \tag{4}$$

The transformation

$$\mathsf{Z}_{(1)}^T \mathsf{P} \mathsf{Z}_{(1)} = \mathsf{P}_{(1)} \tag{5}$$

then regroups the transition matrix $\mathsf{P}$ into $\mathsf{P}_{(1)} \in \mathbb{R}^{k\times k}$ which describes the transition probabilities between the newly found communities, rather than our original boxes. By repeating this clustering and lumping process we finally arrive at a transition probability matrix of desired size. We have

$$\mathsf{Z}_{(k)}^T \cdots \mathsf{Z}_{(1)}^T \mathsf{P} \mathsf{Z}_{(1)} \cdots \mathsf{Z}_{(k)} = \mathsf{P}_{(k)}. \tag{6}$$

It is worth noting that the entries of the matrix $\mathsf{Z}_{(i)}$ depend on the graph structure of the associated matrix $\mathsf{P}_{(i-1)}$; the above formula thus expresses a complex and nonlinear transformation process.

Once we are satisfied with the number of identified communities, we can classify the identified communities with associated flow structures. By relating the communities back to the original snapshots, we can catalogue similar dynamical behaviour. In particular, communities that have a nearly unitary transition probability represent flow structures that occupy and remain (with high probability) in a specific region of phase space. Physically, these structures, while unsteady, show a great degree of persistence in the overall flow. In our case of turbulent flow, we expect the slowly meandering, streamwise elongated structures to fall into this category. On the other hand, communities that show a high transition probability to neighbouring communities signify transitory structures and can be thought of as precursors for a violent (and/or rare) event. These structures are far less observable, but highly controllable.

## 3. Low-dimensional models

While designed to apply to flow field sequences of fully turbulent flows, from numerical or experimental sources, we will demonstrate the key elements of the above methodology on a simplified yet relevant model of the underlying self-sustaining process that maintains turbulent fluid motion. We select the nine-dimensional system of nonlinear ordinary differential equations proposed in [24]. This model has been following earlier attempts at deriving a compact system by projecting the Navier–Stokes equations onto pertinent models that play a critical role in the sustainance of turbulence in generic wall-bounded flows. In our case, the flow is based on a uni-directional, parallel, sinusoidal shear flow, and the flow is considered in a doubly periodic computational domain with one inhomogeneous direction. Modelling the base flow profile, the streak and the downstream vortex, the spanwise modulations of the latter components, normal vortex modes and fully three-dimensional perturbations – all with appropriate spatial scales and unknown amplitudes – we arrive at an evolution problem for the nine coefficients $\{a_1, a_2, \ldots, a_9\}$ that capture the growth, instability, breakdown and regeneration of flow structures. The model [24] has strong similarity to an earlier eight-coefficient model [4]. Previous work has suggested that the self-sustaining process, described by the model equation, constitutes a universal characteristic of wall-bounded shear flow turbulence [3, 4, 25, 26]. The chosen model can be stated in the form of nine quadratically interacting ordinary differential equations according to

$$\begin{aligned}
\dot a_1 + \zeta_1 a_1 &= \zeta_1 + \xi_{11} a_6 a_8 + \xi_{12} a_2 a_3, &(7a)\\
\dot a_2 + \zeta_2 a_2 &= \xi_{21} a_4 a_6 - \xi_{22} a_5 a_7 - \xi_{23} a_5 a_8 - \xi_{24} a_1 a_3 - \xi_{25} a_3 a_9, &(7b)\\
\dot a_3 + \zeta_3 a_3 &= \xi_{31}(a_4 a_7 + a_5 a_6) + \xi_{32} a_4 a_8, &(7c)\\
\dot a_4 + \zeta_4 a_4 &= -\xi_{41} a_1 a_5 - \xi_{42} a_2 a_6 - \xi_{43} a_3 a_7 - \xi_{44} a_3 a_8 - \xi_{45} a_5 a_9, &(7d)\\
\dot a_5 + \zeta_5 a_5 &= \xi_{51} a_1 a_4 + \xi_{52} a_2 a_7 - \xi_{53} a_2 a_8 + \xi_{54} a_4 a_9 + \xi_{55} a_3 a_6, &(7e)\\
\dot a_6 + \zeta_6 a_6 &= \xi_{61} a_1 a_7 + \xi_{62} a_1 a_8 + \xi_{63} a_2 a_4 - \xi_{64} a_3 a_5 + \xi_{65} a_7 a_9 + \xi_{66} a_8 a_9, &(7f)\\
\dot a_7 + \zeta_7 a_7 &= -\xi_{71}(a_1 a_6 + a_6 a_9) + \xi_{72} a_2 a_5 + \xi_{73} a_3 a_4, &(7g)\\
\dot a_8 + \zeta_8 a_8 &= \xi_{81} a_2 a_5 + \xi_{82} a_3 a_4, &(7h)\\
\dot a_9 + \zeta_9 a_9 &= \xi_{91} a_2 a_3 - \xi_{92} a_6 a_8, &(7i)
\end{aligned}$$

where the coefficients in the above equations are given as

$$\zeta_1 = \frac{\beta^2}{Re},\quad \zeta_2 = \frac{1}{Re}\left(\frac{4\beta^2}{3} + \gamma^2\right),\quad \zeta_3 = \frac{\beta^2+\gamma^2}{Re},\quad \zeta_4 = \frac{3\alpha^2+4\beta^2}{3Re},$$
$$\zeta_5 = \frac{\alpha^2+\beta^2}{Re},\quad \zeta_6 = \frac{3\alpha^2+4\beta^2+3\gamma^2}{3Re},\quad \zeta_7 = \zeta_8 = \frac{\alpha^2+\beta^2+\gamma^2}{Re},\quad \zeta_9 = 9\zeta_1 \tag{8}$$

and

$$\xi_{11} = \xi_{62} = \xi_{66} = \xi_{92} = \sqrt{\frac{3}{2}}\frac{\beta\gamma}{\kappa_{\alpha\beta\gamma}},\quad \xi_{12} = \xi_{24} = \xi_{25} = \xi_{91} = \sqrt{\frac{3}{2}}\frac{\beta\gamma}{\kappa_{\beta\gamma}},$$
$$\xi_{21} = \frac{5\sqrt{2}}{3\sqrt{3}}\frac{\gamma^2}{\kappa_{\alpha\gamma}},\quad \xi_{22} = \frac{\gamma^2}{\sqrt{6}\kappa_{\alpha\gamma}},\quad \xi_{23} = \xi_{53} = \frac{1}{\sqrt{6}}\kappa_2,\quad \xi_{31} = \xi_{55} = \frac{2}{\sqrt{6}}\kappa_1,$$
$$\xi_{32} = \frac{\beta^2(3\alpha^2+\gamma^2) - 3\gamma^2(\alpha^2+\gamma^2)}{\sqrt{6}\kappa_{\alpha\gamma}\kappa_{\beta\gamma}\kappa_{\alpha\beta}},\quad \xi_{41} = \xi_{45} = \xi_{51} = \xi_{54} = \xi_{61} = \xi_{65} = \xi_{71} = \frac{\alpha}{\sqrt{6}},$$
$$\xi_{42} = \frac{10\alpha^2}{3\sqrt{6}\kappa_{\alpha\gamma}},\quad \xi_{43} = \sqrt{\frac{3}{2}}\kappa_1,\quad \xi_{44} = \sqrt{\frac{3}{2}}\frac{\alpha^2\beta^2}{\kappa_{\alpha\gamma}\kappa_{\beta\gamma}\kappa_{\alpha\beta\gamma}},\quad \xi_{52} = \frac{\alpha^2}{\sqrt{6}\kappa_{\alpha\gamma}},$$
$$\xi_{63} = \frac{10}{3\sqrt{6}}\frac{\alpha^2-\gamma^2}{\kappa_{\alpha\gamma}},\quad \xi_{64} = 2\sqrt{\frac{2}{3}}\kappa_1,\quad \xi_{72} = \frac{1}{\sqrt{6}}\frac{\gamma^2-\alpha^2}{\kappa_{\alpha\gamma}},\quad \xi_{73} = \frac{1}{\sqrt{6}}\kappa_1,$$
$$\xi_{81} = \frac{2}{\sqrt{6}}\kappa_2,\quad \xi_{82} = \frac{\gamma^2(3\alpha^2-\beta^2+3\gamma^2)}{\sqrt{6}\kappa_{\alpha\gamma}\kappa_{\beta\gamma}\kappa_{\alpha\beta\gamma}} \tag{9}$$

with

$$\kappa_{\alpha\gamma} = \sqrt{\alpha^2+\gamma^2},\qquad \kappa_{\beta\gamma} = \sqrt{\beta^2+\gamma^2},\qquad \kappa_{\alpha\beta\gamma} = \sqrt{\alpha^2+\beta^2+\gamma^2} \tag{10}$$

and

$$\kappa_1 = \frac{\alpha\beta\gamma}{\kappa_{\alpha\gamma}\kappa_{\beta\gamma}},\qquad \kappa_2 = \frac{\alpha\beta\gamma}{\kappa_{\alpha\gamma}\kappa_{\alpha\beta\gamma}}. \tag{11}$$

The remaining, user-specified coefficients are the Reynolds number $Re$, and the streamwise and spanwise wavenumbers $\alpha$ and $\gamma$, respectively. The wavenumber $\beta$, which determines the length scale in the wall-normal direction, is taken as constant and set to $\pi/2$.

The coefficients $\{a_1, a_9\}$ collectively describe the mean flow and shear, the coefficients $\{a_2, a_3, a_4\}$ represent the rolls and streaks, and the coefficients $\{a_5, a_6, a_7, a_8\}$ characterize the burst event. For visualization purposes, we will thus combine the nine coefficients in this manner. We choose the dimension of our minimal unit as $L_x = 4\pi$ and $L_z = 2\pi$ from which we determine $\alpha = 2\pi/L_x$ and $\gamma = 2\pi/L_z$ and the remaining system coefficients. The Reynolds number (based on the half-channel height) is set to $Re = 400$.

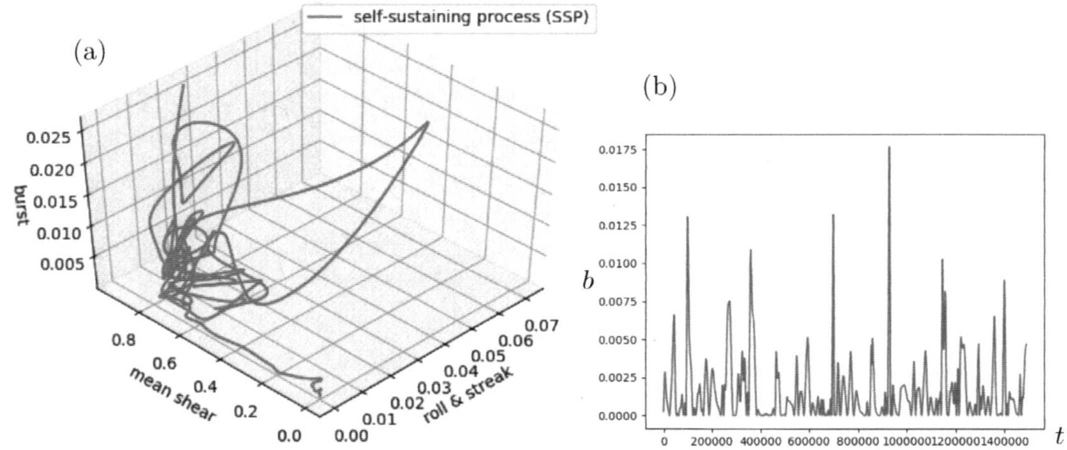


**Figure 1.** (a) Phase space trajectory of a simulation of the nine-dimensional Moehlis-Faisst-Eckhardt model (with $L_x = 4\pi, L_z = 2\pi, Re = 400$). The phase space is visualized by grouping the nine coefficients into groups describing rolls and streaks, mean shear and burst components. (b) Burst component versus time, showing temporal intermittency.

## 4. Results

We integrate the governing equations (7) forward in time and, after an initial transient period, visualize the results in phase space. Due to the high dimensionality of the system, we choose to combine the nine coefficients into three groups characterizing the contributions from (i) the streamwise roll and streak ($\|(a_2, a_3)^T\|$), (ii) the mean shear ($\|1 - a_1\|$), and (iii) the three-dimensional instability structure responsible for the breakdown of the streak ($\|(a_4, a_5)^T\|$). Figure 1 shows a typical phase-space trajectory, visualized by combining the nine coefficients in the manner described above. We observe a clumping of the trajectory in phase space that is interrupted by occasional and violent excursions from the cluster and an eventual and equally rapid return to the cluster. This phase-space behavior is further exemplified by displaying the burst component, consisting of the three-dimensional instability coefficients, as a function of time (see Figure 1(b)). We notice a signal that consists of spikes in amplitude that occur rapidly, intermittently and in varying intensity. This behaviour indicates the type of flow pattern described above, where a quasi-coherent motion is periodically but irregularly disrupted by violent events of a different nature. In our case, these events represent the breakdown of the streamwise roll and streak and its subsequent reconstitution.

The phase-space trajectory, represented by the original nine coefficients $a_{1,\ldots,9}$ over 2000 time units, is then tessellated such that each nine-dimensional hypercube does not contain more than 150 snapshots. From this tessellation the raw transition probability matrix is computed following (2). The fill pattern of this matrix is shown in Figure 2(a): it is clearly diagonally dominant, with few off-diagonal entries, reflecting the initial, sequential processing of the phase-space trajectory. No clustering or categorization of the fluid motion has yet been applied. Interpreting the transition probability matrix $\mathsf{P}$ as the adjacency matrix of an associated directed graph, we can now apply the clustering algorithm and determine graph-communities that show an optimal level of graph modularity according to (3). The identified community allows us to regroup the snapshots and deflate the original matrix $\mathsf{P}$ – describing the motion between

IOP Conf. Series: Journal of Physics: Conf. Series **1001** (2018) 012015 doi:10.1088/1742-6596/1001/1/012015

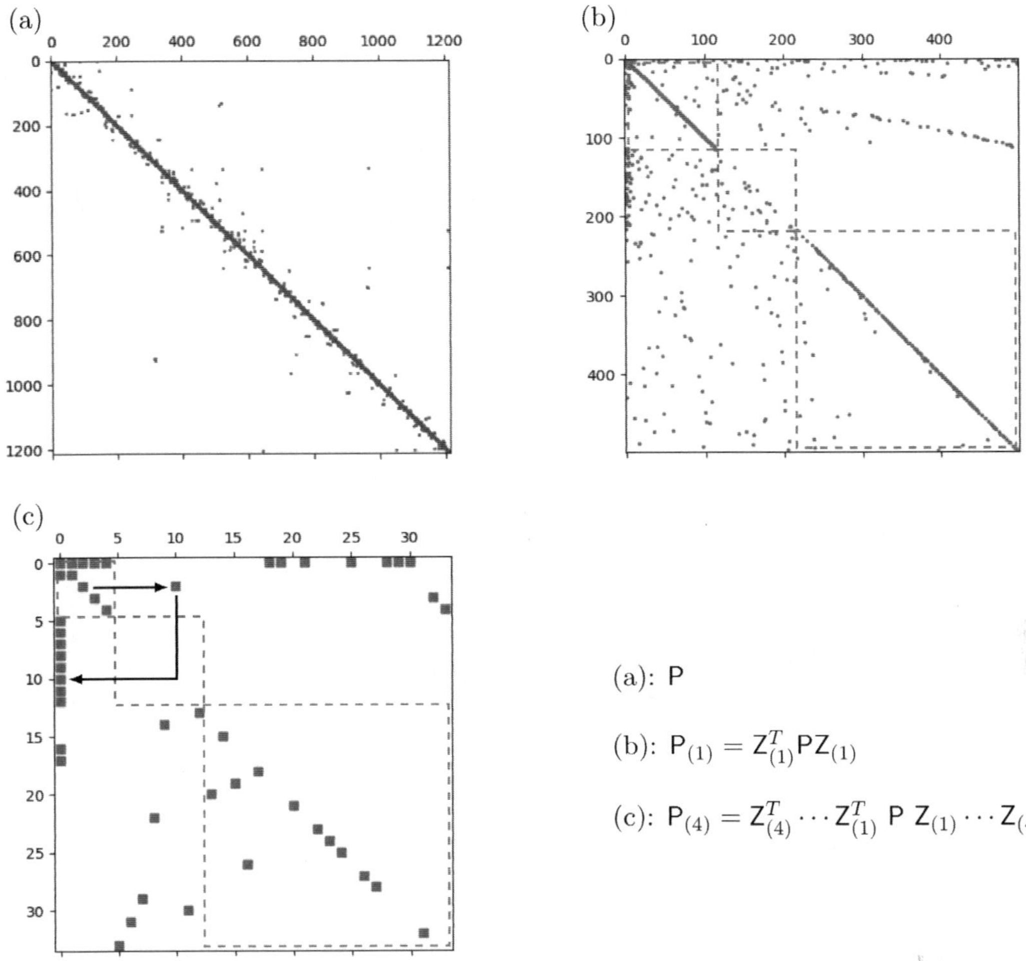


(a): $\mathsf{P}$

(b): $\mathsf{P}_{(1)} = \mathsf{Z}_{(1)}^T \mathsf{P} \mathsf{Z}_{(1)}$

(c): $\mathsf{P}_{(4)} = \mathsf{Z}_{(4)}^T \cdots \mathsf{Z}_{(1)}^T \; \mathsf{P} \; \mathsf{Z}_{(1)} \cdots \mathsf{Z}_{(4)}$

**Figure 2.** Fill pattern of transition probability matrices: (a) original matrix from the tesselation of the phase-space trajectory, (b) matrix after one step of the graph-community algorithm followed by deflation, (c) matrix after five successive community-clustering and deflation steps.

more than 1200 hypercubes – into a transition probability matrix $\mathsf{P}_{(1)}$ between nearly 500 communities. The fill pattern of this latter matrix is shown in Figure 2(b), where each symbol represents a nonzero entry in the matrix. The figure displays a substantially different pattern: the communities can be grouped into more persistent elements (with a strong concentration along the diagonal) and more transitory elements (where the diagonal is absent or sparsely populated). An approximate classification is indicated by dashed boxes in the figure. It appears that after only one application of the graph-community clustering algorithm, a separation of persevering, slow fluid motion and volatile events is already emerging. By repeated application of the clustering algorithm, we can further solidify this feature. Figure 2(c) shows the fill pattern of the transition matrix after four deflation applications; at this point, only 34 different communities are tracked. We notice that the gap along the diagonal consisting of communities $c_{5,\ldots,12}$, which was already notable in Figure 2(b), is now even more pronounced.

The layout of the fill pattern now allows the quantitative and objective description of rare events. This is indicated in one example in Figure 2(c). The probability of resting in community $c_2$ is given by the diagonal entry in $\mathsf{P}_{(4)}$; however, there is also a (smaller) probability of transitioning to community $c_{10}$. The probability of resting in community $c_{10}$ is, however, negligibly small (i.e., below the threshold of marking the entry in the matrix, taken as $10^{-4}$). We thus revert to community $c_0$ and then close the cycle by slowly moving again to community $c_2$. The slow-fast cycle is indicated by arrows in Figure 2(c). Instances of similar behaviour can be extracted and quantified in this manner from the transition probability matrix $\mathsf{P}_{(4)}$. The flow states associated with communities $c_{0,\dots,4}$ can be labelled as quasi-stationary (with rather large residence times and low transition probabilities), while the flow states corresponding to communities $c_{5,\dots,12}$ represent highly transitory states with comparatively low residence time and high transition probabilities. In this manner, cyclical processes that consist of a combination of slow motion and intermittent, fast events can be detected objectively and described quantitatively.

## 5. Summary and conclusions

In this preliminary study, a computational framework for the description and analysis of rare and violent events embedded in a slowly varying fluid motion has been introduced. It consists of the analysis of an embedded phase-space trajectory using a probabilistic approach by estimating the transition probabilities between discretized regions of phase-space traversed by the flow. Employing techniques from graph clustering, the discretized regions are recombined and grouped into communities that describe persistent and highly observable fluid motion as well as transitory and highly controllable processes.

We applied this technique to a reduced-order minimal-channel model of the self-sustaining process (SSP) – a mechanism that is pivotal in sustaining turbulent fluid motion. The framework confirmed the existence of quasi-stationary and transitory states and extracted flow states of either kind objectively and without prior definitions.

In future work, data from direct numerical simulations will be used to detect burst events. Moreover, the above framework will be extended to devise optimized (sparse) control techniques to avoid or attenuate intermittently occurring fluid motion. The application to larger-dimensional systems should not pose significant difficulties, since the chosen data-structure for the phase-space trajectory avoids the curse of dimensionality of a more naïve approach and allows the efficient computation of the transition matrix even for higher-dimensional phase-space embeddings. Results of our analysis of turbulent data-bases will be reported elsewhere.

## Acknowledgements

Support from the European Union under the ERC Advanced Grant "COTURB: Coherent Structures in Wall-bounded Turbulence" is gratefully acknowledged. We are grateful to S. Illingworth for his careful refereeing comment on the manuscript.

## References

[1] Pope S 2000 *Turbulent Flows* (Cambridge University Press)
[2] Jimenez J and Moin P 1991 *J. Fluid Mech.* **225** 213
[3] Hamilton J M, Kim J and Waleffe F 1995 *J. Fluid Mech.* **287** 317
[4] Waleffe F 1995 *Phys. Fluids* **7** 3060
[5] Cossu C and Hwang Y 2017 *Phil. Trans. Roy. Soc. A* **375** 20160088
[6] Antonia R, Danh H and Prabhu A 1976 *Phys. Fluids* **19** 1680
[7] Kim J 1985 *Phys. Fluids* **28** 52
[8] Lumley J 1970 *Stochastic Tools in Turbulence* (Academic Press)
[9] Sirovich L 1987 *Q. Appl. Math.* 561
[10] Berkooz G, Holmes P and Lumley J 1993 *Ann. Rev. Fluid Mech.* **25** 539
[11] Schmid P 2010 *J. Fluid Mech.* **656** 5

IOP Conf. Series: Journal of Physics: Conf. Series **1001** (2018) 012015 doi:10.1088/1742-6596/1001/1/012015

[12] Rowley C, Mezic I, Bagheri S, Schlatter P and Henningson D 2009 *J. Fluid Mech.* **641** 85
[13] Mezic I 2013 *Ann. Rev. Fluid Mech.* **45** 357
[14] McKeon B and Sharma A 2010 *J. Fluid Mech.* **658** 336
[15] Kaiser E, Noack B, Cordier L and Spohn A 2014 *J. Fluid Mech.* **754** 365
[16] Ulam S 1960 *A Collection of Mathematical Problems* (Interscience Publisher NY)
[17] Lasota A and Mackey M 1994 *Chaos, Fractals, and Noise: Stochastic aspects of dynamics* (Springer Verlag)
[18] Chartrand G and Zhang P 2012 *A First Course in Graph Theory* (Dover Publishing)
[19] Girvan M and Newman M 2002 *Proc. Nat. Acad. Sci.* **99** 7821
[20] Fortunato S and Barthelemy M 2007 *Proc. Nat. Acad. Sci.* **104** 36
[21] Blondel V, Guillaume J L, Lambiotte R and Lefebvre E 2008 *J. Stat. Mech.* **10**
[22] Dugué N and Perez A 2015 *Directed Louvain: maximizing modularity in directed networks* Ph.D. thesis Université d'Orléans
[23] Leicht E and Newman M 2008 *Phys. Rev. Lett.* **100** 118703
[24] Moehlis J, Faisst H and Eckhardt B 2004 *New J. Phys.* **6**
[25] Reddy S, Schmid P, Baggett J and Henningson D 1998 *J. Fluid Mech.* **365** 269
[26] Jimenez J and Pinelli A 1999 *J. Fluid Mech.* **389** 335

IOP Conf. Series: Journal of Physics: Conf. Series **1001** (2018) 012016 doi:10.1088/1742-6596/1001/1/012016

# Overdamped large-eddy simulations of turbulent pipe flow up to $Re_\tau = 1500$

**Daniel Feldmann and Marc Avila**
University of Bremen, Center of Applied Space Technology and Microgravity (ZARM), 28359 Bremen, Germany
E-mail: `daniel.feldmann@zarm.uni-bremen.de`

**Abstract.** We present results from large-eddy simulations (LES) of turbulent pipe flow in a computational domain of 42 radii in length. Wide ranges of shear the Reynolds number and Smagorinsky model parameter are covered, $180 \leq Re_\tau \leq 1500$ and $0.05 \leq C_s \leq 1.2$, respectively. The aim is to asses the effect of $C_s$ on the resolved flow field and turbulence statistics as well as to test whether very large scale motions (VLSM) in pipe flow can be isolated from the near-wall cycle by enhancing the dissipative character of the static Smagorinsky model with elevated $C_s$ values. We found that the optimal $C_s$ to achieve best agreement with reference data varies with $Re_\tau$ and further depends on the wall normal location and the quantity of interest. Furthermore, for increasing $Re_\tau$, the optimal $C_s$ for pipe flow LES seems to approach the theoretically optimal value for LES of isotropic turbulence. In agreement with previous studies, we found that for increasing $C_s$ small-scale streaks in simple flow field visualisations are gradually quenched and replaced by much larger smooth streaks. Our analysis of low-order turbulence statistics suggests, that these structures originate from an effective reduction of the Reynolds number and thus represent modified low-Reynolds number near-wall streaks rather than VLSM. We argue that *overdamped* LES with the static Smagorinsky model cannot be used to unambiguously determine the origin and the dynamics of VLSM in pipe flow. The approach might be salvaged by e.g. using more sophisticated LES models accounting for energy flux towards large scales or explicit anisotropic filter kernels.

## 1. Introduction

One of the most conspicuous features of wall-bounded turbulent flows is the prevalence of near-wall streaks, which consist of elongated neighbouring regions of high- and low speed fluid alternating in the spanwise direction [1]. It is well established, that their typical streamwise length and spanwise spacing in viscous units is independent of the Reynolds number ($Re_\tau$) and is about $\mathcal{O}(10^3)$ and $\mathcal{O}(10^2)$, respectively. For the example of streaks in pipe flow, characteristic streamwise and azimuthal (spanwise) wavelengths in the energy spectra are $\lambda_z^+ \approx 800$ and $\lambda_{r\theta}^+ \approx 130$, respectively, e.g. [2], and two-point correlations indicate a radial thickness of roughly 80 viscous units, e.g. [3].

For sufficiently high $Re_\tau \geq \mathcal{O}(10^3)$, there is growing experimental evidence of the rise of very large-scale motions (VLSM) of several outer units in extent. In pipe flow, very long meandering structures carrying a substantial portion of the kinetic energy in streamwise wavelengths exceeding $12R$, where $R$ is the pipe radius, have been reported [4, 5, 6, 7]. In boundary layer flow, Hutchins & Marusic [8, 9] detected superstructures commonly exceeding $20\delta$ in streamwise length, where $\delta$ is the boundary layer thickness. Their spanwise extent is $\mathcal{O}(1)$ in outer units, and so as *Re* increases there is an increasing scale separation between superstructures and near-wall structures. While glimpses of VLSM have been observed [10] already at $Re_\tau$ as low as 380, a distinct outer peak in the azimuthal spectra only emerges for $Re_\tau \geq 1500$ [2]. A distinct outer peak in the streamwise spectra develops at much larger $Re_\tau$ [11, 12]. In addition to these smoothly growing outer peaks in the energy spectra, a sudden increase of the streamwise turbulence intensity in the outer region occurs at $Re_\tau \approx 1500$ and can be traced back to VLSM-relevant length scales [13, 14, 2]. Because there is no universal definition available yet, we hereinafter use the terms superstructures and VLSM interchangeably.

Uncovering the onset and origin of superstructures, analysing their dynamics, and determining their contribution to the global momentum transport are challenging tasks because of the large Reynolds numbers at which they can be unambiguously observed. Numerical studies of wall-bounded turbulence at the highest Reynolds numbers attained to date (see e.g. [15, 16] for channel flow at $Re_\tau = 4200$ and 5200 or [2, 17] for pipe flow at $Re_\tau = 1500$ and 3008) barely reach the onset of superstructures. This limitation is because of the enormous computational costs of resolving all spatial and temporal scales with direct numerical simulation (DNS) of the Navier–Stokes equations. Note that not only $Re_\tau$ must be large, but additionally the computational domain must also be large enough in terms of outer units (e.g. $42R$ in pipe flow) to capture the energetically most relevant length scales associated with VLSM [2].

Large-eddy simulations (LES) [18, 19] with the static Smagorinsky model [20] have been extensively used by Cossu, Hwang and co-workers in an attempt to isolate superstructures from the near wall cycle in turbulent channel [21, 22, 23] and plane Couette [24] flows. Their approach — termed *overdamped* LES (ODL) in [25] — is based on the idea that by increasing the Smagorinsky parameter $C_s$, an increasing range of small-scale structures is damped out (i.e. accounted for by the eddy-viscosity model), whereas large structures may be maintained without loosing valuable resolution [25]. For two reasons, ODL is a very appealing approach to investigate the physical mechanisms underlying superstructures. First, compared to DNS, LES allows for a Reynolds number roughly one order of magnitude larger for a given amount of computational resources. Second, and even more important, the use of ODL reduces the complexity of the resolved flow field substantially because of the absence of small rapidly varying structures. This makes high *Re* flows, and thus superstructures, accessible to dynamical-systems approaches, which have proven very useful in understanding low Reynolds number (transitional) flows [26, 27, 28].

We here apply the ODL approach proposed by Cossu, Hwang and co-workers to pipe flow. We perform LES with the Smagorinsky model in a computational domain of length $42R$ for $180 \leq Re_\tau \leq 1500$ and $0.05 \leq C_s \leq 1.2$. The paper is structured as follows. In §2, we briefly introduce the details of our numerical approach, before comparing our results to DNS reference data sets, as well as former ODL of channel and Couette flows in §3. Finally, in §4 we discuss the effect of $C_s$ on the resolved flow field and turbulence statistics and assess the feasibility of studying superstructures with ODL.

## 2. Numerical approach

We consider an incompressible Newtonian fluid with constant properties confined by a cylindrical wall of diameter $D = 2R$ and length $L$. The fluid is driven in the axial direction with a constant bulk velocity $u_b$. The flow is described through the Navier–Stokes equations, where $t$ denotes the time, $p$ the pressure, and $\boldsymbol{u}$ the velocity vector. Its components $u_r$, $u_\theta$, and $u_z$ point along the cylindrical coordinates in the radial ($r$), azimuthal ($\varphi$), and axial ($z$) directions.

The pressure gradient $\boldsymbol{\nabla} p = \boldsymbol{\nabla} p' + \boldsymbol{P}$ is separated into a fluctuating part $\boldsymbol{\nabla} p'$ and a constant part $\boldsymbol{P}$ driving the flow, such that periodic boundary conditions (BC) can be employed for $p'$ in $\theta$ and in $z$. For $\boldsymbol{u}$, the governing equations are complemented with periodic BC in $\theta$ and $z$ as well as with no-slip and impermeability BC at $r = R$ to model a smooth and rigid pipe wall.

IOP Conf. Series: Journal of Physics: Conf. Series **1001** (2018) 012016 doi:10.1088/1742-6596/1001/1/012016

The only physical control parameter is the bulk Reynolds number

$$Re = \frac{D u_{\mathrm{b}}}{\nu} \tag{1}$$

where $\nu$ is the kinematic viscosity of the fluid.

We use the pseudo-spectral simulation code `openpipeflow.org` [29] to solve the filtered Navier–Stokes equations in the form

$$\boldsymbol{\nabla} \cdot \overline{\boldsymbol{u}} = 0 \quad \text{and} \quad \frac{\partial \overline{\boldsymbol{u}}}{\partial t} + (\overline{\boldsymbol{u}} \cdot \boldsymbol{\nabla})\,\overline{\boldsymbol{u}} + \boldsymbol{\nabla}\overline{p}' - \frac{1}{Re}\boldsymbol{\Delta}\overline{\boldsymbol{u}} - \boldsymbol{\nabla} \cdot \boldsymbol{\tau}^r = \overline{\boldsymbol{P}} \tag{2}$$

in combination with a standard eddy viscosity closure model. The driving pressure gradient $\overline{\boldsymbol{P}}$ is adapted every time step such that a constant mass flux through the pipe domain is maintained. Filtered quantities are denoted by an overbar and $\boldsymbol{\tau}^r$ is the unknown residual stress tensor. To integrate eq. (2) in time, `openpipeflow.org` employs a second-order predictor-corrector scheme with dynamic time-step control. For the spatial discretisation, Fourier expansions are used in the two homogeneous directions $\theta$ and $z$ and finite differences with a seven point stencil are used in $r$. Grid points are distributed in $r$ according to a hyperbolic tangent function in such a way that high wall-normal velocity gradients in the viscous sublayer can be resolved. Additionally, the first few points for $r \approx 0$ are also clustered to preserve the high order of the finite difference scheme across the pipe axis.

The simple but robust linear eddy viscosity model of Smagorinsky [20] is used to relate the unknown residual stresses to the filtered rate of strain tensor ($\overline{\boldsymbol{S}}$) via

$$\boldsymbol{\tau}^r = -2\nu_r \overline{\boldsymbol{S}} \quad \text{with} \quad \nu_r = D l_s^2 \left(2\overline{S}_{ij}\overline{S}_{ij}\right)^{1/2} \quad \text{and} \quad l_s = C_s\overline{\Delta}, \text{ where } \overline{S}_{ij} \equiv \frac{1}{2}\left(\frac{\partial \overline{u}_i}{\partial x_j} + \frac{\partial \overline{u}_j}{\partial x_i}\right) \tag{3}$$

are the components of $\overline{\boldsymbol{S}}$. Through the model parameter $C_s$, the Smagorinsky (mixing) length scale $l_s$ is taken to be proportional to the volumetric mean grid filter width

$$\overline{\Delta} = (\Delta z \cdot r\Delta\theta \cdot \Delta r)^{1/3}, \tag{4}$$

where $\Delta\alpha$ denotes the local grid spacing in the direction $\alpha \in \{\theta, r, z\}$. Alternative formulations (e.g with the hypotenuse form [30]) can also be found in the literature. In contrast to eq. (4), such formulations do not lead to vanishing $\overline{\Delta}$ for $r \to 0$, but also to much higher values of $\overline{\Delta}$ everywhere else (especially for $r \to R$). Since the choice of the definition of $\overline{\Delta}$ is to some extent empirical/arbitrary, we here chose the form given by eq. (4) to be consistent with the ODL studies of Cossu and co-workers [21, 22, 24]. Further, the modelled eddy viscosity $\nu_r$ of the residual motions is modified by a van Driest damping function of the form

$$D(y^+) = \left(1 - e^{-(y^+/26)^2}\right)^2 \tag{5}$$

to enforce the correct behaviour of $l_s$ close to the wall, see e.g. [31, 32, 33, 19] for commonly used formulations. Here, $y^+ = (R - r)Re_\tau$ is the distance from the pipe wall in viscous (wall) units denoted by plus. The relevant viscous length scale is, as in [21, 22, 24], adaptively calculated in the course of the simulation based on the instantaneous shear velocity $\langle u_\tau \rangle_{\theta,z}$, where angle brackets denote averaging with respect to the noted subscripts. Note that as for all eddy viscosity approaches with $\nu_r > 0$, the Smagorinsky model does not allow for backscatter, i.e. no energy can be transferred from unresolved (small) to resolved (large) scales.

A summary of the spatial resolutions used in our LES is provided in tab. 1. For all $Re_\tau$, the

**Table 1.** Parameter settings for our LES. The control parameter is the bulk Reynolds number *Re*. The friction Reynolds number $Re_\tau$ is calculated with Blasius' law to a priori estimate the necessary spatial resolution in wall units (denoted by $^+$). $2K$, $2M$, and $N$ are the number of modes/points in $z$, $\theta$, and $r$, respectively. The total number of degrees of freedom (DOF) can be calculated as $2HN$ with $H = KM - K - M + 1$.

| *Re* (Control parameter) | 5300 | 11 700 | 25 800 | 59 600 |
|---|---|---|---|---|
| $Re_\tau$ (Blasius) | 180.4 | 360.7 | 720.6 | 1499.3 |
| $K \times M \times N$ | $40 \times 18 \times 40$ | $76 \times 36 \times 72$ | $152 \times 72 \times 128$ | $320 \times 144 \times 256$ |
| DOF | $1.1 \times 10^5$ | $7.7 \times 10^5$ | $5.5 \times 10^6$ | $4.7 \times 10^7$ |
| $\Delta z^+$ | 94.7 | 99.7 | 99.6 | 98.4 |
| $\Delta\theta R^+$ | 31.5 | 31.5 | 31.4 | 32.7 |
| $\Delta r^+\vert_{r\approx 0}$ | 9.1 | 11.1 | 12.7 | 13.0 |
| $\Delta r^+\vert_{r=R}$ | 0.8 | 0.7 | 0.7 | 0.7 |
| $N_{y^+<5}$ | 5 | 6 | 6 | 7 |
| $N_{y^+<30}$ | 16 | 22 | 27 | 33 |

resolution is chosen to be roughly the same when scaled in wall units. When $C_s$ is increased, according to the ODL approach proposed by Cossu and co-workers [25] the grid resolution is kept constant, so that $C_s$ is the only free parameter to change the mixing length scale $l_s$ and thus enhance the dissipative character of the closure model. With the resolution listed in tab. 1, eq. (4) evaluates to $\mathcal{O}(3.5) \leq \overline{\Delta}^+ \leq \mathcal{O}(35)$ depending on $r$. As a consequence, $l_s$, which is a uniform length scale representative of the smallest resolved eddies [34], reaches at most values of $\mathcal{O}(4.2) \leq l_s^+ \leq \mathcal{O}(42)$ (depending on $r$) for the highest $C_s$ considered here, see tab. 2. A typical size in $r$, $\theta$, and $z$ of the target structures to be damped out by the ODL approach — i.e. typical near-wall streaks — is roughly $80 \times 130 \times 800$ viscous units in pipe flow, e.g. [2, 3].

For the lowest $Re_\tau \approx 180$ considered here, a statistically steady flow field was generated in a well-resolved DNS and used as initial condition for the LES. For successively higher Reynolds numbers a statistically steady flow field from the LES of the previous $Re_\tau$ was used. For the LES at successively higher $C_s$, a statistically steady flow field from the previous $C_s$ was used. In all cases, turbulence statistics were sampled over at least $375^{D}/_{u_b}$ after a transient phase of at least the same length.

## 3. Large eddy simulation results

### 3.1. Instantaneous flow field

We visualise streaks by plotting iso-surfaces of the streamwise velocity fluctuations $u_z' = u_z - \langle u_z \rangle_{t,\theta,z}$ and focus on low-velocity streaks — i.e. connected regions of fluid moving slower than the local mean velocity $\langle u_z \rangle_{t,\theta,z}$ — defined by a negative threshold of $u_z' = -2.5u_\tau$. Although this choice is to some extent arbitrary, thresholds in the range of $-0.5u_\tau$ to $-2.5u_\tau$ are commonly used in the literature [10, 35, 36, 22]. In what follows, velocities are scaled in units of the actual $u_\tau \equiv \langle u_\tau \rangle_{t,\theta,z}$ resulting from each simulation and not with the nominal friction velocity according to Blasius' law.

The effect of the Smagorinsky parameter $C_s$ on the structure and organisation of low velocity streaks is shown in fig. 1. The top panels show typical turbulent flow fields for two selected nominal $Re_\tau \in \{360, 1500\}$ from LES with a commonly used moderate value of $C_s = 0.10$. They reveal the characteristic elongated streak structures, which decrease in size (when scaled in bulk units) and hence increase in number with growing *Re*. These qualitative observations for the regular $C_s$ values are in excellent agreement with flow field visualisations from pipe flow

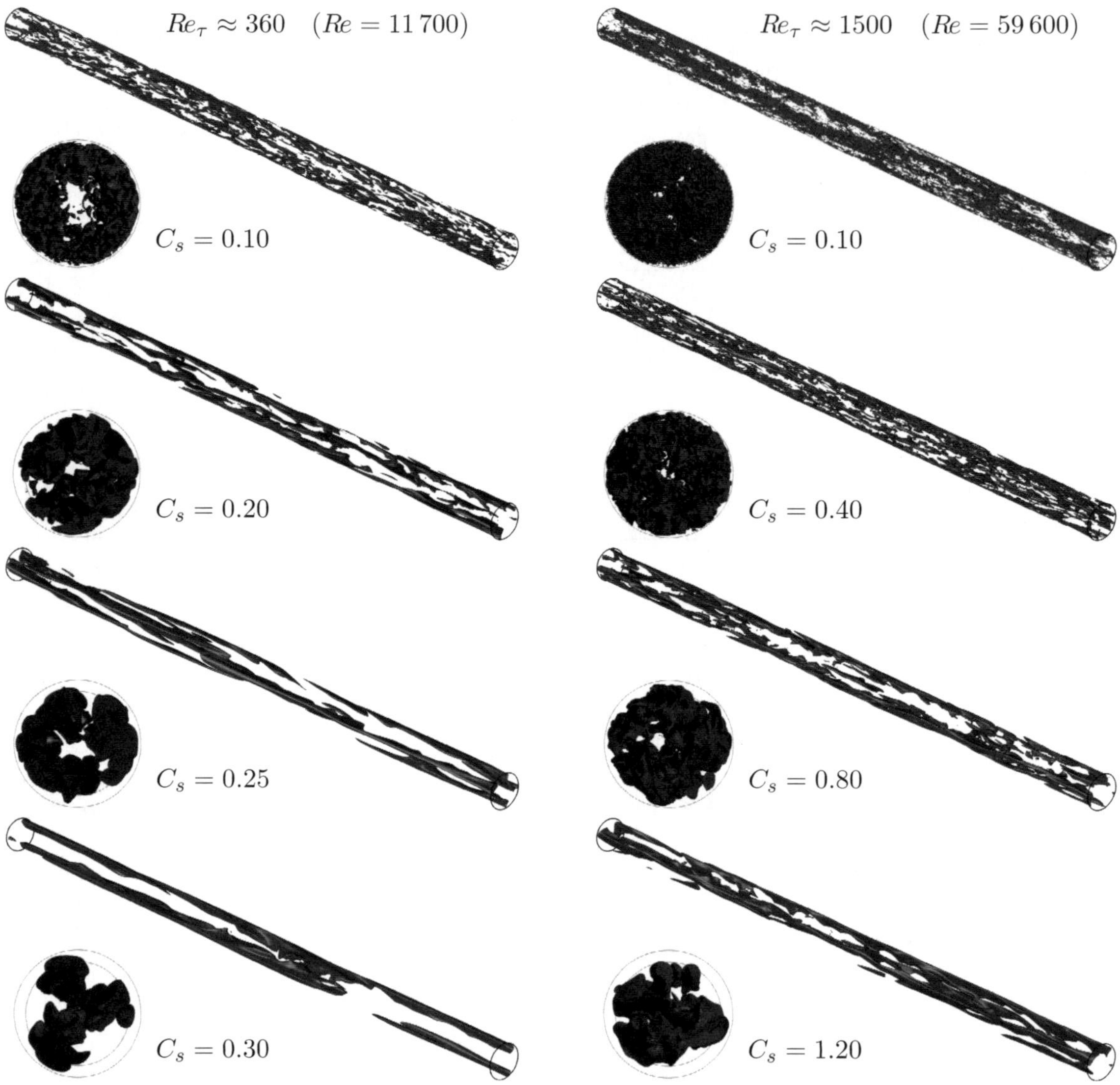


**Figure 1.** Visualisation of low-speed streaks in instantaneous flow field realisations from different LES. Shown are iso-surfaces of $u'_z = -2.5u_\tau$ for nominal $Re_\tau \approx 360$ (left) and $Re_\tau \approx 1500$ (right) and increasing values of the Smagorinsky parameter $C_s$ (top to bottom). The flow is directed from the top left to the bottom right, and out of the plane, respectively.

DNS data, see e.g. [37, 38, 14, 2] amongst many others. In addition to the near-wall streaks, the incipient formation of clusters with a large wavelength modulation can be observed for the highest $Re_\tau \approx 1500$ considered here. This modulated clusters provide qualitative evidence for the rise of very large scale coherent motions [14], which can be more clearly discerned at even higher $Re_\tau$ [8, 11, 12].

As $C_s$ increases, the smallest visible length scales are gradually removed from the picture, while very large smooth streaks appear instead. The resulting three- and fourfold symmetries correspond to azimuthal streak spacings of approximately one to two pipe radii, while their radial extent is slightly less than $R$. This trend — exemplarily shown for the two selected Reynolds

**Table 2.** Measured mean friction Reynolds number $\langle Re_\tau \rangle$ and its relative error $\epsilon_{\text{rel}}$ with respect to the nominal $Re_\tau$ from Blasius' law as a function of the Smagorinsky model parameter $C_s$ at four different bulk Reynolds numbers ($Re$).

| $Re$ (contr. param.) | 5300 | | 11 700 | | 25 800 | | 59 600 | |
|---|---|---|---|---|---|---|---|---|
| $Re_\tau$ (Blasius) | 180.4 | | 360.7 | | 720.6 | | 1499.3 | |
| | $\langle Re_\tau \rangle$ | $\epsilon_{\text{rel}}$ | $\langle Re_\tau \rangle$ | $\epsilon_{\text{rel}}$ | $\langle Re_\tau \rangle$ | $\epsilon_{\text{rel}}$ | $\langle Re_\tau \rangle$ | $\epsilon_{\text{rel}}$ |
| $C_s = 0.05$ | 195.3 | 0.083 | 398.6 | 0.105 | 784.8 | 0.089 | 1626.2 | 0.085 |
| $C_s = 0.10$ | 165.6 | -0.082 | 366.2 | 0.015 | 741.8 | 0.029 | 1586.1 | 0.058 |
| $C_s = 0.12$ | 160.9 | -0.108 | | | | | | |
| $C_s = 0.15$ | laminarised | | 326.0 | -0.096 | 664.6 | -0.078 | 1501.6 | 0.002 |
| $C_s = 0.20$ | | | 320.5 | -0.111 | 622.1 | -0.137 | 1419.7 | -0.053 |
| $C_s = 0.25$ | | | 326.8 | -0.094 | | | | |
| $C_s = 0.30$ | | | 322.7 | -0.105 | 618.8 | -0.141 | | |
| $C_s = 0.32$ | | | laminarised | | | | | |
| $C_s = 0.40$ | | | | | 645.6 | -0.104 | 1516.3 | 0.011 |
| $C_s = 0.55$ | | | | | 679.1 | -0.058 | | |
| $C_s = 0.60$ | | | | | 694.7 | -0.036 | | |
| $C_s = 0.62$ | | | | | 700.4 | -0.028 | | |
| $C_s = 0.65$ | | | | | laminarised | | | |
| $C_s = 0.80$ | | | | | | | 1948.2 | 0.299 |
| $C_s = 1.20$ | | | | | | | 2092.7 | 0.396 |

numbers in fig. 1 — is basically the same for all $Re_\tau$ considered here. See tab. 2 and fig. 2a for a summary of all performed simulations. To obtain qualitatively the same degree of *overdamping* for a four times higher Reynolds number, $C_s$ has to be also increased by a factor of four, see fig. 1. For $Re_\tau = 1500$ — the lowest Reynolds number for which we expect VLSM-related effects [2] — the Smagorinsky parameter has to be increased to $C_s = 1.2$ to observe a degree of small-scale quenching comparable to that observed here for $Re_\tau \approx 360$ or in [21, 22, 24], where $C_s \approx 0.3$ is sufficient. For only moderately increased values of $C_s \lesssim 0.4$, there is almost no reduction of complexity (removal of small scales) at $Re_\tau \approx 1500$. However, the observed clustering of smaller structures has been completely suppressed already at $C_s = 0.4$. Thus, increasing $C_s$ is comparable to reducing the Reynolds number, since the observed clustering emerges first for $Re_\tau \gtrsim \mathcal{O}(1500)$ [14].

### 3.2. *Non-dimensional shear stress*

The effect of $C_s$ on the integral shear stress in the turbulent flow is quantified here by comparing the actual shear Reynolds number

$$\langle Re_\tau \rangle = \frac{R \langle u_\tau \rangle_{t,\theta,z}}{\nu} \tag{6}$$

extracted from each LES to the nominal $Re_\tau = f(Re)$ given by Blasius' law at a fixed $Re$. Note that as $R$ and $\nu$ are fixed via the control parameter $Re$, changes in $\langle Re_\tau \rangle$ are due to changes in the wall shear stress $\tau_{\text{w}} \sim u_\tau^2$, and hence of the total energy input. The relative errors $\epsilon_{\text{rel}} = {}^{\langle Re_\tau \rangle}/_{Re_\tau} - 1$ for four different nominal $Re_\tau$ are given in fig. 2b and tab. 2, where positive/negative values correspond to over-/underestimated shear stress. With increasing $C_s$ the measured shear stress decreases and reaches a minimum (which depends on $Re_\tau$). For the two lowest $Re_\tau$, the flow laminarises completely around this minimum when $C_s$ is further

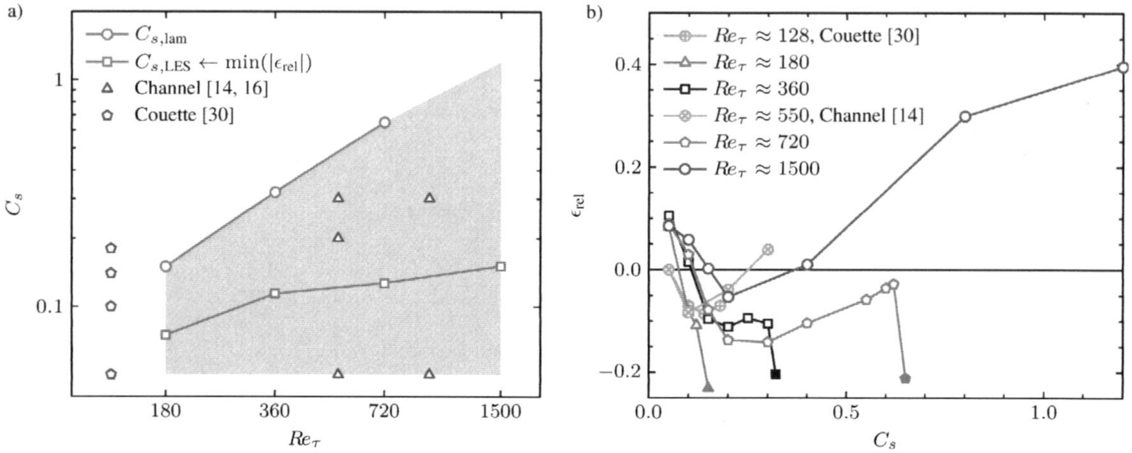


**Figure 2.** The shaded area in panel a) indicates the parameter range considered in the present study. At each $Re_\tau$, the optimal value of the Smagorinsky parameter ($C_{s,\mathrm{LES}}$) is marked with a square. This is defined as the value of $C_s$ for which $\epsilon_{\mathrm{rel}} = 0$. Values leading to complete flow laminarisation are shown as circles. The triangles and pentagons show the parameter values reported in previous ODL studies of turbulent channel [21, 23] and plane Couette [24] flows, respectively. The effect of $C_s$ on $\epsilon_{\mathrm{rel}} = {}^{\langle Re_\tau \rangle}/_{Re_\tau} - 1$ is shown in panel b), where full symbols denote values for a stationary flow after complete laminarisation. Results of Couette and channel flows are normalised with respect to the reference simulations at $C_{s,\mathrm{LES}} = 0.05$ reported in [21, 24].

increased. The shear stress corresponding to this laminar flow state is indicated by full symbols and is much higher compared to the shear stress of Hagen-Poiseuille (HP) flow at the same *Re*, which is further discussed in §3.3. This demonstrates, that a large $\langle Re_\tau \rangle$ does not necessarily imply dynamical/kinematic similarity with a truly high $Re_\tau$ turbulent pipe flow. For the two highest $Re_\tau$, the measured $\langle Re_\tau \rangle$ increases again when $C_s$ is further increased, before ultimate laminarisation takes place. The dependence of the laminarisation threshold $C_{s,\mathrm{lam}}$ on $Re_\tau$ is illustrated in fig. 2a.

Minimizing $|\epsilon_{\mathrm{rel}}|$ by linear interpolation yields an estimate of the optimal $C_{s,\mathrm{LES}}$ for a standard LES in the sense that the measured shear stress is closest to the nominal value from Blasius' law for a fully-developed healthy turbulent flow. Of course this choice does not necessarily optimise one-point statistics, energy spectra, etc. as discussed in the following sections. Interestingly, the value of $C_{s,\mathrm{LES}}$ increases from about 0.075 to 0.151 as $Re_\tau$ is increased (see squares in fig. 2a). To some extent, $C_{s,\mathrm{lam}}$ quantifies the proposed/observed reduction of the effective Reynolds number with increasing Smagorinsky parameter: a two or roughly three times higher $C_s$ is needed to laminarise the flow for a two or three times higher $Re_\tau$, respectively.

### *3.3. Low-order-moment turbulence statistics*

As a next step, we analyse the effect of an elevated Smagorinsky parameter on low-order turbulence statistics. Figure 3 shows the mean and root-mean-square (*RMS*) velocity profiles for $Re_\tau \approx 360$ scaled in inner (a, c, e, g) as well as in outer (b, d, f, h) units. Independent DNS data sets [38, 2] are shown as a reference (with excellent agreement). The best matches to the DNS data are obtained for Smagorinsky parameters roughly between $C_s = 0.05$ and 0.10, depending on the quantity and the wall normal location. Thus, the optimal $C_s$ value estimated by matching first- and second-order turbulence statistics are slightly smaller than that from the (integral) mean shear stress ($C_{s,\mathrm{LES}} = 0.114$).

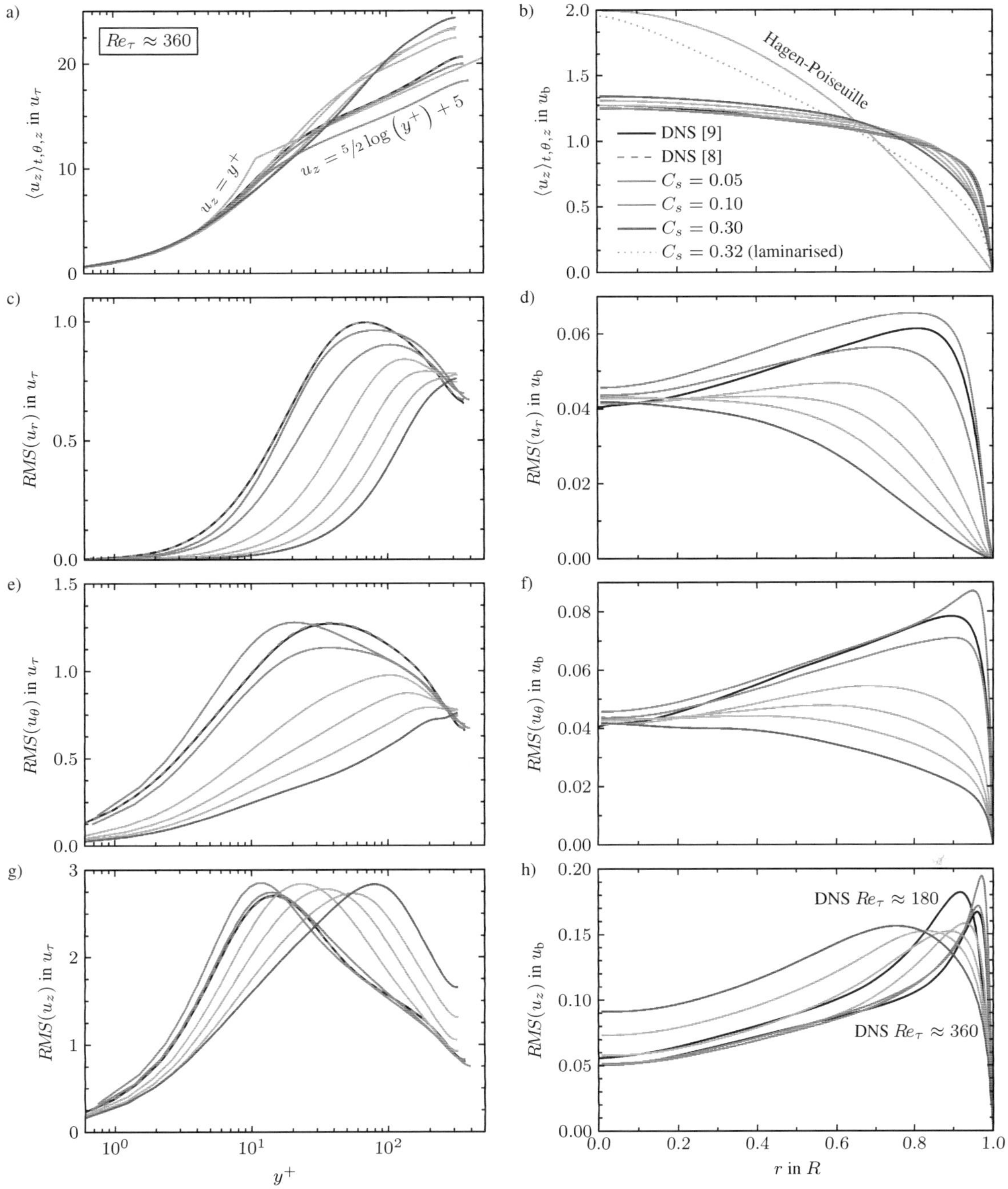


**Figure 3.** Mean (a, b) and *RMS* (c-h) velocity statistics calculated from LES at $Re_\tau \approx 360$ with varying $C_s$ and fixed spatial discretisation. The data is presented in inner (a, c, e, g) and outer (b, d, f, h) units. Independent DNS data [38, 2] as well as the HP solution (b), the linear law (a), and the logarithmic law (a) are shown as references. Red and green lines localise best matches to the DNS data. Unlabelled grey lines correspond to intermediate $C_s$ as listed in tab. 2. Additionally, an instantaneous velocity profile for the stationary flow state after laminarisation is presented in panel b). DNS data for the next lower $Re_\tau \approx 180$ is included in panel h).

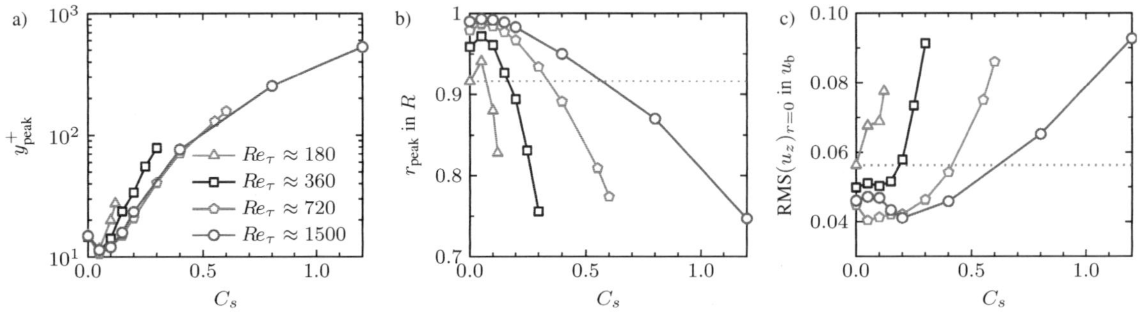


**Figure 4.** Effect of $C_s$ on $RMS(u_z)$. Panels a) and b) show the wall-normal location of the near-wall peak in inner and outer units, respectively. Panel c) shows the turbulence intensity level at the pipe axis ($r = 0$). The DNS reference data are plotted at $C_s = 0$ and marked for $Re_\tau \approx 180$ by a dashed line, to exemplarily quantify an effective reduction of the Reynolds number with increasing $C_s$.

By further increasing $C_s$, the mean velocity profile departs almost everywhere from the reference profile. Figure 3a shows that the mean velocity develops a large wake component already for moderately increased $C_s$. Note that large wake components are characteristic of low Reynolds number (transitional) pipe flow. Figure 3b shows that for $C_s \to C_{s,\text{lam}}$, the HP solution is quickly approached but never reached. The wall-normal mean velocity gradient, and thus the measured $\langle Re_\tau \rangle$, is still much higher despite the flow being laminar ($u'_r = u'_\theta = u'_z = 0$). This is because the implementation of the Smagorinsky model used here and also in [21, 22, 24] leads, incorrectly, to $\boldsymbol{\tau}^r = \mathcal{O}(\overline{\Delta}^2)$ for a laminar flow with $C_s \neq 0$, due to the non-vanishing components of $\overline{S}_{ij}$ related to the mean velocity profile, see pp. 597 in [19]. Alternative implementations based on $\overline{S}_{ij} - \langle \overline{S}_{ij} \rangle$ (e.g. [39, 31]), however, suffer from the disadvantage that an appropriate averaging operation has to be performed during the simulation.

Figures 3c–f reveal that with increasing $C_s$ both cross-stream components of the *RMS* velocity fluctuations are reduced almost everywhere in the radial direction except for a small region close to the centreline, where the level of turbulence seems nearly unaffected by $C_s$. The *RMS* of the streamwise velocity fluctuations shown in fig. 3g–h exhibit a progressive outward shift of the high intensity near-wall peak as $C_s$ increases. As a consequence of the shift, the streamwise turbulence intensity in the outer region increases substantially. While the peak amplitude remains roughly unchanged in wall units, it is gradually reduced for increasing $C_s$ when scaled in outer units. Additionally, fig. 3h includes DNS reference data for the next lower $Re_\tau \approx 180$. This shows that increasing the Smagorinsky parameter for $Re_\tau \approx 360$ by a factor of two (from $C_s = 0.1$ to $C_s = 0.2$), perfectly matches the *RMS* level (for over 80 % of the radial direction) and the *RMS* peak position of DNS reference data for $Re_\tau \approx 180$. These features are common to all simulations performed here. Figure 4 shows the variation (in inner and outer units) of the near-wall peak (level and location) and of the centreline value of $RMS(u_z)$ as a function of $Re_\tau$ and $C_s$. The observed trends suggest a smooth outward displacement of the near-wall cycle as $C_s$ is increased. Hence, the effect of increasing $C_s$ is actually qualitatively similar to reducing the effective Reynolds number of the flow, even though the bulk Reynolds number (i.e. the control parameter) is unchanged and the measured shear stress is roughly similar (see fig. 2b). This suggests, that the measured $\langle Re_\tau \rangle$ cannot be used to infer dynamical similarity for a given *Re*. $Re_\tau$ implies similarity in the near wall region and only has general/global meaning for fully developed, genuinely turbulent pipe flow. It does not guarantee global kinematic similarity in other situations. For example, in oscillatory pipe flow, the instantaneous $\langle Re_\tau \rangle$ can be very high,

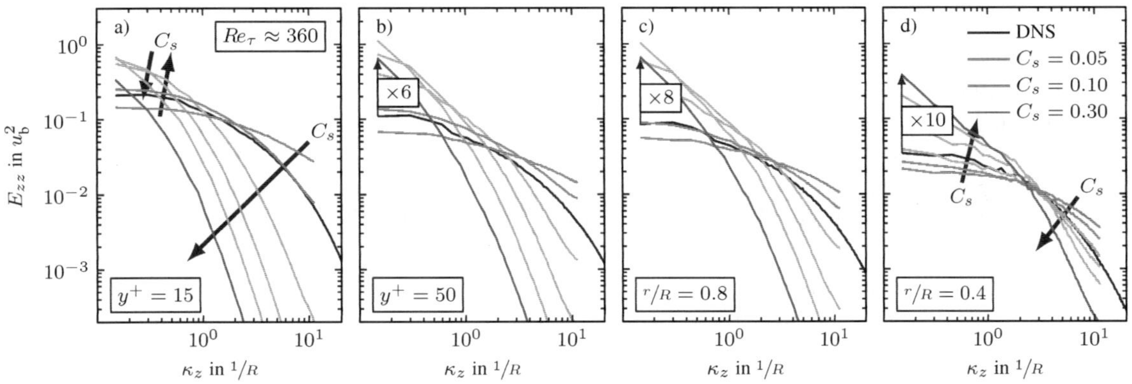


**Figure 5.** Comparison of streamwise energy spectra of $u_z$ as a function of $C_s$ for $Re_\tau \approx 360$ and constant resolution. Panels a) to b) show spectra at different locations ranging from the near-wall ($y^+ = 15$) to the outer region ($r \leq 0.8R$), where VLSM are expected to live.

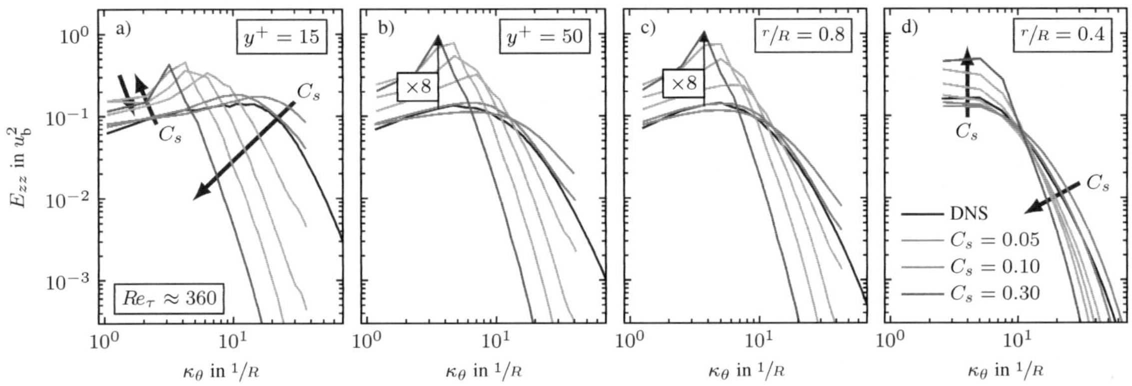


**Figure 6.** Comparison of azimuthal energy spectra of $u_z$ as a function of $C_s$ for $Re_\tau \approx 360$ and constant resolution. Panels a) to b) show spectra at different locations ranging from the near-wall ($y^+ = 15$) to the outer region ($r \leq 0.8R$), where VLSM are expected to live.

even though the flow is completely laminar, see [40]. Or high $\langle Re_\tau \rangle$ pipe flow can be laminarised by destabilising turbulence using appropriate perturbations [41]. Therefore, the turbulent pipe flow computed in an ODL is not necessarily similar to truly turbulent pipe flow at the same $Re$ just because the measured $\langle Re_\tau \rangle$ are similar.

### *3.4. Energy spectra*

In the following, we examine the effect of $C_s$ on the energy distribution across scales at different wall-normal locations. Since VLSM are mostly recognised as long and wide regions of purely positive or negative $u'_z$, we focus on streamwise and azimuthal energy spectra of $u_z$, denoted as $E_{zz}(\kappa_z)$ and $E_{zz}(\kappa_\theta)$, respectively. All spectra are scaled in outer units, since we are mainly concerned with superstructures, which scale in outer units. Figures 5a–d show streamwise spectra at $Re_\tau \approx 360$ for increasing distance from the wall and different values of $C_s$. Additionally, fig. 6a–d show azimuthal spectra for the same parameters. Comparison with the reference DNS data reveals again best matches for $0.05 < C_s < 0.15$, which agrees with the optimal $C_{s,\text{LES}} = 0.114$ obtained by matching the measured dimensionless shear stress with the nominal value from Blasius law (see §3.2). The optimal $C_s$ for the spectra depends slightly on the wavenumber range and also on the wall-normal location. In other words, $C_s$ changes the slope of the spectrum differently strong at different wall-normal locations. At $y^+ = 15$, a moderate value of $C_s = 0.10$ gives best results for all $\kappa_z$. At $r/R = 0.8$, on the other hand, best matches change gradually from $C_s = 0.10$ to 0.15 with increasing $\kappa_z$, while at $r/R = 0.4$ a uniform $C_s = 0.15$ leads to best agreement for all $\kappa_z$. For the entire range of $C_s$, the energy in the lower wavenumbers increases by factors of 8 to 10 in the outer region ($r < 0.8R$). In the buffer region, the energy level also increases with $C_s$, but only by factors of 6 to 8 in the lower wavenumbers, see fig. 5 and 6. In contrast to the lowest $Re_\tau \approx 180$ (not shown), for all $Re_\tau \geq 360$ the highest energy increase does not occur for the strongest possible *overdamping* (i.e. $C_s = 0.3$ in this case), but for moderately elevated $0.12 < C_s < 0.3$. This effect is mainly restricted to the buffer region (black arrows in fig. 5a–b) and absent in the outer region (black arrows in fig. 5d). The azimuthal spectra presented in fig. 6 have similar features and similar effects are observed at all other *Re* considered (hence not shown). The observed trends are consistent with the fact, that the LES filter is local in physical space and thus non-local in Fourier space (the energy spectra change slope everywhere).

### *3.5. Pre-multiplied energy spectra*

We now turn to pre-multiplied energy spectra, which — in contrast to the energy spectra — allow a direct comparison of the energetic contribution of different ranges of wavenumbers. To compensate for the varying amount of energy input, which is proportional to $u_\tau^2 \sim \tau_\mathrm{w}$ for fixed $u_\mathrm{b}$, the energy spectra are scaled in inner units based on the measured $\langle Re_\tau \rangle$. As the effect of $C_s$ on the pre-multiplied spectra is similar at all $Re_\tau$, we here focus on $Re_\tau \approx 1500$, at which incipient VLSM are present. Figures 7 and 8 present streamwise ($\kappa_z E_{zz}(\lambda_z)$) and azimuthal ($\kappa_\theta E_{zz}(\lambda_\theta)$) pre-multiplied energy spectra of the streamwise velocity as a function of the wavelength $\lambda$.

Figures 7a–c demonstrate that at wall-normal locations characteristic of the near-wall cycle ($y^+ \approx 15$) the energy in the range of wavelengths likewise characteristic of the near-wall cycle ($\lambda_z^+ \approx 800$) is strongly damped as $C_s$ increases beyond $C_{s,\mathrm{LES}}$. However, in the outer region (fig. 7d–f) the energy in the range of wavelengths associated with VLSM ($\lambda_z \approx 20R$) substantially grows far beyond the level of energy observed for DNS or standard LES ($C_{s,\mathrm{LES}} \approx 0.15$), as soon as $C_s$ is elevated to values from which on a clear damping of small scales is obvious and only smooth long streaks remain (fig. 1). For the same high $C_s = 1.2$, closer to the wall (fig. 7b–c), the level of energy at $\lambda_z \approx 20R$ drops far below the values observed for standard LES. For only moderate, but still elevated values (e.g. $C_s = 0.4$) on the other hand, the artificial increase of energy in the relevant large and intermediate $\lambda$ can be even larger than for the highest $C_s = 1.2$, e.g. fig. 7b–d.

Similar trends are observed in the azimuthal energy spectra presented in fig. 8. For sufficiently high $C_s$ in the sense of a clear reduction of small scales (fig. 1), energy in wavelengths associated with the azimuthal spacing of near-wall streaks ($\lambda_\theta^+ = 130$) is substantially damped out, whereas it is strongly increased at larger wavelengths associated with the azimuthal spacing of VLSM ($\lambda_\theta = 1R$). In the buffer region, energy is first strongly increased and than damped out as $C_s$ is smoothly elevated, as can be seen from fig. 8b–c. Looking at the series of spectra for smoothly increased $C_s$, this can be explained as follows. The near-wall peak, which is known to appear at $\lambda_\theta^+ \approx 130$ around $y^+ \approx 15$, is well reproduced at $C_s \in [0.10, 0.15]$. As the Smagorinsky parameter is increased by a factor of roughly three ($C_s = 0.4$), the near wall peak is shifted to a roughly three times larger wall distance ($y^+ = 50$) and a roughly six times larger wavelength (fig. 8c). Its near-wall-low-wavelengths flank replaces the near-wall peak at its original location (fig. 8b), while its far-wall-high-wavenumber flank traverses the location of the VLSM (fig. 8d), i.e around $r/R = 0.8$ and $\lambda_\theta = R$ [2]. At a fixed location and wavelength this seemingly appears as an energy amplification followed by damping as the modified near-wall peak passes by. This means, that the original near-wall peak is modified in physical as well as in wavenumber space

IOP Conf. Series: Journal of Physics: Conf. Series **1001** (2018) 012016 doi:10.1088/1742-6596/1001/1/012016

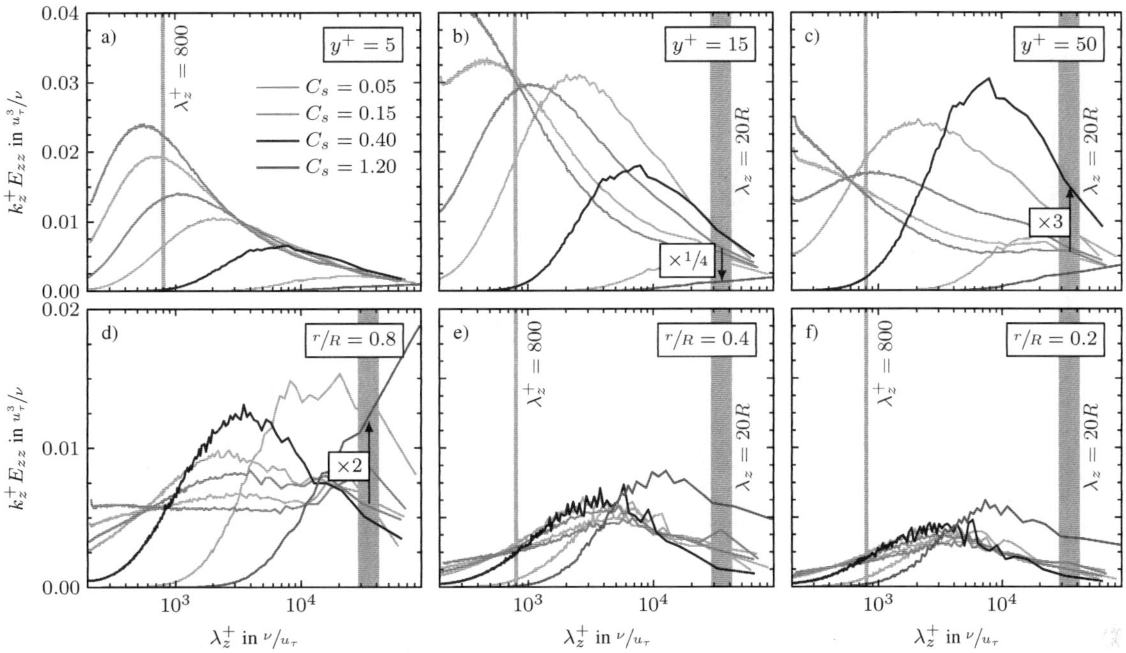


**Figure 7.** Comparison of streamwise pre-multiplied energy spectra for $u_z$ from different LES at $Re_\tau \approx 1500$ with varying $C_s$. Shown are spectra extracted at several wall normal locations from the edge of the viscous sublayer ($y^+ = 5$) up to the outer region ($r \leq 0.8R$). Unlabelled grey lines correspond to further intermediate $C_s$ as listed in tab. 2.

when $C_s$ is elevated, until, eventually, its dynamics are constrained by the pipe wall. In order to verify this claim, an examination of the dynamics of the structures for different $C_s$ should be conducted, as in [42], where various quantitative measures are used to categorise eddies.

## 4. Discussion

We performed an extensive LES study of turbulent pipe flow driven at a constant bulk velocity by using the classic static Smagorinsky model [20]. The computational domain length was fixed to $42R$, allowing the formation of VLSM [2], whereas wide ranges of the Reynolds number and the Smagorinsky model parameter were covered, $180 \leq Re_\tau \leq 1500$ and $0.05 \leq C_s \leq 1.2$, respectively. Overall, our analysis of wall shear stresses, mean velocity profiles, *RMS* velocity fluctuations, and energy spectra confirm the high quality of our LES of turbulent pipe flow for reasonable parameter ranges of $C_s$. The optimal Smagorinsky parameter to achieve best agreement with DNS reference data varies with $Re_\tau$ and further depends on the wall normal location, as well as on the quantity of interest.

We selected an optimal value of the Smagorinsky constant ($C_{s,\text{LES}}$) by minimising the difference between the nominal friction according to Blasius' law and the friction obtained from the LES. For the two lower $Re_\tau \approx 180$ and 360, we found $C_{s,\text{LES}} \approx 0.075$ and 0.114 respectively, which is very similar to those values reported in the literature ($0.065 \leq C_{s,\text{LES}} \leq 0.125$) to adequately reproduce reference data in turbulent pipe [37, 43, 30] and channel [31, 44, 45] flow LES at similar $Re_\tau$. For increasing $Re_\tau$, we found that increasing values of the Smagorinsky parameter produced better results. For the highest $Re_\tau \approx 1500$ investigated, $C_{s,\text{LES}} = 0.151$ approaches values normally used in LES of isotropic turbulence, as in the seminal work of Smagorinsky [20] or in the theoretical derivation by Lilly [46], with $C_s = 0.2$, and 0.17

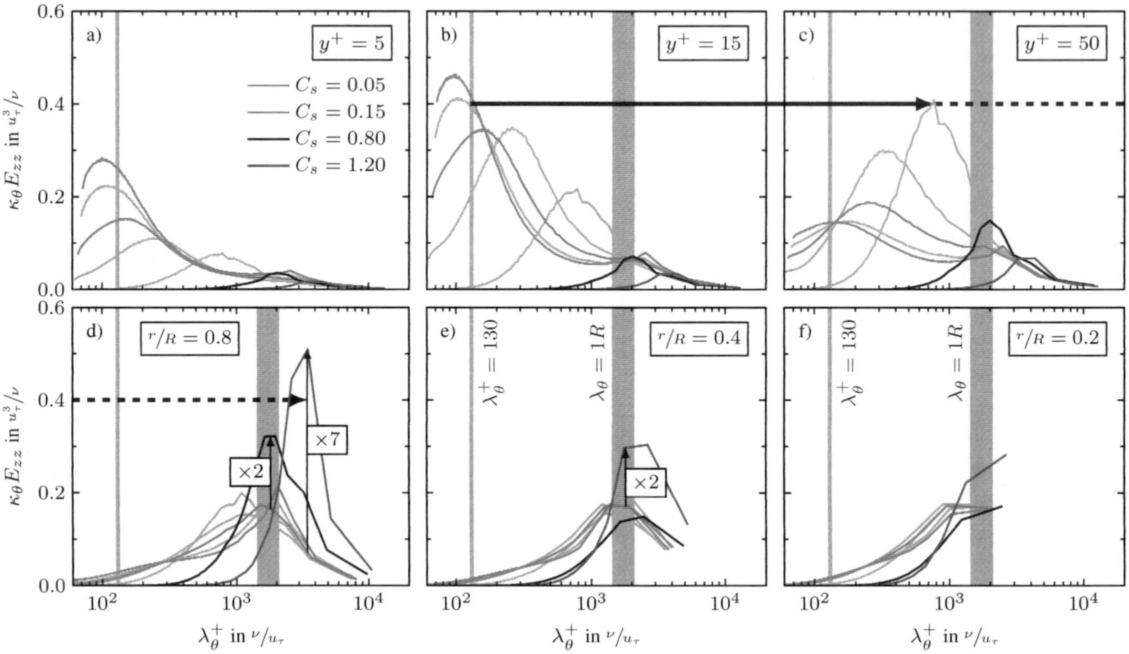


**Figure 8.** Comparison of azimuthal pre-multiplied energy spectra for $u_z$ from different LES at $Re_\tau \approx 1500$ with varying $C_s$. Shown are spectra extracted at several wall normal locations from the edge of the viscous sublayer ($y^+ = 5$) up to the outer region ($r \leq 0.8R$). Unlabelled grey lines correspond to further intermediate $C_s$ as listed in tab. 2.

respectively. This is in agreement with the fact that for increasing $Re_\tau$ the flow gradually becomes more isotropic in a growing portion of the pipe domain. Hence it would be interesting to perform LES at yet larger $Re_\tau$ than here to determine the value at which $C_{s,\text{LES}}$ eventually saturates.

For the highest $Re_\tau \approx 1500$ considered here, the incipient formation of long clusters of streaks leading to VLSM could be observed — in good agreement with recent pipe flow DNS, e.g. [17, 14], and experiments, e.g. [6, 10]. Our long term goal is to understand and model VLSM with dynamical systems approaches, i.e. via exact coherent solutions (ECS) of the governing equations. This requires however to substantially reduce the complexity of high Reynolds number flow fields by isolating VLSM from smaller-scale turbulent motions. We wanted to take this challenge by using *overdamped* LES following the work of Cossu [25] and co-workers. Their idea is that increasing the effective filter width ($l_s$) of the Smagorinsky model via $C_s$, *passivates* an increasing range of small-scale structures without loosing their dissipative effect on the next larger scales. We found in good qualitative agreement with these studies that small-scale streaks in simple flow field visualisations are gradually quenched and replaced by much larger smooth streaks as $C_s$ increases. Surprisingly, for all parameter values reached here, the uniform length scale at which the Smagorinsky model is acting (i.e $l_s^+ = C_s\overline{\Delta}^+ \leq C_s 35 \leq 42$), remains always much smaller than the size of the inhomogeneous/anisotropic target near-wall structures (i.e. $80 \times 130 \times 800$ viscous units). However, our analysis of mean velocities, *RMS* velocity profiles and energy spectra suggest, that these structures originate from an effective reduction of the Reynolds number and thus represent enlarged/modified near-wall streaks rather than superstructures isolated from the original near-wall cycle. As the $C_s$ is increased, the wall cycle progressively moves away from the wall and the associated streaks become longer and wider, when scaled in inner and outer units. Our streak visualisations in §3.1 reveal that for all $Re_\tau$ the remaining structures reach an azimuthal and radial size similar to those of usual near-wall streaks in a turbulent low-Reynolds number ($Re_\tau \approx 90$) pipe flow, i.e. $\Delta\theta = {}^{\Delta\theta^+}/_{Re_\tau} = {}^{130}/_{90} = 1.4R$ and $\Delta r = {}^{\Delta r^+}/_{Re_\tau} = {}^{80}/_{90} = 0.9R$. Just before the flow is forced to relaminarise, connected streaks end up filling our $42R$-long pipe domain. Furthermore, the energy contained in VLSM-related wavelengths is artificially increased by up to factors of seven (inner units) and ten (outer units) for elevated $C_s$, likely because the modified higher-intensity near-wall cycle shifts to outer-scale locations/wavelengths where VLSM are expected to live. The method changes the flow similarly at all investigated $Re_\tau$ regardless of the clear presence of VLSM, which in DNS may only be discerned at the highest Reynolds number investigated here ($Re_\tau \approx 1500$) [2, 14]. These observations are consistent with other LES studies, where $C_s$ is varied more moderately [47, 30, 43], and can be explained by the purely dissipative nature of the Smagorinsky model, whose dissipation is strongly enhanced by increasing $C_s$. The Smagorinsky model exclusively acts as an energy sink at a uniform length scale and, therefore, can neither mimic structural features of the unresolved motions nor account for feedback (e.g. backscatter, cluster formation etc.) to the resolved scales.

In view of these findings, we here suggest that the *overdamped* LES approach based on the static Smagorinsky model is not suitable to unambiguously determine the origin and the dynamics of superstructures in pipe flow. We expect that ECS computed from the ODL equations would not truly represent the nature of superstructures in turbulent pipe flow. Nevertheless, we believe that the proposal of Cossu [25] and co-workers — to eliminate the wall cycle by filtering in order to check whether VLSM are autonomous — is very valuable. Their ODL approach might be salvaged by e.g. using a more sophisticated LES model accounting for backscatter [48] or maybe by applying explicit anisotropic filtering tailored to match the size and orientation of the target structures. Moreover, ODL appears to be a valuable numerical tool to find ECS using $C_s$ as a third continuation parameter (besides $Re$ and $L$), as already described in e.g. [49]. However, eventually $C_s$ has to be reduced back to zero or at least to $C_{s,\mathrm{LES}}$ to converge the intermediate ECS to solutions of the Navier-Stokes or the standard LES equations, so that they can actually be interpreted as exact coherent superstructures.

## Acknowledgments

This work was funded by the priority programme *SPP1881 Turbulent Superstructures* of the Deutsche Forschungsgemeinschaft (DFG) and by the *COTURB* programme of the European Research Council (ERC). Computational resources on *Konrad* provided by North-German Supercomputing Alliance (HLRN) are also gratefully acknowledged. Further, we are grateful to C. Cossu, Y. Hwang, C. Wagner, and A. P. Willis for fruitful discussions and valuable feedback.

## References

[1] Kline S J, Reynolds W C, Schraub F A and Runstadler P W 1967 *J. Fluid Mech.* **30** 741–773
[2] Feldmann D, Bauer C and Wagner C 2018 *J. Turbul.* **in print** 1–22
[3] Feldmann D and Wagner C 2012 *J. Turbul.* **13** 1–28
[4] Kim K C and Adrian R J 1999 *Phys. Fluids* **11** 417–422
[5] Guala M, Hommema S E and Adrian R J 2006 *J. Fluid Mech.* **554** 521–542
[6] Monty J P, Stewart J A, Williams R C and Chong M S 2007 *J. Fluid Mech.* **589** 147–156
[7] Monty J P, Hutchins N, NG H C H, Marusic I and Chong M S 2009 *J. Fluid Mech.* **632** 431–442
[8] Hutchins N and Marusic I 2007 *Philos. T. Roy. Soc. A* **365** 647–664
[9] Hutchins N and Marusic I 2007 *J. Fluid Mech.* **579** 1–28
[10] Hellström L H O, Sinha A and Smits A J 2011 *Phys. Fluids* **23** 011703
[11] Mathis R, Hutchins N and Marusic I 2009 *J. Fluid Mech.* **628** 311–337
[12] Rosenberg B J, Hultmark M, Vallikivi M, Bailey S C C and Smits A J 2013 *J. Fluid Mech.* **731** 46–63
[13] Jiménez J 2007 *Rev. R. Acad. Cien. Serie A. Mat.* **101** 187–203
[14] Bauer C, Feldmann D and Wagner C 2017 *Phys. Fluids* **29** 125105

[15] Lozano-Durán A and Jiménez J 2014 *Phys. Fluids* **26** 011702
[16] Lee M and Moser R D 2015 *J. Fluid Mech.* **774** 395–415
[17] Ahn J, Lee J H, Kang J h and Sung H J 2015 *Phys. Fluids* **27** 065110
[18] Sagaut P 2006 *Large Eddy Simulation for Incompressible Flows: An Introduction* 3rd ed (Springer)
[19] Pope S B 2000 *Turbulent Flows* 7th ed (Cambridge U. Press)
[20] Smagorinsky J 1963 *Month. Weather Rev.* **91** 99–164
[21] Hwang Y and Cossu C 2010 *Phys. Rev. Lett.* **105** 044505
[22] Hwang Y and Cossu C 2011 *Phys. Fluids* **23** 061702
[23] Hwang Y, Willis A P and Cossu C 2016 *J. Fluid Mech.* **802** R1
[24] Rawat S, Cossu C, Hwang Y and Rincon F 2015 *J. Fluid Mech.* **782** 515–540
[25] Cossu C 2017 Exact invariant solutions for coherent turbulent motions in Couette and Poiseuille flows (KITP Talk)
[26] Eckhardt B, Schneider T M, Hof B and Westerweel J 2007 *Ann. Rev. Fluid Mech.* **39** 447–468
[27] Kawahara G, Uhlmann M and van Veen L 2012 *Ann. Rev. Fluid Mech.* **44** 203–225
[28] Cvitanović P 2013 *J. Fluid Mech.* **726** 1–4
[29] Willis A P 2017 *SoftwareX* **6** 124–127
[30] Schmidt S, McIver D M, Blackburn H M, Rudman M and Nathan G J 2001 *14th Australasian Fluid Mech. Conf.* pp 917–920
[31] Moin P and Kim J 1982 *J. Fluid Mech.* **118** 341–376
[32] Piomelli U, Ferziger J H and Moin P 1987 Models for large eddy simulations of turbulent channel flows including transpiration Tech. Rep. TF-32 Stanford University
[33] Kim W W and Menon S 1999 *Int. J. Numer. Meth. Fluids* **31** 983–1017
[34] Piomelli U, Moin P and Ferziger J H 1988 *Phys. Fluids* **31** 1884
[35] Lozano-Durán A, Flores O and Jiménez J 2012 *J. Fluid Mech.* **694** 100–130
[36] Lee J, Ahn J and Sung H J 2015 *Phys. Fluids* **27** 025101
[37] Unger F 1994 *Numerische Simulation turbulenter Rohrströmungen* Dissertation Technische Universität München
[38] El Khoury G K, Schlatter P, Noorani A, Fischer P F, Brethouwer G and Johansson A V 2013 *Flow Turbul. Combust.* **91** 475–495
[39] Schumann U 1975 *J. Comput. Phys.* **18** 376–404
[40] Feldmann D and Wagner C 2016 *Int. J. Heat Fluid Flow* **61** 229–244
[41] Kühnen J, Song B, Scarselli D, Budanur N B, Riedl M, Willis A P, Avila M and Hof B 2018 *Nature Physics*
[42] Jimenez J 2018 *J. Fluid Mech.* **842** P1–100
[43] Rudman M and Blackburn H M 1999 *Second International Conference on CFD in the Minerals and Process Industries* (Melbourne)
[44] Schmitt L, Richter K and Friedrich R 1986 *Direct and Large Eddy Simulation of Turbulence* (*Notes on Numerical Fluid Mechanics* vol 3) ed Schumann U and Friedrich R (Wiesbaden: Vieweg+Teubner Verlag) pp 161–176
[45] Arnal M and Friedrich R 1993 *Turbulent Shear Flows 8* 1987 (Berlin, Heidelberg: Springer Berlin Heidelberg) pp 169–187
[46] Lilly D K 1967 *IBM Scientific Computing Symposium on environmental sciences* pp 195–210
[47] Prinz S, Boeck T and Schumacher J 2018 *Eur. J. Mech. B/Fluids* **submitted**
[48] Cimarelli A and De Angelis E 2014 *Phys. Fluids* **26**
[49] Rawat S, Cossu C and Rincon F 2017 *Procedia IUTAM* **20** 94–98

IOP Conf. Series: Journal of Physics: Conf. Series **1001** (2018) 012017 doi:10.1088/1742-6596/1001/1/012017

# Exploring the large-scale structure of Taylor–Couette turbulence through Large-Eddy Simulations

**Rodolfo Ostilla-Mónico**[1,2], **Xiaojue Zhu**[3], **Roberto Verzicco**[3,4]

[1] School of Engineering and Applied Sciences, Harvard U., Cambridge, MA 02138, USA.
[2] Dept. of Mechanical Engineering, U. of Houston, Houston, Texas 77204, USA.
[3] Physics of Fluids Group, Faculty of Science and Technology, MESA+ Research Institute, and J. M. Burgers Centre for Fluid Dynamics, U. Twente, PO Box 217, 7500 AE Enschede, The Netherlands.
[4] Dipart. Ing. Industriale, U. Rome "Tor Vergata", Via del Politecnico 1, Roma 00133, Italy

E-mail: rostilla@central.uh.edu

**Abstract.** Large eddy simulations (LES) of Taylor-Couette (TC) flow, the flow between two co-axial and independently rotating cylinders are performed in an attempt to explore the large-scale axially-pinned structures seen in experiments and simulations. Both static and dynamic LES models are used. The Reynolds number is kept fixed at $Re = 3.4 \cdot 10^4$, and the radius ratio $\eta = r_i/r_o$ is set to $\eta = 0.909$, limiting the effects of curvature and resulting in frictional Reynolds numbers of around $Re_\tau \approx 500$. Four rotation ratios from $Rot = -0.0909$ to $Rot = 0.3$ are simulated. First, the LES of TC is benchmarked for different rotation ratios. Both the Smagorinsky model with a constant of $c_s = 0.1$ and the dynamic model are found to produce reasonable results for no mean rotation and cyclonic rotation, but deviations increase for increasing rotation. This is attributed to the increasing anisotropic character of the fluctuations. Second, "over-damped" LES, i.e. LES with a large Smagorinsky constant is performed and is shown to reproduce some features of the large-scale structures, even when the near-wall region is not adequately modeled. This shows the potential for using over-damped LES for fast explorations of the parameter space where large-scale structures are found.

## 1. Introduction

Numerical simulation is a powerful tool for turbulence research because it provides us with the ability to reproduce virtually any quantity, including those which are unavailable in experiments. Direct numerical simulation (DNS) is certainly the best choice from a modeling perspective as it solves all the flow scales up to the smallest scale. However, DNS is computationally very expensive at high Reynolds number which prohibits simulations of practical applications of turbulence. Large eddy simulation (LES) is a less expensive alternative in which the fundamental idea is to reduce the computational cost by using a coarser discretization, and accounting for the small eddies, which are not resolved, by a subgrid-scale (SGS) model. In the last two decades, LES has been successfully implemented in many paradigmatic wall-bounded turbulent flows, such as channel flow [1], pipe flow [2] and boundary layer flow [3]. For additional information on the general topic of LES, we refer the readers to the book by Sagaut [4] or the review by Meneveau and Katz [5].

Here we focus on another paradigmatic flow, Taylor-Couette (TC) flow. TC flow is the flow between two co-axial and independently rotating cylinders, and it has attracted a lot of attention in recent years [6]. It is a popular toolbox for the new development of concepts in fluid dynamics, from instabilities, nonlinear dynamics, pattern formation- and turbulence, the subject of the present study. There are two distinct differences between TC turbulence and the previously mentioned paradigmatic wall-bounded flows: the first is the curvature of the walls and the second is the presence of a considerable mean rotation imposed on the mean shear flow. It is expected that these two effects will have huge impacts on the turbulence properties.

TC turbulence can be used as a benchmark for LES as it has two opposing curvatures, convex and concave at both cylinders, and also the amount of mean rotation can be easily varied using the differential rotation of the cylinders. Due to the high amount of symmetries, and natural confinement, TC is easily realized experimentally up to high Reynolds numbers: velocity profiles have been measured up to $Re \sim \mathcal{O}(10^6)$ [7], so we are not limited by the Reynolds numbers we can reach in experiments for validation.

In TC turbulence, persistent and spatially stable large scale known as turbulent Taylor vortices or Taylor rolls are known to exist. Understanding their origin is very important as it is one of the key mechanisms by which angular velocity is transported in TC flow [8, 9]. Recently, in turbulent channel and Couette flows, Hwang & Cossu [10, 11] showed that large-scale motions can exist even in the absence of small scales of turbulence. This was done by a method known as over-damped LES, which consists of using an LES filter without energy backscatter and artificially increasing its dissipation characteristics. In this way, the small scale motions are suppressed while the dissipation associated to them is kept.

However, different from pipe, channel or Couette flows, where large-scale motions contain only a significant amount of energy for sufficiently high Reynolds number [12], in TC turbulence the turbulent Taylor vortices are suspected to be simply a continuation of the laminar Taylor rolls, i.e. low Reynolds number structures originating from centrifugal instability. Therefore over-damped LES should always show these motions regardless of Reynolds number. Nevertheless, even at low Reynolds numbers Taylor rolls show substantial azimuthal modulation before the transition to chaos [13]. Overdamped LES can then help us elucidate if long-wavelength azimuthal modulations of the turbulent Taylor rolls exist, for which indications were found in Ref. [14].

In this study, we set out with two goals. First, we attempt to validate reference cases for LES with static Smagorinsky and dynamic SGS models in TC turbulence to test whether LES can work in the conditions of wall curvature and mean rotation. This sets a basis for the second step, namely performing simulations of overdamped LES, focusing on the modulations of turbulent Taylor vortices.

## 2. Numerical details

A second-order, centered finite difference code is used to solve the incompressible Navier-Stokes equations in cylindrical coordinates [15]. The resulting discretized system is solved by a fractional-step method: the time advancement is realized through a hybrid low-storage third-order Runge-Kutta approach. The convective terms are discretized by the second-order-explicit Adams-Bashforth scheme and the viscous terms are discretized by the second-order-implicit Crank-Nicholson scheme. One-dimensional MPI directives are adopted for parallelization, i.e. what is commonly known as a slab decomposition. In recent years, the code has been extensively validated and used for the DNS of TC turbulence [8, 16].

As stated before, two SGS models are implemented for LES. First, we use the static Smagorinsky model, including a van Driest damping function,

$$\nu_t = (c_s\Delta)^2 D(r^+)|S| = (c_s\Delta)^2 D(r^+)\sqrt{2S_{ij}S_{ij}} \tag{1}$$

with $S_{ij} = \frac{1}{2}\left(\frac{\partial u_i}{\partial x_j} + \frac{\partial u_j}{\partial x_i}\right)$ the resolved-scale strain rate tensor, the "filter" length scale $\Delta$ is calculated as $\Delta = (\Delta x_1 \Delta x_2 \Delta x_3)^{1/3}$, where $\Delta x_i$ is the grid width in the $i-$direction and $c_s$ is the Smagorinsky constant. The van Driest damping function $D$ is defined as

$$D(r^+) = 1 - \exp(r^+/25)^\alpha, \tag{2}$$

where $r^+$ is the wall-distance in viscous (inner) units, and $\alpha = 3$ controls the asymptotic behaviour close to the wall. As the asymptotic near-wall behaviour of the mixing length has been hypothesized to behave differently in TC flow [17], we simulate selected cases using an alternative van Driest model, denoted by $D'$, which uses $\alpha = 2$.

For additional benchmarking, we use the dynamic LES model [18] to control the possible effects not captured by the static model. The dynamic model has been used successfully by Piomelli *et al.* [19] to simulate rotating channel flow, and has been shown to overcome some of the shortcomings of the static model, which we will discuss later. For numerical stability, we average the resulting (space- and time-dependent) $C_s$ in the azimuthal coordinate.

The geometry considered is a Taylor-Couette system with a radius ratio $\eta = r_i/r_o$ of $\eta = 0.909$, where $r_i$ ($r_o$) is the inner (outer) cylinder radius. This value of $\eta$ limits the effects of curvature. Axially periodic boundary conditions, with a non-dimensional period of $\Gamma = L_z/(r_o - r_i) = 4.66$ are used. This is large enough to fit two Taylor roll pairs. A rotational symmetry order 10 is imposed on the flow, which results in a streamwise (azimuthal) extent of the domain of $2\pi$ gaps at the inner cylinder. A convective reference frame is used, such that the velocities of both cylinders are $\pm U/2$, separating differential rotation (shear) from mean rotation. The shear Reynolds number $Re_s = U(r_o - r_i)/\nu$, with $\nu$ the kinematic viscosity is fixed as $Re_s = 3.56 \cdot 10^4$. Four rotation numbers $Rot = 2\Omega_{rf}(r_o - r_i)/U$, with $\Omega_{rf}$ the rotation rate of the reference frame, are simulated: $Rot = 0$ (no mean rotation), $Rot = 0.0909$ (pure inner cylinder rotation), $Rot = 0.3$ (co-rotating cylinders) and $Rot = -0.0909$ (strongly counter-rotating cylinders) We note that positive values of $Rot$ denote *co*-rotating cylinders which have *anti*-cyclonic mean rotation, while counter-rotating cylinders correspond to cyclonic rotation, and that the rotation number is simply the inverse of the Rossby number $Rot = 1/Ro$.

These parameters result in frictional Reynolds numbers $Re_\tau = u_\tau(r_o - r_i)/(2\nu)$ of $Re_\tau \approx 500$, where $u_\tau$ is the frictional velocity $u_\tau = \sqrt{\tau_w/\rho}$, $\tau_w$ the mean shear at the wall and $\rho$ the fluid density. The shear at the outer cylinder has a geometrical factor $\eta$ with respect to the inner cylinder normalizations, neglected unless otherwise mentioned. The viscous length $\delta_\nu$ is defined as $\delta_\nu = \nu/u_\tau$. For convenience we also define the "outer" normalized radius $\tilde{r} = (r - r_i)/(r_o - r_i)$. The LES resolutions used are $N_\theta \times N_r \times N_z = 256 \times 128 \times 288$, which in viscous units correspond to $r_i\Delta\theta^+ \approx 24.5$, $\Delta r^+ \in (0.9, 11.0)$ and $\Delta z^+ \approx 16$. These resolutions are approximately twice as fine as the resolutions of Hwang & Cossu [10] (validated by Härtel and Kleiser [20]), as we use finite differences, in place of spectral methods.

Using the static model, we run a series of cases with $c_s \in [0, 0.6]$, going from $c_s = 0$ which switches off the SGS model, and is used for benchmarking purposes, to artificially high values of $c_s$. These are benchmarked against well-resolved DNS reference cases, which have half the domain size of the LES simulations in both streamwise and axial directions. This does not considerably affect mean and r.m.s. statistics [21]. The reference cases have resolutions of $N_\theta \times N_r \times N_z = 384 \times 512 \times 768$, which in non-dimensional terms is $r_i\Delta\theta^+ \approx 8.2$, $\Delta r^+ \in (0.2, 2.9)$ and $\Delta z^+ \approx 3$. Statistics for all simulations are collected until the radial flux of angular momentum $J^\omega = r^3(\langle u_r\omega\rangle_{t,\theta,z} - \nu\partial_r\langle\omega\rangle_{t,\theta,z})$ is constant within one percent, where $\langle ... \rangle_i$ denotes averaging with respect to variable $i$. The radial flux of angular momentum (torque at the cylinders) is non-dimensionalized as a quasi-Nusselt number $Nu_\omega = J^\omega/J^\omega_{pa}$ where $J^\omega_{pa}$ is the angular current for the laminar, purely azimuthal flow, $J^\omega_{pa} = (2\nu r_i^2 r_o^2[\omega_i - \omega_o])/(r_o^2 - r_i^2)$.

## 3. LES of Taylor-Couette flow with no mean rotation

First, we focus on TC flow with no mean rotation, i.e. $Rot = 0$. Table 1 shows $Nu_\omega$ for the different cases ran, both LES with the static and dynamic models, as well as the reference cases. The under-resolved reference case ($c_s = 0$) overestimates the torque. Increasing $c_s$ activates the SGS models and the torque decreases. This can be reasoned by noting that increasing $c_s$ increases the turbulent viscosity (stress), and as the stresses increase, the strains decrease as the LES model self-adjusts [22]. And as a consequence of the deceased strains, a lower torque results.

For the static model with $c_s = 0.1$ a good agreement is achieved with the reference data, regardless of the van Driest model used. The simulations with a static $c_s$ larger than 0.2, correspond to "overdamped" LES. These show torques which are more than 10% under the reference value. For $c_s \geq 0.4$, the underlying small-scale flows are almost completely the damped, and $Nu_\omega$ shows a drastic decrease for $c_s = 0.6$. These cases will be discussed in Section 5, for the time being, we focus on cases ran with conventional values of $c_s$ and use the simulations to assess the merits of LES of TC flow. The dynamic model also produces reasonable results within temporal averaging errors.

| $c_s$ | $Nu_\omega$ | Error (%) |
|---|---|---|
| DNS | 22.53 | – |
| 0 | 23.80 | 5.7 |
| 0.05 | 23.32 | 3.5 |
| 0.1 | 22.32 | 0.9 |
| 0.1 ($D'$) | 22.15 | 1.71 |
| 0.2 | 19.96 | 11.4 |
| 0.3 | 18.93 | 16.0 |
| 0.4 | 19.05 | 15.6 |
| 0.6 | 14.71 | 34.4 |
| Dynamic | 21.96 | 2.5 |

**Table 1.** Nusselt number comparison for the simulations ran with $Rot = 0$, and the reference DNS case.

The LES models also improve the agreement of the azimuthal (streamwise) mean velocity profiles, shown in outer units for selected cases in Figure 1. Unless otherwise mentioned, the averaging is always done in the azimuthal and spanwise directions, as well as in time. The velocity in the bulk close to the outer cylinder is slightly lower, and this can be attributed to centrifugal stabilization.

Figure 2 shows the azimuthal (streamwise) mean velocity in inner units for both the inner and the outer cylinder. A reasonable collapse with the DNS reference is observed for the static model with $c_s = 0.1$ and for the dynamic model. The case with $c_s = 0.05$ lies slightly below the reference, while the case with $c_s = 0.2$ lies slightly above. This can be easily explained as the additional turbulent viscosity close to the wall suppresses turbulent mixing and delays the transition to the buffer and logarithmic layers at slightly higher values of $r^+$. The underresolved case (not shown here for the sake of conciseness) would lie below the $c_s = 0.05$. The data for the $c_s = 0.1$ case with the alternate van Driest model collapses onto that of the normal model case at the same $c_s$ and is not shown. The profiles at the outer cylinder show a better match for a slightly lower value of $c_s$, due to the effect of centrifugal stabilization, but the differences are not substantial. This shows further proof that the LES self-adjusts when turbulent stresses are increased.

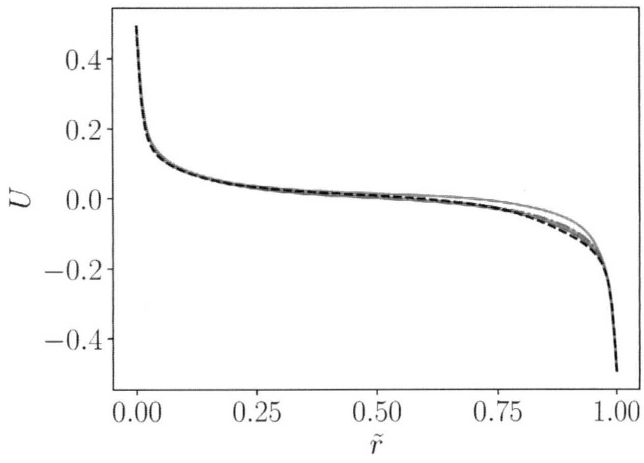


**Figure 1.** Mean azimuthal (streamwise) velocity profiles in outer units for selected cases including DNS reference for $Rot = 0$.

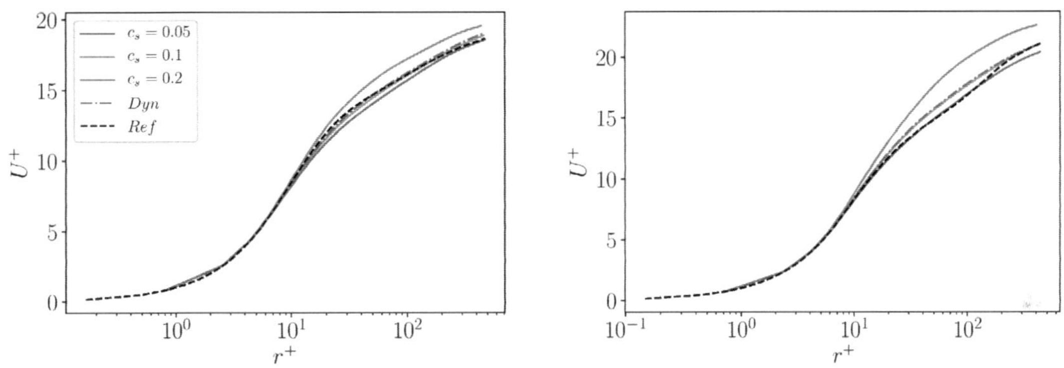


**Figure 2.** Mean azimuthal (streamwise) velocity profiles in wall units and for the inner cylinder (left) and outer cylinder (right) for selected cases including DNS reference for $Rot = 0$.

The effect of SGS models on the velocity fluctuations is shown in Figure 3, which shows the r.m.s. fluctuations for all three velocity components. From now on, we show only the inner cylinder as the profiles in the outer cylinder behave in a similar manner. The match between LES and the DNS reference is not as good as for the mean velocity, but could still be considered within acceptable realms. As expected, increasing $c_s$ reduces the level of fluctuations due to the additional turbulent viscosity. The magnitude of the $u'_\theta$ peak is slightly higher for the LES models, while the $u'_r$ fluctuations are clearly underestimated. This will later be shown to have an effect when $Rot \neq 0$. The axial (spanwise) velocity fluctuations also captured well.

These results show that for TC flow with $Rot = 0$, both the dynamic model and the static model with $c_s = 0.1$ (regardless of van Driest damping) show a reasonable match with the reference data. It could appear that the dynamic model produces no significant improvement over a static model with a calibrated constant, however, dynamic models are predictive while static models require calibration. We can directly look at the turbulent viscosity to explore

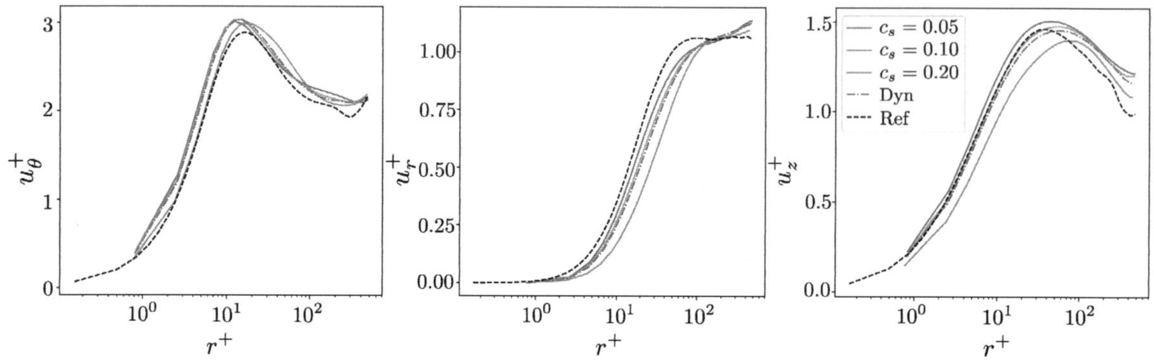


**Figure 3.** R.m.s. velocity fluctuation profiles in inner units for selected cases including DNS reference for $Rot = 0$.

the reasons for these effects. The left panel of Figure 4 shows the near-wall profiles of the averaged turbulent viscosity, while the other two panels show the axial and radial structure of the averaged turbulent viscosity for the static and dynamic models. The first thing to notice is that the $D'$ simulation (with a van Driest model with $\alpha = 2$) only shows a difference for low values of $r^+$, where the turbulent viscosity is negligible with respect to the molecular viscosity. The value of $\alpha$ in $D(r)$ controls the asymptotic behavior of the turbulent viscosity ($\sim r^{+\alpha}$), but the values close to the wall are very small anyway for the grid resolution used, and thus the choice of damping model does not seem to matter.

The choice of $c_s$ varies the turbulent viscosity by a constant factor and does not significantly change its spatial dependence. Both the static model with $c_s = 0.1$ and the dynamic model have similar effective turbulent viscosities, and both models capture the axial inhomogeneities of the flow: the footprint of the rolls can be clearly seen in the two right panels of Figure 4 which show the distribution of the azimuthally- and temporally-averaged turbulent viscosity. We note that both models only produce turbulent viscosities which are at most 30% of the molecular viscosity, and this could also explain the relatively small variance between the static and dynamic models.

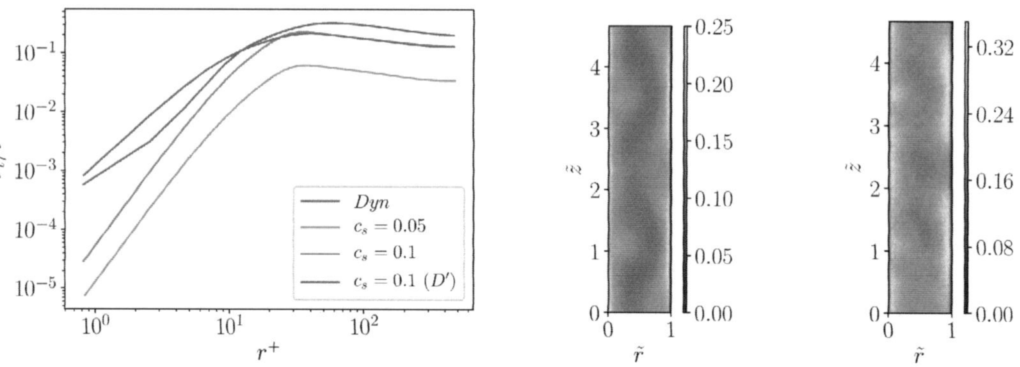


**Figure 4.** Left: Mean turbulent viscosity for selected cases. Middle and right: pseudocolor visualization of the azimuthally- and temporally averaged turbulent viscosity for the static model with $c_s = 0.1$ (middle) and the dynamic model (right).

## 4. The effect of mean rotation on Taylor-Couette LES

We now consider cases where mean rotation is present, i.e. $Rot \neq 0$. Table 2 shows the non-dimensional torque for the cases with rotation, including the reference datasets. Once again, the underresolved reference case ($c_s = 0$) overestimates the torque. The base errors are larger than in the previous case, where $Rot = 0$. Furthermore, with increasing $c_s$ the results do not converge towards the resolved reference case. For $Rot = 0.0909$ (i.e. pure inner cylinder rotation), some improvement is seen, especially for $c_s = 0.2 - 0.3$, as the same trend of decreasing torque for increasing $c_s$ is seen, but the error still remains around 8%, before increasing again for the "overdamped" cases. For $Rot = 0.3$ no significant decrease in the torque value is seen for any of the sub-grid stress models, as the LES model fails to self-adjust.

For the negative $Rot$ case, significant reductions in $Nu_\omega$ are seen for all cases when compared to the cases with positive or zero $Rot$. Strong cyclonic rotation stabilizes the flow, and, if curved walls are present, it enhances the asymmetry between inner and outer boundary layers due to centrifugal effects. The torque is overestimated by the case with no SGS model and the static model with $c_s = 0.05$, while the static case with $c_s = 0.1$ and the dynamic model produce reasonably accurate results, similar to the $Rot = 0$ cases. For the "overdamped" cases, the torque is found to be smaller than the laminar torque, again indicating that increased stresses lead to decreased strains.

| $c_s$ | $Nu_\omega$ for $Rot = -0.0909$ | $Nu_\omega$ for $Rot = 0.0909$ | $Nu_\omega$ for $Rot = 0.3$ |
|---|---|---|---|
| DNS | 9.73 | 23.75 | 23.94 |
| 0 | 9.97 | 32.05 | 31.90 |
| 0.05 | 9.96 | 31.32 | 31.82 |
| 0.1 | 9.76 | 29.20 | 31.53 |
| 0.1 ($D'$) | 9.74 | 29.13 | 31.68 |
| 0.2 | 7.59 | 25.51 | 32.03 |
| 0.3 | 0.97 | 25.09 | 32.76 |
| 0.4 | 0.32 | 26.55 | 30.60 |
| 0.6 | 6.21 | 30.90 | 31.68 |
| Dynamic | 9.58 | 29.55 | 31.16 |

**Table 2.** Nusselt number comparison for the simulations ran with $Rot \neq 0$, and the reference DNS case.

The mean azimuthal (streamwise) velocity profiles also deviate substantially from the DNS references. Figure 5 shows these profiles at the inner cylinder for all $Rot$ considered. For $Rot = -0.0909$, and small values of $c_s$, the velocity profiles have a weak dependence on $c_s$, and are close to the reference case. For $c_s = 0.3$ and $c_s = 0.4$, dispersive errors close to the wall cause anomalous torque measurements (measured by the wall-normal velocity gradient). A substantial amount of "wiggles" in the velocity field in the axial direction due to dispersive errors are also seen for the positive $Rot$ cases. Dispersive errors appear to increase with increasing $Rot$. This will be revisited in a later section.

For $Rot = 0.0909$, large deviations are seen for the base case. Increasing $c_s$ moves the curves towards the reference. For $c_s = 0.2$, which was found to produce the less erroneous torque, the deviations are smallest, even if they are still large on the graph. The deviations for $Rot = 0.3$ are even larger, and no improvement is seen modifying $c_s$. Neither the dynamic model nor any static model is able to correctly approximate the reference.

For now we focus on the cases with $Rot > 0$ to understand the deviations. Further indications of these deviations can be seen when looking at the velocity fluctuation profiles, shown in Figure 6. We only show $Rot = 0.3$ as it shows the largest deviations. The static model cases all seem to

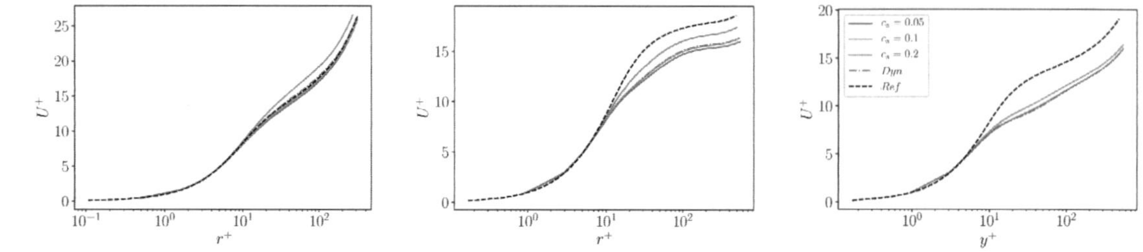


**Figure 5.** Averaged azimuthal (streamwise) velocity in inner units for (left) $Rot = -0.0909$, (middle) $Rot = 0.0909$ (pure IC rotation) and (right) $Rot = 0.3$ for selected cases including DNS reference.

overestimate the $u'_\theta$ (streamwise) velocity peak and locate it closer to the wall, while the dynamic case captures better its magnitude. The inverse happens for the $u'_r$ (wall-normal) fluctuation profile, which is underestimated by all cases, and most considerably by the dynamic model. For the $u'_z$ (spanwise) fluctuations, the static model captures the behaviour reasonably well for the low $c_s$ while the dynamic model again underestimates the fluctuations.

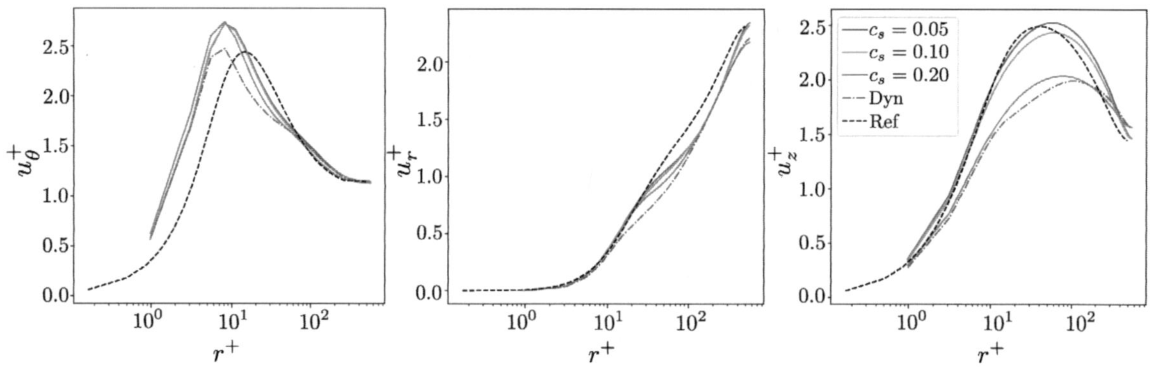


**Figure 6.** R.m.s. profiles of the velocity fluctuations for $Rot = 0.3$ for selected cases including DNS reference.

We can elucidate the reasons for the failure of the sub-grid stress models by looking at the turbulent viscosity. The left panel of Figure 7 shows the turbulent viscosity for the static model with $c_s = 0.1$, while the right panel shows the turbulent viscosity for the dynamic model. All static models produce a relatively similar turbulent viscosity regardless of *Rot*, and the cyclonic rotation tends to slightly decrease $\nu_t$ due to the smaller strains. For the static model, only the strain can change the turbulent viscosity, as the mixing length $(c_s^2\Delta^2)$ is kept constant. The averaged strain rate $|\mathcal{S}|$ is shown in the left panel of figure 8, and indeed it appears to be relatively independent of background rotation for $Rot \geq 0$. Therefore, this does not seem to be adequate to reproduce the dependence of the mean velocity profiles on *Rot*.

The dynamic model does show a dramatic increase for $\nu_T$ for $Rot = 0.3$, explaining why the velocity fluctuations for the dynamic model seen in 6 were much smaller than those for the static case. This adequately corrects for the magnitude of the peak of $u'_\theta$, but this also substantially constrains the radial velocity fluctuations, leading to their underestimation as seen in figure 6. The ordering of the curves is different with respect to the static model, and here the cyclonic case has the lowest turbulent viscosity, which is in line with what we expect from the smaller reference torques.

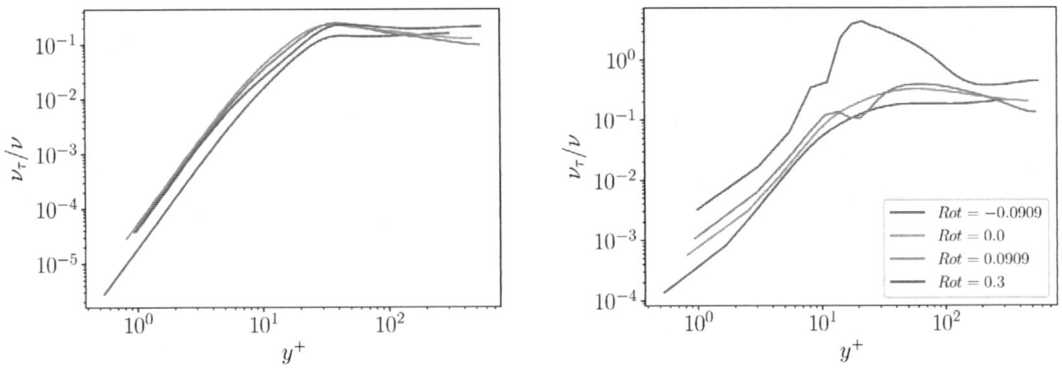


**Figure 7.** Turbulent viscosity for the static case with $c_s = 0.1$ (left) and the dynamic case (right) for the four rotation parameters.

The right panel of figure 8 shows the variations of the mixing length for the reference cases. This is a Reynolds averaged-like mixing length for the mean azimuthal velocity profile given by:

$$l_m = \frac{r_i u_\tau}{r^2}\left(\frac{\partial(U_\theta/r)}{\partial r}\right). \tag{3}$$

While this mixing length is a RANS-type mixing length resulting from a complete averaging of fluctuations, and thus cannot be directly compared to the LES mixing length, it can give an indication of why the LES subgrid models fail in rotating flows. The right panel of figure 8 shows a strong variation of the mixing length of almost of an order of magnitude, caused by the background anti-cyclonic rotation. The inset shows a zoom-in of the near-wall region, with the mixing length compensated so the differences can be clearly seen.

The mixing length is a proxy for fluctuations, both for the total fluctuations in the reference case and for the SGS fluctuations in the LES. The enhanced SGS fluctuations, modeled in the static LES by the constant mixing length $c_s^2\Delta^2$ multiplied by the average strain $|S|$ rate are not captured. The average strain, the only varying term in the turbulent viscosity equation, remains approximately the same in the near-wall region.

However, the dynamic model also fails. The dynamic model can reproduce aspects of the rotating flow, as the enhanced mixing length (cf. figure 7*b*)), but still yields unsatisfactory results. This is because the turbulent viscosity approximation assumes a degree of isotropy, but there is a clear anisotropy in the fluctuations, enhanced by rotation. The SGS cannot be adequately represented by an isotropic molecular viscosity. The stress-strain correlation, which is already low without rotation, could potentially be even worse with anti-cyclonic rotation: the failure of the LES models to decrease torque (strain) also points in this direction.

This is evident in the gross underestimation of radial and streamwise fluctuations in Figure 6 by the dynamic model. The two different directions of rotation act in opposite ways on the SGS. As cyclonic rotation tends to reduce the level of overall fluctuations, the SGS models become less important, and it can be seen that even the case with no SGS model has a small error for $Nu_\omega$ for $Rot = -0.0909$, but a large error for $Rot = 0.0909$. This reduced level of fluctuations can also be seen by the small mixing length of figure 8.

To show evidence of the anisotropy of the SGS, we filter a single DNS reference snapshot using a box-filter, by downsampling the resolution by a factor two in every direction. We then compute the values of the SGS for the single snapshot. Figure 9 shows two of these stresses: the radial and axial stresses of the azimuthal velocity. Aside from the saw-tooth oscillations of the

IOP Conf. Series: Journal of Physics: Conf. Series **1001** (2018) 012017 doi:10.1088/1742-6596/1001/1/012017

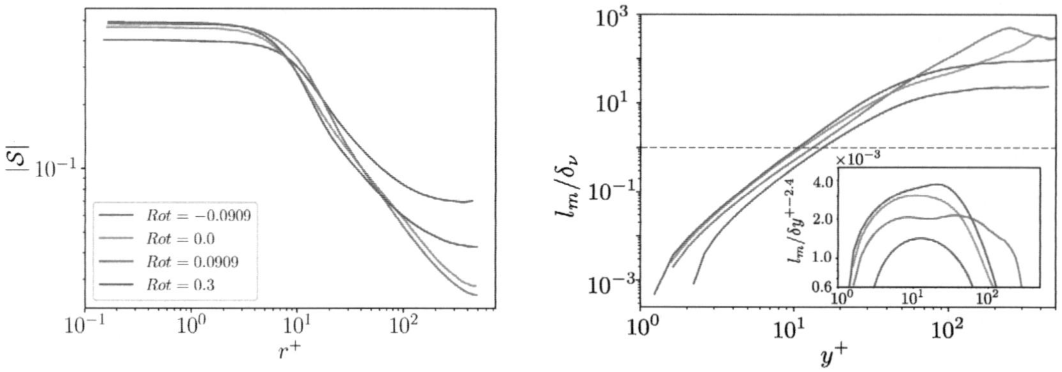


**Figure 8.** Left: Average strain rate for the reference cases and the three values of *Rot*. Right: mixing length for reference cases. Dashed line shows $l_m/\delta_\nu = 1$, separating the regions where the molecular viscosity is either larger or smaller than the equivalent turbulent viscosity. The inset shows a compensated $l_m/\delta_\nu$ highlighting the variation with *Rot*.

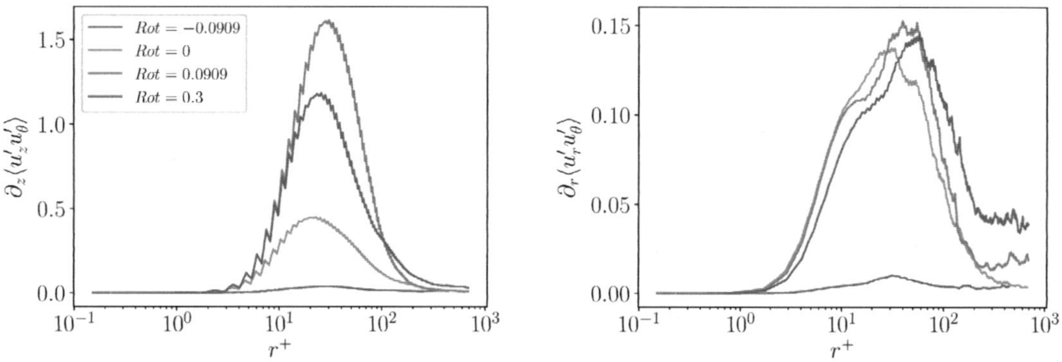


**Figure 9.** Two components of the SGS for all rotation ratios. The degree of anisotropy increases with increasing rotation.

graphs, due to the discrete nature of the box-size filter, two things can be noticed here: firstly, the $Rot = -0.0909$ case has extremely small SGS values, even if the average strain was seen in the left panel of 8 to be comparable to the other cases. This could explain the low values of $Nu_\omega$ seen for high $c_s$. Secondly, the values for the radial SGS in the azimuthal velocity are more or less constant with increasing *Rot*, but the values of the axial SGS depend substantially on *Rot*, and are almost five times larger for $Rot = 0.3$ than for the case with no rotation. Together, these graphs show that indeed, the anisotropy of the SGS increases with rotation. This provide a rationale for the failure of the LES, and also the high dispersive errors in the axial direction seen for anti-cyclonic rotation.

For completeness, we explore the effect of removing the mean Coriolis force and applying the rotation directly as a boundary condition. In principle, this reformulates the sub-grid Coriolis force fluctuations, a possible source of discrepancy, as part of the non-linear term. However, simulations in an inertial frame also incur in additional losses of accuracy. It has been shown that finite-differences are not Gallilean invariant, and that dispersive errors for second order finite differences are smallest when the mean velocity is close to zero [23]. Thus, by running in the inertial frame we amplify dispersive errors.

To check the balance of effects, we ran two additional cases for pure inner cylinder rotation in an inertial frame both with either a static Smagorinsky model and damping constants $c_s = 0.05$ and $c_s = 0.1$, as well as an underresolved reference case with no sub-grid model. The Nusselt numbers for the inertial frame and the convective frame cases are compared in Table 3. These can be benchmarked against the reference $Nu_\omega = 23.5$. In both cases, the errors are large, but we note that they are larger for the inertial frame as the increased dispersive errors dominates any possible effect which could have come from reformulating the Coriolis forces. Finally, we mention that preliminary results have shown that the axial (spanwise) resolution is especially critical for obtaining accurate results: this direction shows the highest dispersion in the velocity fields, especially for the large *Rot* cases.

| $c_s$ | $Nu_\omega$ (Convective frame) | $Nu_\omega$ (Inertial frame) |
|---|---|---|
| 0. | 32.1 | 34.0 |
| 0.05 | 31.3 | 33.2 |
| 0.1 | 29.4 | 31.1 |

**Table 3.** Nusselt number comparison for pure inner cylinder rotation in both convective and inertial reference frames.

## 5. Summary and outlook

In this report we have shown the possibility of using large eddy simulations to obtain reasonable results for Taylor-Couette flow at resolutions below what would be considered well resolved for a direct numerical simulation. LES was found to give good estimates of not only the torque and mean velocity profiles but also the fluctuations for cases with no mean rotation. Increasing anti-cyclonic rotation destroys the validity of the LES sub-grid models, as the linear relationship between sub-grid fluctuations and strain of the static model, as well as the isotropic turbulent viscosity hypothesis of both static and dynamic models come into question. The SGS models tend to severely underestimate the torque increases caused by anticyclonic rotation. The LES of cyclonic rotation is more accurate when compared to the DNS reference, as this rotation stabilizes the flow, and overall reduces the fluctuation level, so is less demanding on SGS models.

However, the main sources of discrepancy occur close to the wall, and could potentially not affect the behaviour of the large-scale motions, i.e. the Taylor rolls far from the wall. Therefore, overdamped simulations of Taylor-Couette flow which reveal the underlying large-scale structure appear to be possible.

In Figure 10 we show visualizations of the flow at the mid-gap, for both the resolved reference DNS (at a reduced domain) and the "overdamped" LES with the static model and $c_s = 0.6$ for three of the four values of *Rot* simulated. The figure shows that the underlying axial signatures of the large-scale motion are preserved at the mid-gap if they exist for the reference case. For $Rot = 0.0909$ where the signature is the strongest, the overdamped LES also preserves a very clear axially-pinned and quasi-axisymmetric structure, with certain spaced out blobs that could correspond to the wavyness found in Ref. [14].

For $Rot = 0$, the large-scale motion shows a larger degree of azimuthal fluctuations, which are captured by the overdamped LES. For the largest *Rot* simulated, no predominant large-scale motion is seen both in the reference case and in the overdamped LES. Overdamped LES appears to be a promising tool for rapidly exploring the parameter space to determine where axially pinned large-scale rolls are found in a faster manner than DNS.

IOP Conf. Series: Journal of Physics: Conf. Series **1001** (2018) 012017 doi:10.1088/1742-6596/1001/1/012017

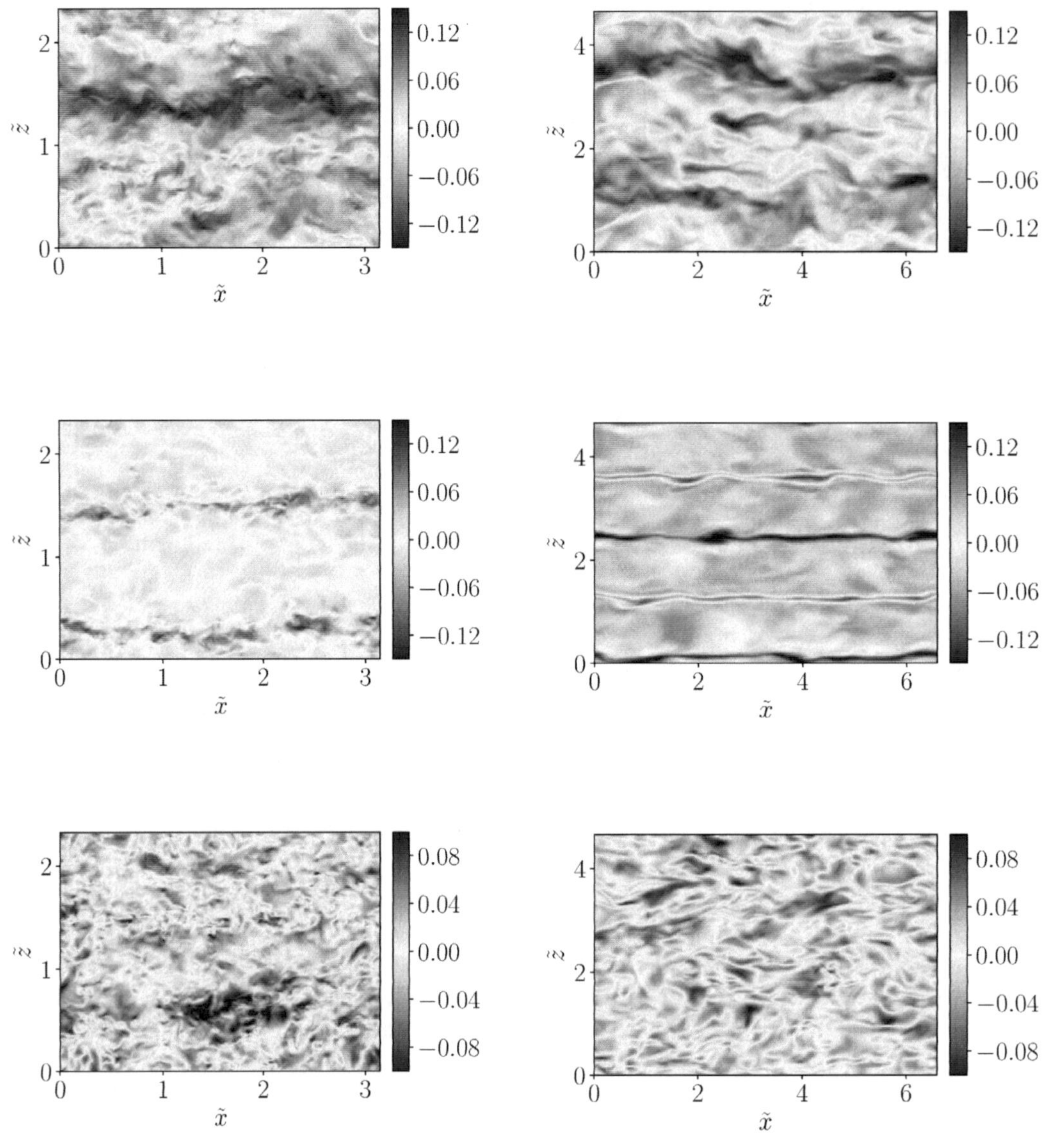


**Figure 10.** Pseudocolor visualization of $u_\theta$ at the mid-gap for the reference DNS cases (left) and the LES cases with the static model and $c_s = 0.6$ (right) for the three rotation numbers $Rot = 0$ (top), $Rot = 0.0909$ (middle) and $Rot = 0.3$ (bottom). The DNS domains are half the length in both spanwise and streamwise directions. The streamwise coordinate $\tilde{x}$ is defined as $\tilde{x} = r\theta/d$.

By this, we note that overdamped LES can reproduce some features of the large-scale motions seen in the mid-gap, and can be a useful tool for a fast exploration of the TC parameter space to check for the presence of axially pinned large-scale structures.

**Acknowledgments**
We thank the participants of the Third Madrid Turbulence Workshop, and especially T. Kwon and D. Feldmann for their comments on the manuscript. We also thank R. J. A. M. Stevens for valuable discussions. Funded in part by the Coturb program of the European Research Council. We also gratefully acknowledge computational time for the simulations provided by SURFsara on resource Cartesius through a NWO grant.

## References

[1] Moin P and Kim J 1982 *J. Fluid Mech.* **118** 341–377
[2] Rütten F, Schröder W and Meinke M 2005 *Phys. Fluids* **17** 035107
[3] Kosovic B 1997 *J. Fluid Mech.* **336** 151–182
[4] Sagaut P 1998 *Large Eddy Simulation for incompressible flows: An Introduction* (Berlin: Elsevier)
[5] Meneveau C and Katz J 2000 *Ann. Rev. Fluid Mech.* **32** 1–32
[6] Grossmann S, Lohse D and Sun C 2016 *Ann. Rev. Fluid Mech.* **48** 53–80
[7] Huisman S G, Scharnowski S, Cierpka C, Kähler C, Lohse D and Sun C 2013 *Phys. Rev. Lett.* **110** 264501
[8] Ostilla-Mónico R, Verzicco R, Grossmann S and Lohse D 2016 *J. Fluid Mech.* **768** 95–117
[9] Spandan V A, Ostilla-Mónico R, Lohse D and Verzicco R 2016 *J. Phys: Conf. Series* **708** 012006
[10] Hwang Y and Cossu C 2010 *Phys. Rev. Lett.* **105** 044505
[11] Hwang Y and Cossu C 2011 *Phys. Fluids* **23** 061702
[12] Smits A J, McKeon B J and Marusic I 2011 *Ann. Rev. Fluid. Mech.* **43** 353–375
[13] Andereck C D, Liu S S and Swinney H L 1986 *J. Fluid Mech.* **164** 155–183
[14] Ostilla-Mónico R, Verzicco R and Lohse D 2015 *Phys. Fluids* **27** 025110
[15] Verzicco R and Orlandi P 1996 *J. Comput. Phys.* **123** 402–413
[16] van der Poel E P, Ostilla-Monico R, Donners J and Verzicco R 2015 *Comput. Fluids* **116** 10–16
[17] Singh H, Suazo C A T and Liné A 2016 *Phys. Rev. E* **94** 053120
[18] Germano M, Piomelli U, Moin P and Cabot W H 1991 *Phys. Fluids A* **3** 1760–65
[19] Piomelli U and Liu J 1994 *AGARD Conf. Proc.* **551** 3
[20] Haertel C and Kleiser L 1998 *J. Fluid Mech.* **356** 327–352
[21] Ostilla-Mónico R, Lohse D and Verzicco R 2016 *Phys. Rev. Flu.* **1** 054402
[22] Jiménez J and Moser R D 2000 *AIAA J.* **38** 605–612
[23] Bernardini M, Pirozzoli S, Quadrio M and Orlandi P 2013 *J. Comput. Phys.* **232** 1–6

# Towards the computation of time-periodic inertial range dynamics

**L van Veen**[1], **A Vela-Martín**[2], **G Kawahara**[3]

[1] Faculty of Science, University of Ontario Institute of Technology, Oshawa, Ontario, Canada
[2] School of Aeronautics, Universidad Politécnica de Madrid, Madrid, Spain
[3] Graduate School of Engineering Science, Osaka University, Toyonaka, Osaka, Japan

E-mail: lennaert.vanveen@uoit.ca

**Abstract.** We explore the possibility of computing simple invariant solutions, like travelling waves or periodic orbits, in Large Eddy Simulation (LES) on a periodic domain with constant external forcing. The absence of material boundaries and the simple forcing mechanism make this system a comparatively simple target for the study of turbulent dynamics through invariant solutions. We show, that in spite of the application of eddy viscosity the computations are still rather challenging and must be performed on GPU cards rather than conventional coupled CPUs. We investigate the onset of turbulence in this system by means of bifurcation analysis, and present a long-period, large-amplitude unstable periodic orbit that is filtered from a turbulent time series. Although this orbit is computed on a coarse grid, with only a small separation between the integral scale and the LES filter length, the periodic dynamics seem to capture a regeneration process of the large-scale vortices.

## 1. Introduction

Starting with the pioneering work of Nagata on Couette flow [1], and the parallel development of numerical techniques by Cliffe and various collaborators (e.g. [2]), computational dynamical systems theory has made a number of valuable contributions to turbulence research. Perhaps the best example of the impact of this approach is found in the transition to turbulence in parallel shear flows in minimal flow units. This transition has been demystified by the computation and subsequent study of equilibrium and time-periodic flows that can co-exist with both the laminar flow and long-lasting turbulent transients. An overview of results in this vein was given by Kawahara *et al.* in 2012 [3]. Around the time this paper appeared, the focus of much research in this area shifted to the more challenging problem of transitional flows on larger or open domains. Duguet *et al.*, for instance, computed spatially localized solutions in Couette flow [4] and, more recently, Khapko *et al.* obtained similar results for the asymptotic suction boundary layer [5]. One conclusion that can be drawn from the latter works is that, when the number of Degrees Of Freedom (DOF) in the simulation is in the order of millions it get hard, if not impossible, to perform the dynamical systems-based computations with the same amount of rigour as in the less demanding case of minimal domains and marginal Reynolds numbers.

We can demonstrate the difficulty of computing, for instance, time-periodic orbits on large domains by combining some estimates for Homogeneous Isotropic Turbulence (HIT), in which the separation of spatial scales can be expressed asymptotically in terms of the Reynolds number. The commonly used estimate is that the ratio of the integral scale to the Kolmogorov scale behaves as $L/\eta \propto Re^{3/4} \propto Re_\lambda^{3/2}$, where $Re$ is a geometric Reynolds number and $Re_\lambda$ is Taylor's microscale Reynolds number [6]. In order to resolve the spatial scales of the fluid motion, we must thus let the linear grid size scale as $Re_\lambda^{3/2}$. As a consequence, the computation time of a single time step, using standard pseudo-spectral methods, will scale as $Re_\lambda^{9/2}$ up to a logarithmic correction. At the same time, the size of a single time step should decrease at least in proportion to the Kolmogorov time scale, namely as $Re_\lambda^{-1}$. The computational complexity of a simulation over a fixed time interval then scales as $Re_\lambda^{11/2}$.

In order to compute invariant solutions, we use Newton-Raphson iteration, or some variation thereof with an enlarged radius of convergence. To compute one such iteration, we need to solve a dense system of linear equations. Currently, the only known way to do this is by the use of a Krylov subspace method. Each Krylov iteration requires the simulation of the flow along the invariant solution, at a computational cost scaling as $Re_\lambda^{11/2}$. The number of Krylov iterations necessary to obtain convergence is largely determined by the spectral properties of the matrix of derivatives of the finite-time flow. The more clustered its eigenvalues, the faster the convergence, and clustering is a consequence of viscous damping [7]. As the Reynolds number increases, an increasing number of eigenvalues gets scattered across the complex plane, and it seems reasonable to assume that their numbers grows as $(L/\eta)^3 \propto Re_\lambda^{9/2}$.

The overall estimate is then that the computational complexity of a single Newton-like iteration, edging us closer to a time-periodic Navier-Stokes flow, will grow as $Re_\lambda^{10}$. The gap between computations in marginally turbulent flows, with $Re_\lambda = \mathrm{O}(10)$, and fully turbulent flows, with $Re_\lambda = \mathrm{O}(100)$ seems impossible to bridge. Moreover, there is no guarantee that the convergence properties of the Krylov and Newton iteration persist at higher Reynolds numbers. The accumulation of round-off error on various levels of the computation may lead to stagnation of the Krylov residual and linear, rather than quadratic convergence for Newton iteration.

In conclusion, it seems that we need to strike a compromise in order to study turbulent flow with a wide separation of length scales with the tools of dynamical systems. One possibility is to compute "shadows" of invariant solutions, e.g. time segments that are periodic up to a residual that is small compared to the variations along the orbit but large compared to that of the result of a convergent Newton-Krylov iteration. This approach was indeed taken by Duguet [4], Khapko [5], and others in the works cited above. In contrast to truly invariant solutions, the definition of a ghost is subjective. The former can be identified by estimating the minimal achievable residual given the finite-accuracy arithmetic and measuring the rate of convergence of the Newton iteration. The definition of the latter is necessarily based on some threshold value of the residual. A further disadvantage of this approach is that the persistence of ghosts when varying parameters is not guaranteed. While a truly periodic orbit can be tracked uniquely in the Reynolds number or other parameters of the flow, ghosts may simply disappear.

A different possibility is to compute invariant solutions in a model flow with fewer DOF. In particular, one can apply Large Eddy Simulation (LES) for this purpose. In LES, we introduce a filter length scale beyond which the dynamics is assumed to be strongly dissipative and well-represented by an effective eddy viscosity. The number of DOF in LES is thus greatly reduced as compared to Navier-Stokes flow with a comparable separation of length scales. Although the LES introduces extra nonlinearities into the equations of motion, the net effect is to reduce the time it takes to simulate the flow. Although LES does not alleviate the bounds on the time step size or the number of Krylov iterations, this reduction may just be sufficient to compute invariant solutions in fully turbulent flow.

This possibility was recently explored by Sasaki *at al.* [8] and by Hwang *at el.* [9] in channel flow. The presence of walls in this geometry brings a complication with it. Near the wall, where the rate of strain is high, the molecular viscosity is important. Thus, we must consider both molecular and eddy viscosity. Usually the regions where the two are dominant are glued together by an ad-hoc transition function. One can avoid these complications by considering a fluid on a periodic domain. Sekimoto and Jiménez [10] applied a homogeneous shear to such a spatially periodic fluid and managed to compute several equilibrium solutions and travelling waves. In their computations, the filter length is several time greater than the grid spacing, and not small enough compared to the integral scale to observe a power-law energy spectrum. Nonetheless, the invariant solutions exhibit interesting spatial structures and we can consider their results a proof-of-principle that interesting invariant solutions can be computed for a spatially periodic LES fluid.

The geometry we chose is simpler than the homogeneous shear layer. Instead of shear, we impose an external force that is constant in time and excites four counter-rotating vortex columns. The setup is identical to that of Yasuda *et al.* [11]. In their study, Yasuda *et al.* showed that the there are quasi-cyclic, large-amplitude oscillations in aggregate quantities like the energy and energy dissipation rate. These large-scale oscillations hint at the existence of a regeneration cycle for vortical structures in this flow, which could explain elements of inertial range dynamics. An obvious goal for the computation of simple invariant solutions in this flow is to capture the regeneration cycle with periodic orbits.

In the current paper we lay out the governing equations, briefly discuss earlier work on Navier-Stokes and LES flow in this geometry and discuss the behaviour of the LES fluid on a grid of $32^3$ points. We present one large-amplitude, long-period orbit that seems to capture the regeneration cycle to some extent, although it is computed in a regime with little scale separation. Finally, we speculate on the numerical requirements for reproducing similar invariant solutions on larger grids and in more turbulent flows.

## 2. Governing equations

The incompressible Navier-Stokes equation with periodic boundary conditions and body forcing is given by

$$\begin{aligned} \boldsymbol{u}_t + \boldsymbol{u}\cdot\nabla\boldsymbol{u} + \frac{1}{\rho}\nabla p - \nu\Delta\boldsymbol{u} &= \gamma\boldsymbol{f} \\ \nabla\cdot\boldsymbol{u} &= 0 \\ \boldsymbol{u}(x+L,y,z,t) = \boldsymbol{u}(x,y+L,z,t) = \boldsymbol{u}(x,y,z+L,t) &= \boldsymbol{u}(x,y,z,t), \end{aligned} \tag{1}$$

where $\rho$ is the fluid density, assumed to be constant, $\nu$ is the kinematic viscosity and $\gamma$ is the amplitude of the forcing.

We will numerically approximate solutions to system (1) on a regular grid of $n^3$ points. In order to reduce the number of degrees of freedom in the numerical simulations, we adopt the strategy proposed by Smagorinsky [12] to model the model the effect of motion on the smallest scales with an effective eddy viscosity. This approach can be interpreted as the application of a spatial filter given by a convolution

$$\bar{\boldsymbol{u}} = G * \boldsymbol{u}. \tag{2}$$

The kernel $G$ can be thought of as a low-pass filter with width $l_{\rm f}$. Since the spatial filter commutes with all spatial derivatives, the treatment of the linear terms in system (1) is straightforward. Filtering the advection term we obtain

$$\overline{u_i\partial_i u_j} = \partial_i\overline{u_i u_j} = \partial_i(\bar{u}_i\bar{u}_j) + \partial_i(\overline{u_i u_j} - \bar{u}_i\bar{u}_j) \equiv \bar{u}_i\partial_i\bar{u}_j + \partial_i\tau^{(\rm s)}_{ij}, \tag{3}$$

where summation over repeated indices is implied and the last equation defines the *subgrid stress* tensor $\boldsymbol{\tau}^{(\rm s)}$. It is split into a diagonal component, that modifies the pressure, and a deviatoric component, according to

$$\tau^{(\rm s)}_{ij} = \frac{1}{3}\Pi\delta_{ij} + 2\nu_T\bar{S}_{ij}, \tag{4}$$

where $\bar{S}_{ij} = (\partial_i \bar{u}_j + \partial_j \bar{u}_i)/2$ is the filtered rate-of-strain tensor, $\boldsymbol{\delta}$ is the Kronecker delta symbol and the *eddy viscosity* $\nu_T$ is defined by

$$\nu_T = (C_S \Delta)^2 \sqrt{2\bar{S}_{ij}\bar{S}_{ij}}. \tag{5}$$

This expression for $\boldsymbol{\tau}^{(s)}$, which contains a filtered term quadratic in $\boldsymbol{u}$, in terms of the filtered rate of strain closes the system of equations and constitutes the Smagorinsky model. The closure introduces a dimensionless parameter, $C_S$, called the *Smagorinky parameter.*

The resulting equations are

$$\partial_t \bar{\boldsymbol{u}} + \bar{\boldsymbol{u}}\cdot\nabla\bar{\boldsymbol{u}} + \nabla\left(\frac{\bar{p}}{\rho} + \frac{1}{3}\Pi\right) - 2\nabla([\nu_T + \nu]\bar{\boldsymbol{S}}) = \gamma\bar{\boldsymbol{f}} \tag{6}$$

$$\nabla\cdot\bar{\boldsymbol{u}} = 0. \tag{7}$$

We fix $\bar{\boldsymbol{f}} = (-\sin(k_f x)\cos(k_f y), \cos(k_f x)\sin(k_f y), 0)^t$ with $k_f = 2\pi/L$. This body force induces a flow with four counter-rotating vortex columns which, up to a slight deformation due to the eddy viscosity, is one of a family of flows studied by Taylor and Green[13].

In the following, we nondimensionalize the equations according to

$$\boldsymbol{x}' = k_f\,\boldsymbol{x} \qquad t' = \sqrt{L\gamma k_f^2}\,t \qquad \Pi' = \frac{1}{L\gamma}\Pi \tag{8}$$

$$\bar{\boldsymbol{u}}' = \sqrt{\frac{1}{L\gamma}}\,\bar{\boldsymbol{u}} \qquad \bar{p}' = \frac{1}{\rho L\gamma}\,\bar{p} \tag{9}$$

and drop both the over bars for spatially filtered quantities and the primes for non dimensional quantities. The resulting momentum balance equation is

$$\boldsymbol{u}_t + \boldsymbol{u}\cdot\nabla\boldsymbol{u} + \nabla\left(p + \frac{1}{3}\Pi\right) - 2\nabla\left(\left[\frac{1}{R^{3/2}}\nu_T + \frac{1}{Re}\right]\boldsymbol{S}\right) = \frac{1}{\alpha}\boldsymbol{f} \tag{10}$$

with non dimensional parameters

$$R = \left(\frac{1}{C_S \Delta k_f}\right)^{4/3} \qquad Re = \frac{\sqrt{L\gamma}}{\nu k_f} \qquad \alpha = L k_f = 2\pi. \tag{11}$$

Here, $R$ is the 'LES Reynolds number', which measures the ratio of the forcing length scale, $k_f^{-1}$, to $C_S\Delta$, which is the scale below which the eddy viscosity strongly damps motion. The effect of molecular viscosity is represented by the geometric Reynolds number $Re$. In the following we will take the limit of the latter to infinity or, equivalently, of vanishing molecular viscosity, and use $R$ as the control parameter. We will refer to the resulting system as governing an *LES fluid.* The other limit, in which $C_S = 0$ and $\nu > 0$, will be referred to as the DNS case.

Two quantities important for the analysis of the flow are the energy and enstrophy, which can be considered as the norm of velocity and vorticity, $\boldsymbol{\omega} = \nabla \times \boldsymbol{u}$, respectively:

$$E = \frac{1}{\alpha^3}\int \frac{1}{2}\|\boldsymbol{u}\|^2\,\mathrm{d}\boldsymbol{x} \qquad Q = \frac{1}{\alpha^3}\int \frac{1}{2}\|\boldsymbol{\omega}\|^2\,\mathrm{d}\boldsymbol{x}, \tag{12}$$

where the integral is taken over the entire domain and $\|.\|$ denotes the standard vector norm. In addition, we compute the rate of energy input,

$$e = \frac{1}{\alpha^4}\int \mathbf{u}\cdot\mathbf{f}\,\mathrm{d}\boldsymbol{x}, \tag{13}$$

and dissipation, or transfer to sub-filter scales,

$$\epsilon = \frac{2}{\alpha^3}\int \left[\frac{1}{R^{3/2}}\nu_T + \frac{1}{Re}\right] S_{ij}S_{ij}\,\mathrm{d}\boldsymbol{x}. \tag{14}$$

### *2.1. Symmetries*

It is straightforward to verify that equation (10) is equivariant under a group of symmetries generated by the following transformations:

- Translation over any distance $d$ in the vertical direction, $T_d$.
- Reflection in the $x$-direction, $S_x$.
- Reflection in the $y$-direction, $S_y$.
- Reflection in the $z$-direction, $S_z$.
- Rotation by $\pi/2$ about $x = y = 0$, followed by a shift over $L/2$ in the $x$-direction, $R$.
- A shift over $L/2$ in both the $x$ and $y$ directions, $D$.

In addition, we define the shift in time along a periodic orbit of period $P$ as $Q_\delta$ where $0 \leq \delta < P$. We will not consider Galilean boosts and in the simulation code the net flux in the vertical direction is identically equal to zero.

### *2.2. Numerical considerations*

For the purpose of numerical simulation the LES system is written on the Fourier basis

$$\boldsymbol{u} = \sum_{\boldsymbol{k}} \hat{\boldsymbol{u}}(\boldsymbol{k}) e^{i\boldsymbol{k}\cdot\boldsymbol{x}}, \tag{15}$$

where each sum is taken from $-n/2 + 1$ to $n/2$ for a resolution of $n^3$ grid points. The pressure is eliminated by formulating the system in terms of the vorticity. Two components of vorticity are time-stepped with a pseudo-spectral code using a fourth-order accurate Runge-Kutta-Gill scheme with a time step of 0.05 in non dimensional units. For de-aliasing we use the phase-shift method introduced by Patterson and Orszag[14]. The largest resolved wave number is $2\sqrt{2}/3 \times n/2 \approx 0.94 \times n/2$, where $n/2$ is the Nyquist wave number. The total number of degrees of freedom in a simulation is 28484 for $n = 32$, 230240 for $n = 64$ and 1839283 for $n = 128$. The exact same methods are used to propagate perturbations to the vorticity field and the system parameters using the tangent linear model.

Using the time-steppers for the vorticity and its linear perturbations we can solve boundary value problems in time to find equilibria and periodic orbits relative to the shift $T_d$. For this end, we use Newton-Krylov continuation. An review of this method was recently presented by Sánchez and Net[15], and we follow their notes, using, in addition, Viswanath's method to handle the translation symmetry [16].

The most time-consuming steps of the algorithm are the Fast Fourier Transforms (FFTs) and the reconstruction of the velocity from the vorticity. When implemented with process and thread parallel processing, the scaling is nearly linear up to $n/2$ cores. For instance, for $n = 64$ a single time step took 1.8(s) on 8 cores, 1(s) on 16 cores and 0.6(s) on 32 cores, using 3.3GH Intel Xeon processors. When implemented on a GPU architecture, the wall time per step was only 0.25(s) on a NVidia GeForce Titan X card. The large-amplitude, long-period orbit shown below required about 2000 time steps per integration, 100 Krylov iterations per Newton-hook step, and around 200 Newton-hook steps, for convergence. The GPU implementation enabled their computation in the course of four months.

## 3. Overview of qualitative dynamics

The governing equations in (10) constitute a dynamical system with two parameters, $Re$ and $R$. If either is small, the motion of the LES fluid will be laminar, closely resembling the external forcing pattern of counter-rotating vortex columns. If both are large, say of order O(100), the motion is turbulent. Figure 1 gives a schematic overview of the transition from laminar to

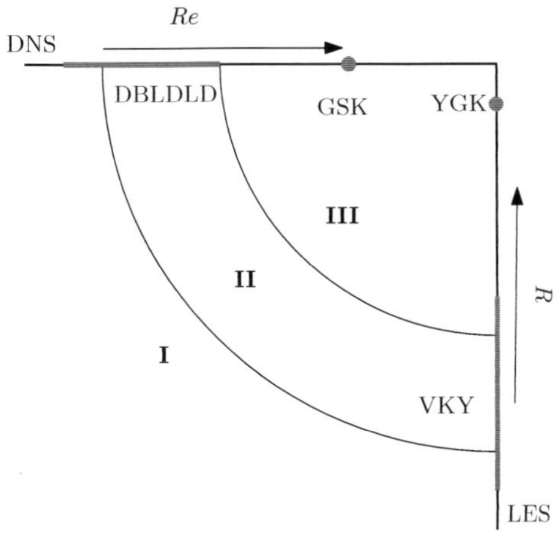


**Figure 1.** Schematic overview of the dynamics of four-vortex flow and related publications. The three zones are labeled **I** for laminar flow, **II** for periodic, homoclinic or heteroclinic motion or low-dimensional chaos and **III** for turbulence. The publications are abbreviated DBLDLD for Dubrulle *et al.* [17], GSK for Goto *et al.* [19], VKY for van Veen *et al.* [18] and YGK for Yasuda *et al.* [11].

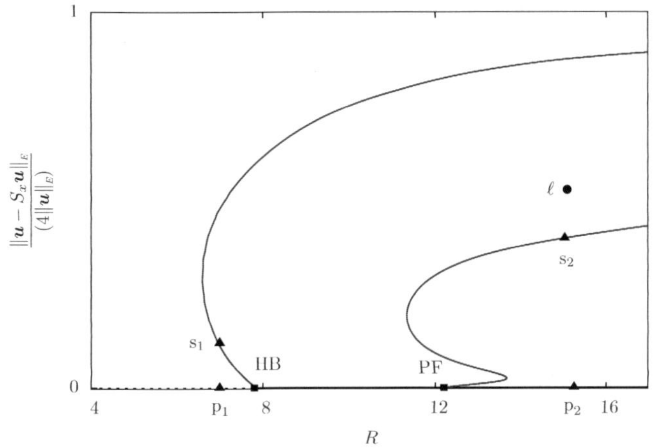


**Figure 2.** Partial bifurcation diagram of the LES fluid computed on $32^3$ grid points. Shown on the vertical axis is the deviation from reflection symmetry, normalized by its maximum value. Dashed lines denote stable solutions and solid lines denote unstable solutions. The squares denote instabilities of the laminar flow, the first being a Hopf bifurcation (labeled HB) and the second a pitchfork (labeled PF). The labels $p_{1,2}$ on the primary branch and $s_{1,2}$ on the secondary branches correspond to the physical space portraits in figures 3-5. The label $\ell$ corresponds to the long-period, large-amplitude unstable periodic solution portrayed in figure 7.

turbulent motion in parameter space. In red, we have indicated parameter values or ranges that have been covered in earlier publications. For the DNS case, Dubrulle *et al.* [17] got an overview of the transition by analysing time series [17]. Their forcing mechanism is different from ours as they keep the Fourier coefficients corresponding to forcing function $\boldsymbol{f}$ constant in time rather than applying a body force. Nevertheless, the qualitative dynamics seems to be similar: they observe a number of equilibrium and time-periodic solutions that are marginally unstable and

IOP Conf. Series: Journal of Physics: Conf. Series **1001** (2018) 012018 doi:10.1088/1742-6596/1001/1/012018

have a smooth spatial structure [17]. The state of the fluid visits several such invariant solutions in long transients. For higher $Re$, chaotic states are observed. Similar behaviour was discovered in the LES case by van Veen *et al.*, who performed a numerical bifurcation analysis. A detailed report can be found in reference [18], while a brief overview is presented in the next section.

For higher Reynolds numbers, say $Re > 50$ or $R > 50$, the fluid motion is chaotic and gets progressively more complicated as the separation between the integral scale and the dissipation scale, set by the Kolmogorov length in the DNS case and by $C_S\Delta$ in the LES case, grows. Two studies investigated the turbulent dynamics. Goto *et al.* focused on the dynamics of vortical structures across spatial scales [19]. The mechanism of energy transfer they highlight is characterized by anti-parallel vortex pairs, created in the strain field of vortex pairs on a scale two to eight times larger. In principle, this process can be repeated to transfer energy to progressively smaller scales across the inertial range. The other study considers the LES case and focuses on the long-period, large-amplitude oscillations exhibited by the energy and energy dissipation rate [11]. These oscillations correspond to a regeneration cycle of the largest-scale vortices and hints at the possibility of capturing elements of the turbulence by unstable periodic orbits. The ultimate goal of the study of invariant solutions in the LES case is to identify one or more solutions that capture the large-amplitude oscillations and vortex dynamics.

## 4. Examples of simple invariant solutions

Up to $R = 7.45$, the laminar flow is stable. A portrait of the stable laminar flow, which is invariant under the symmetry operations listed in §2.1, is shown in figure 3. At $R = 7.45$, this equilibrium becomes unstable in a Hopf bifurcation that produces a family of unstable periodic orbits that is translation invariant in the vertical direction. Two snap shots of a representative orbit are shown in figure 4. These periodic orbits are invariant under the symmetries when combined with a shift in time over half the period. To be precise, they are invariant under $T_d$, $S_z$, $S_x \circ Q_{P/2}$, $S_y \circ Q_{P/2}$, $R \circ Q_{P/4}$ and $D \circ Q_{P/2}$.

The second instability of the laminar equilibrium is a pitchfork bifurcation that gives rise to a family of equilibria with a three-dimensional structure. A portrait taken at $R = 15.25$ is shown in figure 5. The large-scale vortices are bent into a corkscrew-like shape with dominant wave number two in the vertical direction. In addition, pairs of smaller, counter-rotating vortices have appeared in the perpendicular directions. The latter are reminiscent of so-called *ribs* that have previously been identified in Taylor-Green flow as a consequence of an instability concentrated at the hyperbolic stagnation points $(\pi/4 \pm \pi/4,\ \pi/4 \pm \pi/4,\ z)$ [20]. The three-dimensional equilibria are invariant under the transformations $S_x \circ S_y \circ T_{L/4}$, $D$ and $R \circ T_{L/8}$.

The periodic orbits and equilibria that branch off the laminar flow do not capture the large fluctuations the the rate of energy input and transfer to sub-grid scales observed in weakly turbulent motion. We computed the Probability Density Function (PDF) for turbulence at $R = 15.25$. It is shown in figure 6, along with the laminar and three-dimensional equilibrium. The former has high rates of energy input and transfer, while the latter has low rates. In time-series, large oscillations that approach these states intermittently can be observed. We used one such spontaneous, large-amplitude oscillation to compute an unstable periodic orbit, shown in figure 6 in red, with dots evenly spaced in time. Its period is about 46 large-eddy turnover times, the latter computed from the root-mean-square velocity of turbulent motion and the domain size. In figure 7, we have extracted four snapshots, labeled **a**–**d**. At label **a**, where the rate of energy input exceeds the rate of transfer, the vertical vortex columns are growing in strength under the influence of the forcing. At **b**, the vortex columns have gained strength, but they are not stable. Pairs of smaller vortices are growing in the perpendicular directions, extracting energy from the large-scale ones. This process accelerates in the third phase, around snapshot **c**, when the rate of energy transfer is relatively high as small-scale vortices are damped through eddy viscosity. Finally, near the minimum of the energy input rate, the flow approaches the three-dimensional

IOP Conf. Series: Journal of Physics: Conf. Series **1001** (2018) 012018 doi:10.1088/1742-6596/1001/1/012018

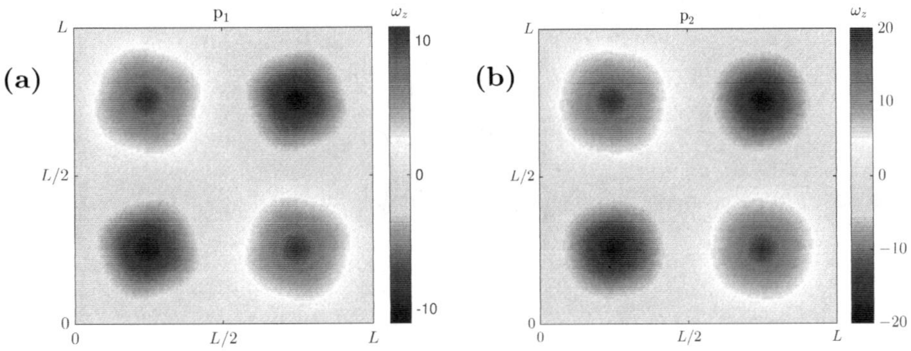


**Figure 3.** Physical space portrait of the viscous equilibrium at: (a) $R = 7$; (b) $R = 15.25$. Shown is the vertical vorticity in a constant-$z$ plane. The labels correspond to those in figure 2.

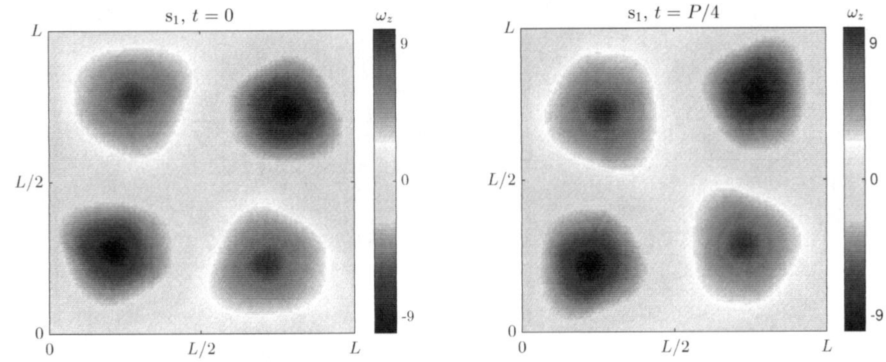


**Figure 4.** Physical space portraits of the secondary branch consisting of two-dimensional periodic orbits. Shown is the vertical vorticity in a plane of constant height $z$. Two snapshots are shown, separated by a quarter of the period. The label corresponds to the one in figure 2.

equilibrium, which is also unstable. Both the large-scale and the smaller-scale vortices are weak and the external force is starting to restore the large-scale structure.

Perhaps the most interesting segment of the periodic orbit is that between snapshots **b** and **d**. Here, the instability of the large-scale vortices induces the growth of pairs of smaller vortices, much like described by Goto *et al.* [19]. Of course, at this low LES Reynolds number, we can only see one step of this process. In order to see inertial range dynamics, we would need to increase it to $R \approx 70$ on the grid of $32^3$ points. However, in spite of the small grid this will be a difficult task. The computation of the periodic orbit presented here took 200 Newton-hook iterations with, on average, 100 Krylov iterations. Since it takes about 2000 time steps to compute one Krylov iteration, the total computation time was close to four months.

## 5. Outlook

We have presented an overview of the behaviour of the LES of four-vortex flow on a grid of $32^3$ points and at low LES Reynolds numbers. The "derivative solutions", i.e. the solutions that bifurcate off the laminar flow, could easily be computed on up 16 CPUs. In contrast, the long-period, large-amplitude solution, which seems well embedded in the ambient turbulence, took about 20,000 simulations over 2000 time steps to iteratively refine. Only by using GPU parallelism were we able to compute it in the course of months rather than years. Assuming the the number of Krylov iterations per Newton (or Newton-hook) step will grow as $[L/(C_S\Delta)]^3$, a

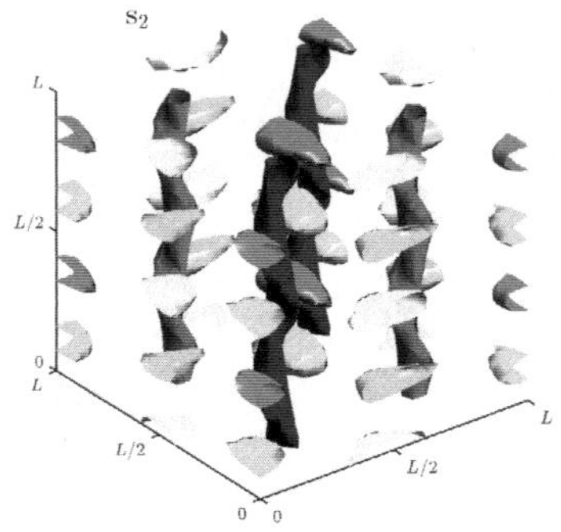


**Figure 5.** Physical space portrait of the three-dimensional equilibrium labeled $s_2$ in figure 2. Shown are the isosurfaces of $z$-vorticity (red and blue for positive and negative), $x$-vorticity (green and yellow for positive and negative) and $y$-vorticity (cyan and magenta for positive and negative) at 75% of the maximal and minimal values.

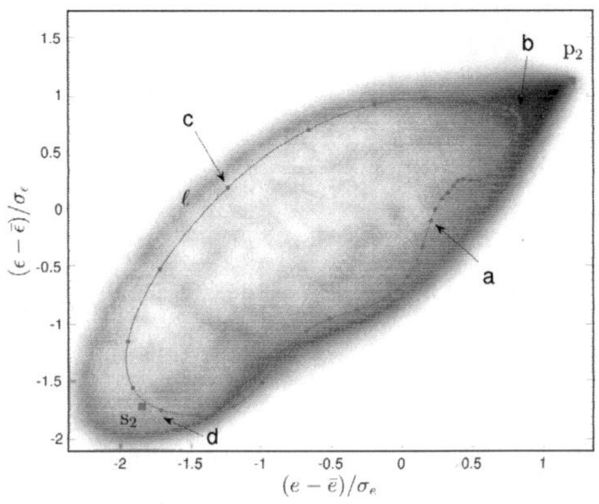


**Figure 6.** Projection of the PDF of weak turbulence, the large-amplitude periodic orbit $\ell$, the laminar equilibrium $p_2$ and the three-dimensional equilibrium $s_2$ on the rate of energy input and transfer to sub filter scales. Shown is the deviation from their respective time mean values, normalized by their standard deviation. The red dots on the periodic orbit are drawn at fixed time intervals to show the slowing down of the phase point in the vicinity of the equilibria. The PDF is shown on a logarithmic grey scale.

similar computation in the regime of developed turbulence would take about 30 times longer, which is taxing. Moving up to a grid of $64^3$ points seems out of the realm of possibility. The estimate of the number of Krylov iterations is, however, probably overly pessimistic. Also, there may be preconditioning techniques that can be applied to alleviate this problem.

Our current approach is to compute invariant solutions on the small grid with $32^3$ points at a LES Reynolds number as high as possible, and then recomputing it on a finer grid with the same value of $C_S\Delta$. Subsequently, the filter length must be reduced by continuation. Whether this approach will finally lead to an invariant solution with properties of developed turbulence remains to be seen.

**Acknowledgements**
This work was funded by the Coturb program of the European Research Council. We are grateful to Dr. Atsushi Sekimoto for his helpful comments on the manuscript.

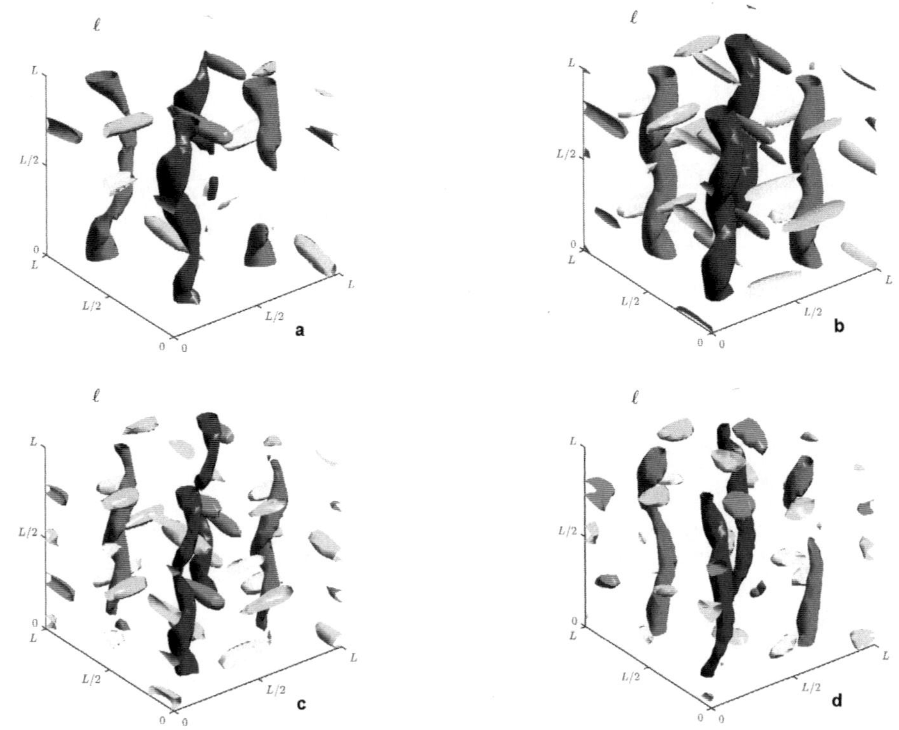


**Figure 7.** Physical space portrait of the long-period, large-amplitude unstable periodic solution labeled $\ell$ in figure 2. The labels **a**–**d** correspond to those in figure 6. The isosurfaces are drawn as in figure 5.

## References


[1] Nagata M 1990 *J. Fluid Mech.* **217** 519–527
[2] Mullin T, Cliffe K A and Pfister G 1987 *Phys. Rev. Lett.* **58** 2212–2215
[3] Kawahara G, Uhlmann M and van Veen L 2012 *Ann. Rev. Fluid Mech.* **44** 203–225
[4] Duguet Y, Schlatter P and Henningson D S 2009 *Phys. Fluids* **21** 111701
[5] Khapko T, Kreilos T, Schlatter P, Duguet Y, Eckhardt B and Henningson D S 2013 *J. Fluid Mech.* **717** R6
[6] Constantin P, Foias P, Manley O P and Temam R 1985 *J. Fluid Mech.* **150** 427–440
[7] Sánchez Umbría J, Net M, Garcia-Archilla B and Simó C 2004 *J. Comput. Phys.* **201** 13–33
[8] E Sasaki G Kawahara A S and Jiménez J 2016 *J. Phys. Conf. Ser.* **708** 012003
[9] Hwang Y, Willis A P and Cossu C 2016 *J. Fluid Mech.* **802** R1
[10] Sekimoto A and Jiménez J 2017 *J. Fluid Mech.* **827** 225–249
[11] Yasuda T, Goto S and Kawahara G 2014 *Fluid Dyn. Res.* **46** 061413
[12] Smagorinsky J 1963 *Mon. Weather Rev.* **91** 99–164
[13] Taylor G I and Green A E 1937 *P. Roy. Soc. Lond. A Mat.* **158** 499–521
[14] Patterson G S and Orszag S A 1972 *Phys. Fluids* **14** 2538–2541
[15] Sánchez Umbría J and Net M 2016 *Eur. Phys. J.-Spec. Top.* **225** 2465
[16] Viswanath D 2007 *J. Fluid Mech.* **580** 339–358
[17] Dubrulle B, Blaineau P, O M Lopes O, F Daviaud F, Laval J P and Dolganov R 2007 *New J. Phys.* **9** 308
[18] van Veen L, Kawahara G and Yasuda T 2017 Transitions in large eddy simulation of box turbulence URL `http://arxiv.org/abs/1711.02289`
[19] Goto S, Saito Y and Kawahara G 2017 *Phys. Pev. Fluids* **2** 064603
[20] Leblanc S and Godeferd F S 1999 *Phys. Fluids* **11** 497–499

IOP Conf. Series: Journal of Physics: Conf. Series **1001** (2018) 012019 doi:10.1088/1742-6596/1001/1/012019

# Linearised dynamics and non-modal instability analysis of an impinging under-expanded supersonic jet

**Shahram Karami[1], Paul C. Stegeman[1], Vassilis Theofilis[2], Peter J. Schmid[3] , Julio Soria[1,4]**

[1]Department of Mechanical and Aerospace Engineering, Monash University, Melbourne, Australia
[2]School of Engineering, University of Liverpool, UK
[3]Department of Mathematics, Imperial College London, London, UK
[4]Department of Aeronautical Engineering, King Abdulaziz University, Jeddah, Kingdom of Saudi Arabia

E-mail: shahram.karami@monash.edu

**Abstract.** Non-modal instability analysis of the shear layer near the nozzle of a supersonic under-expanded impinging jet is studied. The shear layer instability is considered to be one of the main components of the feedback loop in supersonic jets. The feedback loop is observed in instantaneous visualisations of the density field where it is noted that acoustic waves scattered by the nozzle lip internalise as shear layer instabilities.

A modal analysis describes the asymptotic limit of the instability disturbances and fails to capture short-time responses. Therefore, a non-modal analysis which allows the quantitative description of the short-time amplification or decay of a disturbance is performed by means of a local far-field pressure pulse. An impulse response analysis is performed which allows a wide range of frequencies to be excited. The temporal and spatial growths of the disturbances in the shear layer near the nozzle are studied by decomposing the response using dynamic mode decomposition and Hilbert transform analysis.

The short-time response shows that disturbances with non-dimensionalised temporal frequencies in the range of 1 to 4 have positive growth rates in the shear layer. The Hilbert transform analysis shows that high non-dimensionalised temporal frequencies ($>4$) are dampened immediately, whereas low non-dimensionalised temporal frequencies ($<1$) are neutral. Both dynamic mode decomposition and Hilbert transform analysis show that spatial frequencies between 1 and 3 have positive spatial growth rates. Finally, the envelope of the streamwise velocity disturbances reveals the presence of a convective instability.

## 1. Introduction

The scenario of a supersonic jet impingement on a ground surface is inevitable in new aircraft technologies where short distance and vertical take-offs and landings are desirable. The intense pressure waves in such flows result in numerous challenges. The high level of oscillations on the landing surface damages the surface structure and can lead to fatigue of the aircraft structure as the pressure waves are reflected by the surrounding surfaces. Environmental and safety problems need to be addressed because the resulting noise and pressure waves are intense. The impingement creates an extremely complicated flow field with a number of interconnected flow

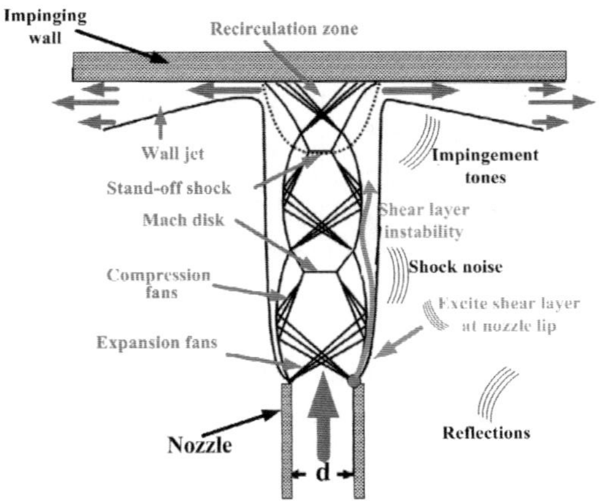


Figure 1: Schematic of a impinging jet.

physics involved. Specifically, development of shear layer instabilities that evolve into vortices, vortex-shock interaction, as well as an axisymmetric stagnation flow and a wall jet are present in this flow.

An important phenomenon is the instability waves that are the result of acoustic waves being scattered by the nozzle lip and grow as they propagate in the shear layer of the jet which is sketched in figure 1. The interactions of the sound and the jet form a feedback loop. The shear-layer disturbances develop in the form of a Kelvin-Helmholtz instability and interact with the Mach disk and stand-off shock which produces shock movements that alter the flow characteristics. This pulsatile motion creates high intensity acoustic waves which travel upstream and perturb the shear layer, closing the feedback loop. Considerable evidence for the feedback loop being the mechanism responsible for the self-sustained oscillations has been accumulated in a number of experimental studies of supersonic impinging jets [1, 2, 3, 4, 5, 6]. However, our understanding of the feedback loop is not complete as the process is sensitive to jet inlet conditions and geometrical boundary conditions. The flow field and hence, the level of pressure oscillations depend on the nozzle pressure ratio (NPR)[1] and the distance between the nozzle exit and the wall. The instability is intensified when the nozzle exit to wall distance (h/d) is shorter than the potential core [2] [7, 8]. In experimental studies of under-expanded supersonic impinging jets with NPR $\approx 4$, three tone-producing regimes are observed. Symmetrical oscillations occur for $x/d < 2.7$, a zone of silence takes place for $2.7 < x/d < 3.5$, and normal tone-producing regimes are active for $3.5 < x/d < 3.8$ [7].

The propagation of the instabilities that were internalised due to the acoustic waves being scattered by the nozzle lip is one of the main components of the feedback loop. The time development of these instabilities can be modeled by linearised Navier Stokes equations. Depending on the research questions two main approaches have been developed to study the process. The first class of the linear stability analysis which deals with the response behaviour of the base flow to disturbances as time approaches to infinity is called modal analysis [9]. The second method is interested in the short-time behaviour of disturbances known as non-modal analysis. Comprehensive reviews of the modal and non-modal methods are presented in the review papers by Theofilis [10] and Schmid [11], respectively. The modal instability analysis

[1] The nozzle pressure ratio (NPR) is here defined as the ratio between the stagnation pressure measured in the jet plenum and the ambient pressure.

[2] The potential core of the jet is the distance from the nozzle where the jet centerline velocity is relatively constant.

leads to a generalized large-scale eigenvalue problem. A solution of this sparse eigenvalue problem provides the modal frequencies and amplification/damping rates. Therefore, the modal instability analysis describes the asymptotic fate of the perturbation. It appears that this asymptotic solution and the shape of the least stable mode may not describe the overall perturbation dynamics accurately and analysis of the instability growth and decay in a finite-time horizon is necessary [11]. Therefore, the subsequent non-modal shear layer instability, which explains the short-time responses of the shear layer to external disturbances, is the principle gap that the present paper tries to address.

## 2. Governing equations

The compressible conservation equations of mass, momentum, and total energy in cylindrical coordinates are the governing equations that apply to supersonic under-expanded impinging jets. These equations are non-dimensionalised with respect to the ambient conditions. The non-dimensional variables are defined as follows,

$$\mathbf{x} = \frac{\mathbf{x}_i^*}{d}, \quad \mathbf{u} = \frac{\mathbf{u}^*}{a_o^*}, \quad \rho = \frac{\rho^*}{\rho_o^*}, \quad t = \frac{t^* a_o^*}{d}, \quad p = \frac{p^*}{\rho_o^* a_o^{*2}},$$
$$T = \frac{T^*}{T_o^*}, \quad \mu = \frac{\mu^*}{\mathrm{Re}\mu_o^*}, \quad \mathrm{Pr} = \frac{\upsilon^*}{\alpha^*}, \quad M_o = \frac{a_o^*}{\sqrt{RT_o^*}} \quad \mathrm{Re} = \frac{\rho_o^* a_o^* d}{\mu_o^*}. \tag{1}$$

Here, the superscripts '*' and 'o' represent the dimensional values and ambient conditions, respectively. **x** represents the spatial vector; **u** the velocity vector; $d$ the jet diameter (see figure 3); $a$ the speed of sound; $\rho$ the density; $T$ the temperature; $t$ the time; $p$ the pressure; and $\mu$ the dynamic viscosity, which is given by $\mu/\mu_o = (T/T_o)^{0.7}$ [12]. The variable Re represents the Reynolds number and Pr the Prandtl number. The non-dimensional forms of the equations are

$$\begin{aligned}
\frac{\partial \rho}{\partial t} &= -\nabla \cdot (\rho \mathbf{u}), \\
\frac{\partial \rho \mathbf{u}}{\partial t} &= -\nabla \cdot (\rho \mathbf{u} \otimes \mathbf{u} + p\mathbb{I} - \mu \mathbb{T}), \\
\frac{\partial \rho e}{\partial t} &= -\nabla \cdot [\mathbf{u}(\rho e + p) - \mathbf{q} - \mu \mathbb{T} \cdot \mathbf{u}].
\end{aligned} \tag{2}$$

where the deviatoric stress tensor, $\mathbb{T}$, and heat flux, **Q** are defined as

$$\begin{aligned}
\mathbb{T} &= \left[\nabla \mathbf{u} + (\nabla \mathbf{u})^T + \left(\frac{\mu_b}{\mu} - \frac{2}{3}\right)(\nabla \cdot \mathbf{u})\mathbb{I}\right], \\
\mathbf{Q} &= \frac{\mu \gamma}{\gamma - 1} \frac{1}{M_o^2 Pr} \nabla T.
\end{aligned} \tag{3}$$

Here,$\gamma$ =1.4 is the heat capacity ratio of air, $e$ is the sensible energy, and the density weighted sensible energy is given by

$$\rho e = \frac{1}{2}\rho \mathbf{u} \cdot \mathbf{u} + \frac{\rho T}{M_o^2(\gamma - 1)} = \frac{1}{2}\rho \mathbf{u} \cdot \mathbf{u} + \frac{p}{\gamma - 1}. \tag{4}$$

The ideal gas assumption is used in this study and the equation of state is $p = \rho T/M_o^2$. Further details of the equations in cylindrical coordinates are found in Ref. [13]. Equation (2) is commonly presented as an evolution equation in the following compact form

$$\frac{\partial \mathbf{Q}}{\partial t} = \mathbf{F}(\mathbf{q}), \tag{5}$$

IOP Conf. Series: Journal of Physics: Conf. Series **1001** (2018) 012019 doi:10.1088/1742-6596/1001/1/012019

From which the linearised Navier Stokes equations (LNSE) are derived by superimposing small amplitude disturbances on a mean flow. Denoting the mean flow variables as $\overline{(...)}$ and perturbations as $(...)'$, the linearised Navier Stokes equations are given as

$$\begin{aligned}\frac{\partial \rho'}{\partial t} &= -\nabla\cdot(\rho\mathbf{u})', \qquad (6)\\ \frac{\partial(\rho\mathbf{u})'}{\partial t} &= -\nabla\cdot\left(\overline{\rho\mathbf{u}}\otimes\mathbf{u}' + (\rho\mathbf{u})'\otimes\overline{\mathbf{u}}\right) - \nabla p' + \overline{\mu}\nabla\cdot\mathbb{T}',\\ \frac{\partial(\rho e)'}{\partial t} &= -\nabla\cdot\left(\overline{(\rho e+p)}\mathbf{u}' + (\rho e+p)'\overline{\mathbf{u}}\right) + \overline{k}\nabla^2 T' + \overline{\mu}V_{KE},\end{aligned}$$

where

$$\begin{aligned}V_{KE} &= \left((\nabla\cdot\mathbb{T}')\cdot\overline{\mathbf{u}} + (\nabla\cdot\mathbb{T}')\cdot\mathbf{u}' + \mathbb{T}':\nabla\overline{\mathbf{u}} + \overline{\mathbb{T}}:\nabla\mathbf{u}'\right), \qquad (7)\\ \mathbb{T} &= \nabla\mathbf{u} + (\nabla\mathbf{u})^T + \left(\frac{\mu_B}{\mu} - \frac{2}{3}\right)(\nabla\cdot\mathbf{u})\,\mathbb{I},\\ (\rho e)' &= \frac{1}{2}\rho\|\mathbf{u}'\|^2 + \frac{p'}{\gamma-1},\end{aligned}$$

and the linearised version of the equation of state is

$$p' = \frac{\overline{\rho}T' + \rho'\overline{T}}{M_0^2}. \qquad (8)$$

Finally, the viscosity and thermal conductivity are given by

$$\overline{\mu} = \frac{\overline{T}^{0.76}}{Re} \qquad \text{and} \qquad \overline{k} = \overline{\mu}\frac{\gamma}{\gamma-1}\frac{1}{M_0^2 Pr}. \qquad (9)$$

These equations are the LNSE, commonly presented in the concise matrix form

$$\frac{\partial q'}{\partial t} = \mathbb{A}q', \qquad (10)$$

where $q'$ is the perturbations $(\rho', u_x', u_r', u_\theta', e')$ , and $\mathbb{A}$ is the linear operator advancing the small perturbation in time.

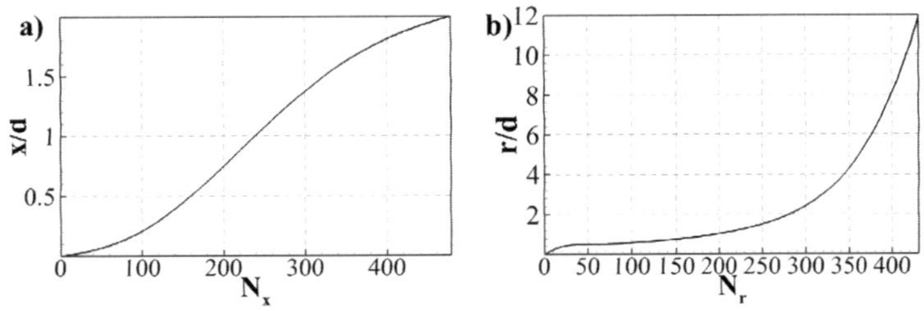


Figure 2: Location of the computational grid points in (a) axial and (b) radial directions.

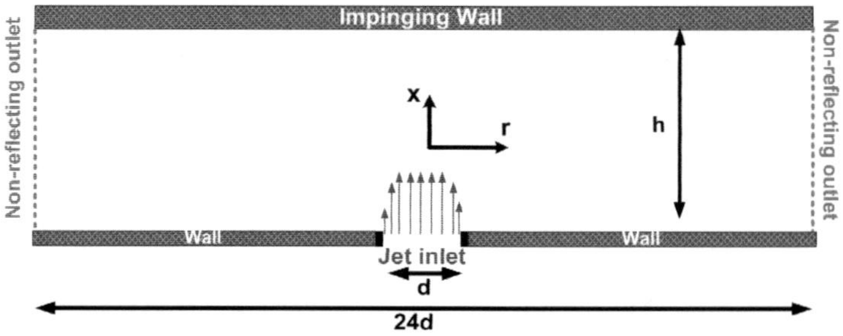


Figure 3: Schematic of the domain and the configuration of this study.

### *2.1. Configuration and numerical methods*

The configuration is an under-expanded impinging jet with a close nozzle–to–wall distance of $h/d$=2 and a radial domain of 12d (figure 3). To obtain the mean flow field (q), an in-house developed high-fidelity LES parallel code that has been tested and validated in previous studies [14, 15, 16, 17] is used to solve the filtered partial differential equations presented in equation (2) with the subgrid scale terms being computed using Germano's dynamic model with the adjustment proposed by Lilly [18]. A sixth-order central finite difference method is applied in the smooth regions in the spatial directions while a fifth-order weighted essentially non-oscillating scheme with local Lax-Friedrichs flux splitting is used in the discontinuous regions. The temporal integration is performed using a fourth-order five-step Runge-Kutta scheme [19, 20]. There are $N_x \times N_r \times N_\theta = 480 \times 400 \times 96$, computational grid points in the LES. A uniform grid is employed in the azimuthal direction. In the axial direction, stretch grid points are used with two stretching rates. A fine grid is used near the nozzle and the impinging wall. A fine grid point of approximately $0.001d$ is employed in the mixing layer with a polynomial stretch of the grids towards the centre of the jet and far-field. The locations of the radial and axial grids are shown in the figure 2 for the sake of completeness. The locally one-dimensional inviscid compressible boundary condition proposed by Poinsot [21] is employed at the outflow boundaries. In addition, the optimised sponge region of Mani [22] is employed near the outflow boundaries to minimise the reflections from the outflow boundaries. In the sponge regions, the flow field is forced to a self-similar solution which was determined a priori. The mean inlet axial velocity $U_{in}$ is specified using the hyperbolic-tangent function of Ref. [23] given by

$$\frac{U_{in}}{U_j} = \frac{1}{2}\left(1 - \tanh\left[\frac{1}{4\delta_{in}}\left(\frac{r}{r_j} - \frac{r_j}{r}\right)\right]\right), \tag{11}$$

where $U_j$ is the jet inlet velocity, $r_j$ the jet inlet radius and $\delta_{in}$ the inlet momentum (mixing layer) thickness. The inlet momentum thickness is equal to $0.04d_j$ in this simulation. The inlet velocity is free of any synthetic turbulence. The Reynolds number is 50,000, the Mach number is 1.0, and the ratio between the stagnation pressure measured in the jet plenum and the ambient pressure (NPR) is 3.4.

The linearised Navier Stokes equations (equation 6) are solved with the same LES resolution in a smaller domain in the radial direction. A sixth-order central finite difference method and fourth-order five-step Runge-Kutta are used for the spatial discretisation and temporal integration, respectively [14]. The mean field obtained from the LES by an ensemble averaging of the data over 38 jet times with the jet time is defined as $h/U_j$, where $h$ is the impingement wall distance from the nozzle and $U_j$ is the jet bulk velocity in the inlet.

#### *2.1.1. Dynamic Mode Decomposition (DMD)*

Dynamic mode decomposition (DMD) is a powerful diagnostic technique which can reveal the spatial and temporal dynamics of coherent structures in fluid flows. The technique relies on sequential snapshots of the flow without any information about the underlying governing equations. Therefore, it can be used in both experimental and numerical studies. Since proposed by Schmid [24], it has been used to study a wide range of fluid flow problems [24, 25, 26, 27]. The concept is similar to the Arnoldi algorithm where a low-dimensional approximation of the high-dimension linear operator is obtained through orthogonalisation techniques. The flow field variables with underlying phenomena of interest are collected by sampling either from a numerical simulation or an experimental measurement in a form of sequential snapshots. These snapshots form a matrix

$$\Psi = \psi_1, \psi_2, ..., \psi_n, \tag{12}$$

where the snapshots, $\psi_i$ have a same time interval $\Delta t$. The snapshots are assumed to be linearly related

$$\psi_{i+1} = A\psi_i, \tag{13}$$

Splitting $\Psi$ into two subspaces of $\Psi_1 = \psi_1, \psi_2, ..., \psi_{n-1}$ and $\Psi_2 = \psi_2, \psi_3, ..., \psi_n$, the linear mapping implies that

$$\Psi_2 = \mathbf{A}\Psi_1 + \mathbf{r}, \tag{14}$$

where $\mathbf{r}$ is the residual and represents the deviation of the system from the linear assumption. In this study, DMD applies to the time series obtained from the solution of the linearised Navier Stokes equations; therefore, the residual is considered to be zero. The SVD-based approach is used in this study [24]. The aim is the construction of a low-dimension matrix $S$ through the similarity transform of $A$. The procedure is as follows. Assume that the SVD of $\Psi_1$ is given as $U\Sigma W^T$. Then,

$$\Psi_2 = \mathbf{A}\Psi_1 = \mathbf{A}U\Sigma W^T, \tag{15}$$

Pre-multiplying both sides of equation (15) by $U^T$ and post-multiply by $W\Sigma^{-1}$ noting that the columns of $U$ are the left singular vectors and the rows of $W^T$ are the right singular vectors, yields

$$U^T AU = U^T \Psi_2 W\Sigma^{-1} \equiv S. \tag{16}$$

Matrix $S$ has the same eigenvalues as $A$ because it is obtained through similarity transformation and the eigenvectors are related by a factor of $U$. The DMD approach is used in this study to obtain both temporal and spatial growth rates in the shear layer. It should be noted that the disturbances are considered to have a temporal form given by

$$q' = \hat{q}(x, r)e^{(\omega_r + i\omega_i)t}e^{im\theta}, \tag{17}$$

for the temporal instability analysis, while they have a spatial form given by

$$q' = \hat{q}(r, t)e^{(\alpha_r + i\alpha_i)x}e^{im\theta}, \tag{18}$$

for the spatial instability analysis. Here, $\omega_i$ is the temporal frequency, $\omega_r$ the temporal growth rate, $\alpha_i$ the spatial frequency and $\alpha_r$ the spatial growth rate. The azimuthal mode number, m, is considered to be zero in the present study.

#### *2.1.2. Hilbert transform analysis*

The Hilbert transform is used to create an analytic signal. For signal $u(t)$, given its Hilbert transform $\hat{u}(t)$, the analytical signal is defined as the composition,

$$u_a(t) = u(t) + i\hat{u}(t), \tag{19}$$

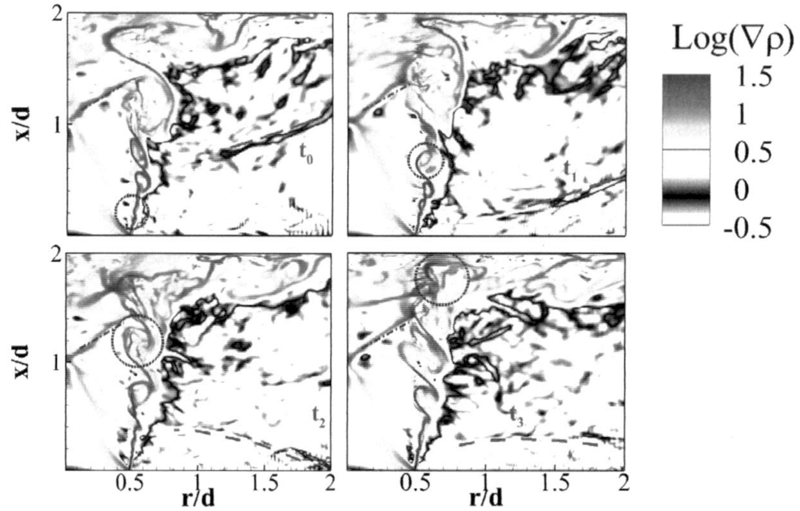


Figure 4: Instantaneous contour plots of logarithm of the density gradient magnitude (Colorbar adjusted to separate the strong gradient regions (hydrodynamic driven gradient) from low gradient regions (acoustic driven gradients), the dash-dot blue lines have fixed locations, the propagating acoustic waves are highlighted by dashed red poly-lines.

which can be expressed in exponential notation as $u_a(t) = K(t)e^{i\phi(t)}$, where $K(t)$ is the instantaneous amplitude or envelop of the signal and $\phi(t)$ is the instantaneous phase. The instantaneous frequency of the signal is also available given by,

$$f(t) = \frac{1}{2\pi}\frac{d\phi}{dt}. \tag{20}$$

The Hilbert transform analysis in this study is applied to answer two questions. First, what are the trailing edge, leading edge and group velocities of the wave packets? Second, how does the energy of the waves with different frequencies evolve temporally and spatially?

## 3. Results and Discussion

### *3.1. Large eddy simulation*

The instantaneous flow fields are presented in this section to illustrate the complexity present in the under-expanded supersonic impinging jet. Figure 4 shows the contour plots of instantaneous density gradient for the sequences of $t_0$ to $t_3$. The contour level is in a log scale with two color scales to facilitate the visualisation of the processes. The regions with high gradients are in red/yellow color while the low gradient regions are in black/grey color. The high gradients represent the Kelvin-Helmholtz structures as well as the shock cells. As can be seen, an inviscid instability of the Kelvin–Helmholtz type is initiated as marked by a green circle in the first instant and a vortex is created and convected as growing in the shear layer (next two time sequences) which is similar to previous observations in the experimental and numerical studies of free and impinging jets [28, 29, 30, 3] and jets in crossflow [31, 32]. This process is followed by an interaction of the vortex with the stand-off shock which has been observed in experimental and numerical studies of impinging jets [28, 29, 30, 3]. The low gradient regions are associated with propagations of acoustic waves. The acoustic waves of impingement propagates towards the inlet as in the first two instants in figure 4 and then reflects from the boundaries (last two instants in figure 4). The self-sustained process of a Kelvin–Helmholtz instability development in the shear layer of the jet followed by acoustic wave generation and propagation towards the

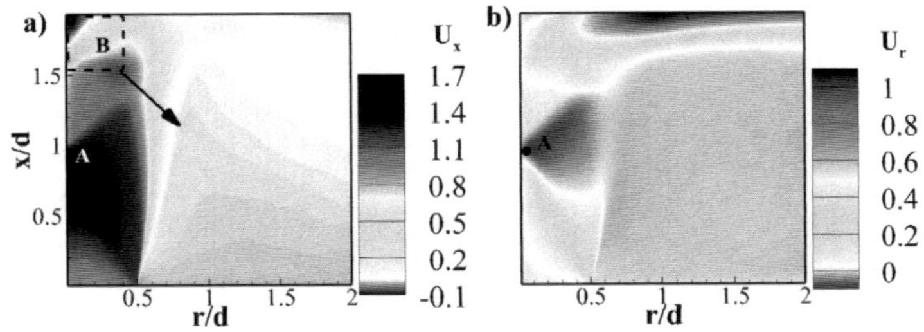


Figure 5: The ensemble averages of (a) streamwise and (b) radial velocity fields (with a zoom of the velocity streamlines in the recirculation region.).

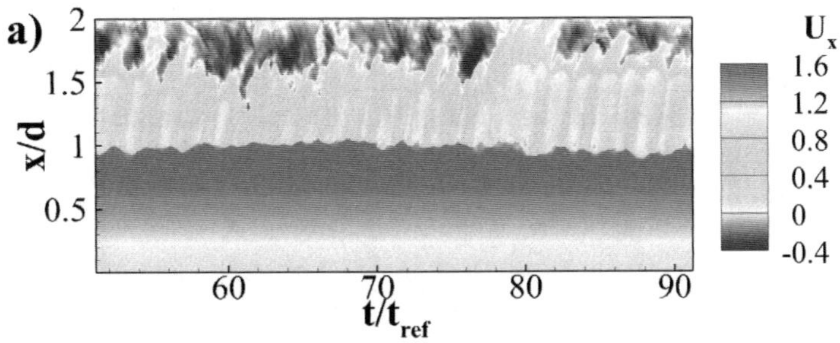


Figure 6: The history of the centreline streamwise velocity.

nozzle has been a topic of research for a few decades and is commonly explained by the feedback loop mechanism [33, 30, 2, 3]. However, the process of completing the feedback loop is still a subject of debate. It is commonly accepted that excitation of the shear layer instability by acoustic waves is one of the main component of the loop. Therefore, the focus of the next sections is on the short-time response of the shear layer to external acoustic waves through the solution of the linearised Navier Stokes equations with an initial acoustic pulse. A base flow is required for this purpose which we will consider to be the temporal average of the LES results.

*3.1.1. Mean flows*

Figure 5 presents the ensemble averages of streamwise and radial velocity fields. The ensemble averages are computed as simultaneous temporal and azimuthal direction means. For the sake of completeness the time history of the centreline streamwise velocity is presented in figure 6 which clearly shows the solution is reached the steady state condition. A sudden drop in streamwise velocity at x/d≈1 is observed in figure 5, marked as 'A' and indicates the Mach disk. The streamwise velocity also shows a local increase near to the wall, which is associated with the bow or stand-off shock. A recirculation region exists between the bow shock and the impinging wall. The streamlines in this region marked as 'B' is also shown in figure 5a where the circulation is clearly observable. Figure 5b shows the mean radial velocity. The formation of a wall jet flow is also observed as the radial velocity is positive in the region away from the stagnation point.

*3.2. Linearised impulse response of the base flow*

The response of the mean flow field to an axisymmetric acoustic pressure pulse in the form of an initial value problem is performed in this section. The linearised Navier-Stokes equations given by equation 6 are solved with an initial pressure pulse described by,

$$p'_{@(t=0)} = A\, exp\left(-\frac{(x-x_o)^2}{2\sigma_x^2} - \frac{(r-r_o)^2}{2\sigma_r^2}\right), \tag{21}$$

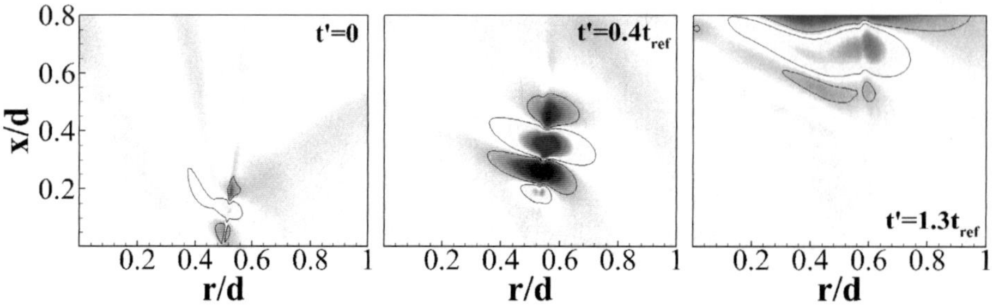


Figure 7: Instantaneous streamwise component of the velocity disturbance.

where 'A' is the amplitude of the pulse; $x_o$ and $r_o$ are the streamwise and radial location of the centre of the pulse, respectively; $\sigma_x$ and $\sigma_r$ are the radii of the pulse in the streamwise and radial directions, respectively. The pulse excites a broad range of frequencies and both the temporal and spatial growth rates as a function of frequency can be recovered by DMD and the Hilbert transform analysis. The Hilbert transform analysis is valid in this context as the signals are generated by a linear mapping and have mean values of zero.

The mean velocity field (figure 5) shows that the Mach disk is located at x/d≈1.0. Therefore, the region and time of the DMD and Hilbert transform analysis should be bounded to avoid nonlinearity effects of the Mach disk on the results. To facilitate this, data for spatio-temporal linear stability is collected in the region from the nozzle lip to 0.8d downstream of the nozzle lip just before the Mach disk location. The time is also limited to the time frame between the time disturbance hits the nozzle lip and the time at which it passes x=0.8d. The time evolution of the disturbance is presented in figure 7, where $t'$ is zero when the disturbance forms at the nozzle lip. The time period used in this analysis is more comprehensible in the time history of the velocity disturbance on the shear layer of the jet, as shown in figure 8. The data used for the DMD and Hilbert transform analysis are considered in the time frame between $t_0$ and $t_n$ in figure 8.

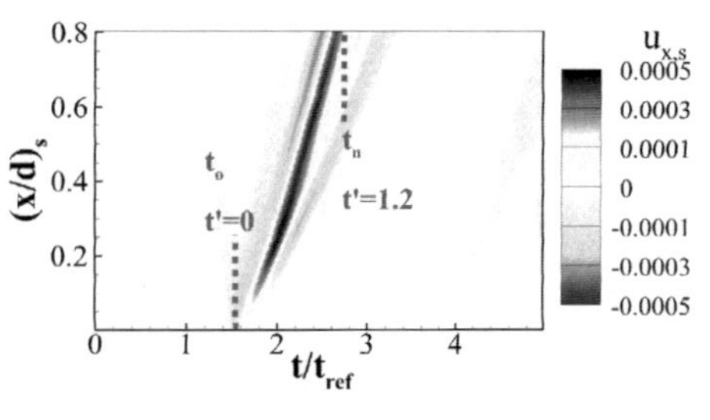


Figure 8: Time history of the streamwsie velocity disturbances on the shear layer of the jet.

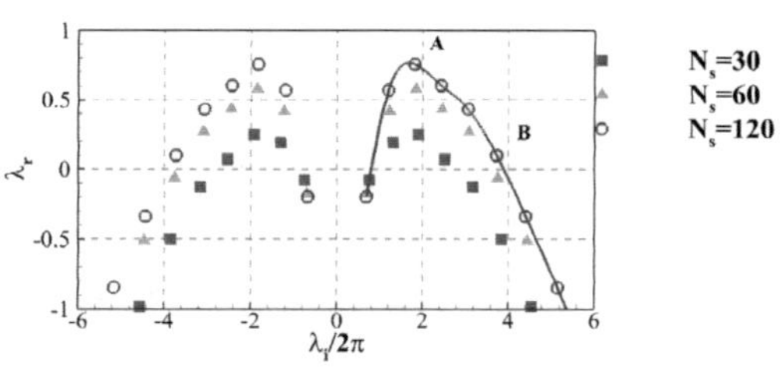


Figure 9: Temporal growth rate obtained from dynamic mode decomposition of linearised responses to pressure pulse.

### *3.2.1. Dynamic mode decomposition*

Dynamic mode decomposition is used for both temporal and spatial linear stability analysis. Temporal decomposition is discussed first, where sequences of streamwise velocity fluctuation snapshots are used as the base for the dynamic mode decomposition. It is expected that by increasing the number of flow fields in our sequence, eigenvalues of the reduced order matrix $S$ converge towards eigenvalues of the linear operator [24]. The convergence behaviour of the dynamic mode decomposition as the number of snapshots increases (i.e. interval of the snapshots decreases) is presented in Figure 9, where the temporal growth rates are shown as a function

IOP Conf. Series: Journal of Physics: Conf. Series **1001** (2018) 012019 doi:10.1088/1742-6596/1001/1/012019

of temporal frequencies. It is clear that an increase in the number of snapshots improves the convergences and that 120 snapshots are sufficient for the purpose of this study. Before we discuss the findings, a caveat is that this analysis has been performed for the short period of time between the formation of the shear layer instability and appearance of the developed disturbance at the Mach disk location. Therefore, the findings are limited to this short period and should not be compared to the *global modes*. Figure 9 shows that disturbances within the non-dimensionalised frequency band of 1–4 grow exponentially, while other frequencies are damped. This result highlights the natural responses of the shear layer of this under-expanded supersonic jet during the evolution of disturbances in the shear layer. The most remarkable finding is that only the disturbance within the non-dimensionalised frequency band of 1–4 grow in this short period. Figures 10 and 11 show the real part of the dynamic modes 'A' and 'B' in Figure 9. The amplitude of the low frequency mode 'A' is stronger and more developed throughout the shear layer, whereas the amplitude of the high frequency mode 'B' is much weaker and waves initially grow in the shear layer and then half way through after reaching a maximum amplitude, they decay.

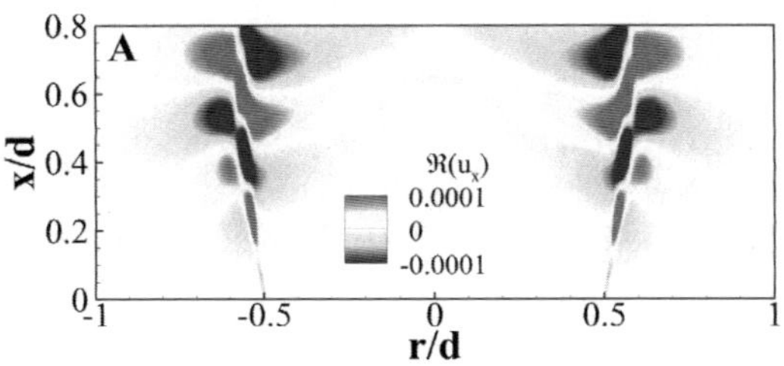

Figure 10: The real part of the dynamic mode 'A' in figure 9.

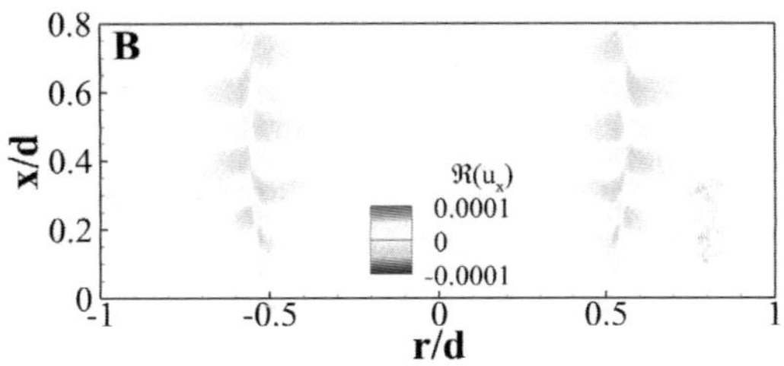

Figure 11: The real part of the dynamic mode 'B' in figure 9.

The spatial growth rate obtained using DMD, where the snapshots are in radial-time coordinate and ordered incrementally in the streamwise direction, is shown in figure 12 as a function of spatial frequencies. The waves with low spatial frequencies (i.e. long wavelengths) have the highest growth rates. On the other hand, waves with high spatial frequencies (i.e. short wavelengths) have low growth rates. This behaviour is expected as the large structures have a high tendency to grow in the direction of the flow motion before the vortex breakdown process which is usually after the potential core. This agrees also with your findings in figure 4, where it is possible to see that the vortex is growing as it is convected.

The summary of the findings of the dynamic mode decomposition are as follows: the waves with streamwise wavelengths ($2\pi/\alpha_i$) of more than approximately 1/8 will grow in the streamwise direction if they have non-dimensionalised temporal frequency in the range of approximately 1 to 4. The overlap is for the non-dimensionalised temporal frequency approximately less than

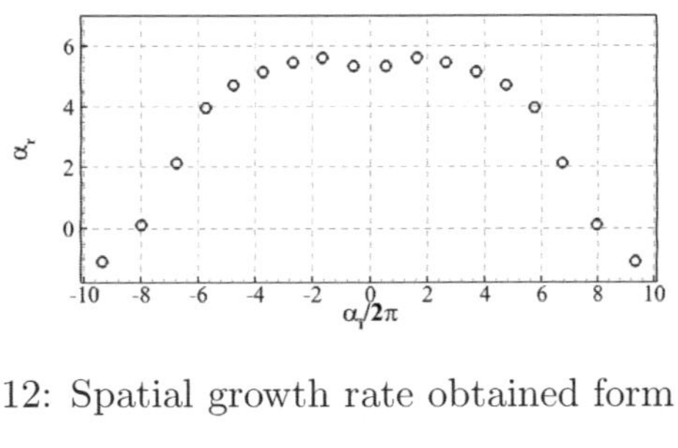

Figure 12: Spatial growth rate obtained form DMD.

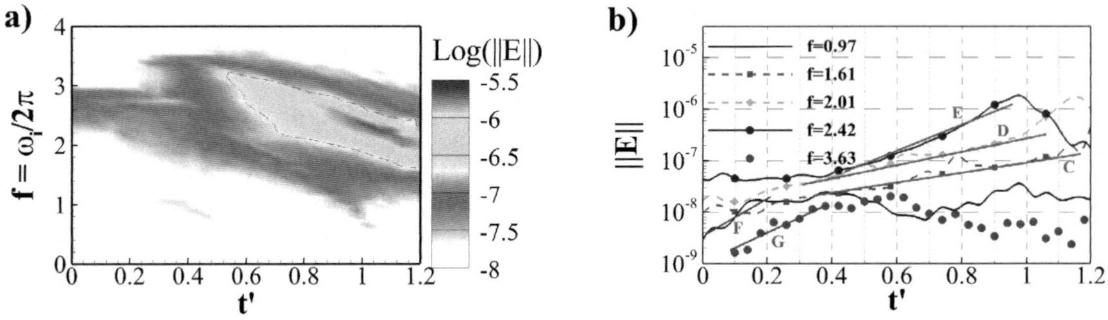


Figure 13: a) Energy as function of time and frequency and b) Time evolution of energy of different frequencies obtained from Hilbert transform analysis.

1. This conveys that the low spatial wavelengths may grow as they progress in the shear layer; however, they decay as time elapses.

*3.2.2. Hilbert transform analysis*

The dynamic mode decomposition analysis presented in the previous section provides interesting information about spatio-temporal growth rates. This section of the paper tries to further clarify the spatio-temporal non-modal instability by using the Hilbert transform analysis of streamwise velocity perturbation where both time and frequency appear as independent variables. The streamwise velocity perturbation is a function of time and space. To perform the analysis, Hilbert transform was applied to the time dependent signal at each grid point in the domain of interest. The analytical signal is obtained for each spacial location. This analytical signal is a function of time, frequency and space. The energy of the signal is defined as the square of the amplitude of the signal. Depending on whether the analysis performed for the spatial or the temporal instability, the energy is conditionally accumulated in time or space for each frequency.

The contour plot of energy of the perturbation conditionally accumulated on frequency and time is presented in Figure 13a. The waves with non-dimensional frequencies in the range of 1.5–3.2 contain 90% of the total energy. This confirms the dominant roles of these frequencies obtained by DMD (see Figure 9). Temporal evolutions of the conditionally accumulated energy for a few non-dimensional frequencies are presented in Figure 13b which shows that the non-dimensional frequencies denoted as 'C', 'D', and 'E' (in the range of 1.5–3.2) have positive growth rates. It is interesting to note that non-dimensional frequencies less than approximately 1 and higher than approximately 3.2 (marked by 'F' and 'G'), show positive growth in the beginning of the process ($t' < 0.4$) and thereafter they are either damped for the high frequencies or neutral for the low frequencies.

The contour plot of energy of the perturbation conditionally accumulated over spatial frequencies and streamwise locations is presented in Figure 14. This figure shows that high spatial frequencies (i.e. short wavelengths) are formed near the nozzle. These waves are damped as they propagate downstream. The disturbances with the wavelengths in the range of 1/3 to 1 contain most of the energy. The energy of these waves grows in streamwise direction. The spatial frequencies less than 0.5 (i.e. wavelengths longer than 1) do not contain any energy. The findings of the Hilbert transform analysis analysis are summarised as follows: the disturbances with streamwise wavelengths ($2\pi/\alpha_i$) between 1/3 and 1 grow in the streamwise direction and the disturbances with non-dimensionalised temporal frequencies in the range of approximately 1.5 to 3.2 grow by time.

It is informative to present the space-time properties of the envelope of the streamwise

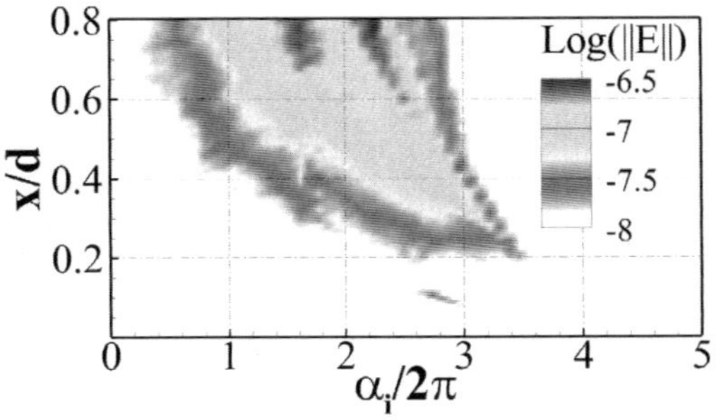


Figure 14: Energy as function of spatial frequency and streamwise location obtained from Hilbert transform analysis.

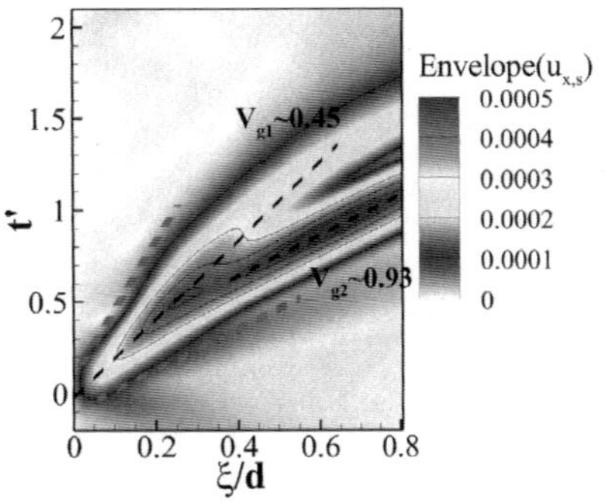


Figure 15: Envelop of the streamwise velocity along the shear layer trace.

velocity perturbation to get a better picture of the spatio-temporal growth. The envelope of the streamwise velocity perturbation along the shear layer of the jet [3] obtained from the Hilbert transform analysis is presented in Figure 15. The leading-edge and trailing-edge of the wave have positive velocities which indicate a convective instability in the shear layer which was also observed in our previous study [14] for a different nozzle–to–wall distance. It is notable that the group velocity, defined as the slope of the curve tracking the maximum value of the envelope, changes along the shear layer. The wave packet seems to split as it progresses in the shear layer, and the high amplitude trace is convected with velocity close to the speed of the sound while the low amplitude trace travels with velocity close to the mean convective velocity of the shear layer. This clearly shows the dispersive characteristic of the shear layer and also indicates that the wave packets are hydrodynamic waves in the region close to the nozzle lip while they split downstream into both acoustic and hydrodynamic waves.

## 4. Conclusions

The spatio-temporal characteristic of the shear layer of an under-expanded supersonic impinging jet has been studied.

A non-modal instability analysis is conducted for the shear layer near the nozzle. To avoid the non-linear effects of shocks, the analysis is limited to a domain between the nozzle lip and and 0.8d downstream of the nozzle lip, as well as the time period in which these disturbances evolve within the shear layer. Both DMD and Hilbert transform analysis show that the frequency band of the shear layer response is in the range of non-dimensional frequencies between approximately 1 and 4. This temporal analysis suggests that if the mean field is perturbed by selective frequencies with the same amplitudes, waves within this range of frequencies will grow exponentially in the shear layer near the nozzle and for the time period of their evolution in the shear layer before reaching the Mach disk location. The spatial DMD analysis reveals that the disturbances with non-dimensionalised wavelengths less than 1/8 will be dampened immediately while other spatial-waves grow in the streamwise direction. The results obtained from the Hilbert transform analysis show that the spatial frequencies between 1 and 3.5 (i.e. non-dimensionalised wavelengths between 0.28 and 1) contain most of the energy of the disturbances. Finally, the envelope of the streamwise velocity disturbances obtain from Hilbert transform at the shear layer confirms the presence of a convective instability.

[3] The axis $\zeta$ is tangential to the shear layer of the jet.

## Acknowledgements
This research was funded in part by the Coturb program of the European Research Council and was undertaken with the assistance of resources from the National Computational Infrastructure (NCI) through NCMAS, which is supported by the Australian Government. The support of the Australian Research Council (ARC) is also gratefully acknowledged through a discovery project.

## References
[1] Henderson B, Bridges J and Wernet M 2005 *J. Fluid Mech.* **542** 115–137
[2] Henderson B and Powell A 1993 *J. Sound Vib.* **168** 307–326
[3] Edgington-Mitchell D, Honnery D R and Soria J 2012 *J. Visual.-Japan* **15** 333–341
[4] Mason-Smith N, Edgington-Mitchell D, Buchmann N A, Honnery D R and Soria J 2015 *Shock Waves* **25** 611–622
[5] Weightman J, Amili O, Honnery D, Edgington-Mitchell D and Soria J 2015 *Australian Conf. Laser Diagnostics in Fluid Mech. and Combustion* (Monash Univ. Pub.) pp 167–172
[6] Weightman J, Amili O, Honnery D, Edgington-Mitchell D and Soria J 2017 *23rd AIAA/CEAS Aeroacoustics Conf.* p 3031
[7] Dauptain A, Cuenot B and M Gicquel L 2010 *AIAA J.* **48** 2325–2338
[8] Ho C M and Nosseir N S 1981 *J. Fluid Mech.* **105** 119–142
[9] Pierrehumbert R and Widnall S 1982 *J. Fluid Mech.* **114** 59–82
[10] Theofilis V 2011 *Annu. Rev. Fluid Mech.* **43** 319–352
[11] Schmid P J 2007 *Annu. Rev. Fluid Mech.* **39** 129–162
[12] Karami S, Hawkes E R, Talei M and Chen J H 2015 *J. Fluid Mech.* **777** 633–689
[13] Constantinescu G S and Lele S K 2002 *J. Comput. Phys.* **183** 165–186
[14] Stegeman P C, Pérez J M, Soria J and Theofilis V 2016 *J. Phys. Conf. Ser.* **708** 012015
[15] Stegeman P C, Soria J and Ooi A 2016 *Fluid-Structure-Sound Interactions and Control* (Springer) pp 241–245
[16] Stegeman P C, Ooi A and Soria J 2015 *Instability and Control of Massively Separated Flows* (Springer) pp 85–90
[17] Stegeman P C, Soria J and Ooi A 2014 *19th Australasian Fluid Mech. Conf.*
[18] Lilly D K 1992 *Phys. Fluids A* **4** 633–635
[19] Kennedy C A and Carpenter M H 1994 *Appl. Numer. Math.* **14** 397 – 433
[20] Kennedy C A, Carpenter M H and Lewis R M 2000 *Appl. Numer. Math.* **35** 177 – 219
[21] Poinsot T and Lele S K 1992 *J. Comput. Phys.* **101** 104 – 129
[22] Mani A 2012 *J. Comput. Phys.* **231** 704–716
[23] Bodony D J and Lele S K 2005 *Phys. Fluids* **17** 085103
[24] Schmid P J 2010 *J. Fluid Mech.* **656** 5–28
[25] Muld T W, Efraimsson G and Henningson D S 2012 *Comput. Fluids* **57** 87–97
[26] Duke D, Honnery D and Soria J 2012 *J. Fluid Mech.* **691** 594–604
[27] Grilli M, Schmid P J, Hickel S and Adams N A 2012 *J. Fluid Mech.* **700** 16–28
[28] Edgington-Mitchell D, Oberleithner K, Honnery D R and Soria J 2014 *J. Fluid Mech.* **748** 822–847
[29] Krothapalli A, Rajkuperan E, Alvi F and Lourenco L 1999 *4th AIAA/CEAS Aeroacoustics Conf.* p 2239
[30] Elavarasan R, Krothapalli A, Venkatakrishnan L and Lourenco L 2001 *AIAA J.* **39** 2366–2373
[31] Bagheri S, Schlatter P, Schmid P J and Henningson D S 2009 *J. Fluid Mech.* **624** 33–44
[32] Meyer K E, Pedersen J M and Özcan O 2007 *J. Fluid Mech.* **583** 199–227
[33] Alvi F, Ladd J and Bower W 2002 *AIAA J.* **40** 599–609

IOP Conf. Series: Journal of Physics: Conf. Series **1001** (2018) 012020 doi:10.1088/1742-6596/1001/1/012020

# Characterisation of minimal-span plane Couette turbulence with pressure gradients

**Atsushi Sekimoto[1,†], Callum Atkinson[1], Julio Soria[1,2]**

[1] Laboratory for Turbulence Research in Aerospace and Combustion, Department of Mechanical and Aerospace Engineering, Monash University, Clayton 3800, AUSTRALIA

[2] Department of Aeronautical Engineering, King Abdulaziz University Jeddah 21589, KINGDOM OF SAUDI ARABIA

[†] Present address: Deparment of Materials Engineering Science, Osaka University, Osaka 560-8531, Japan

E-mail: `asekimoto@cheng.es.osaka-u.ac.jp`

**Abstract.** The turbulence statistics and dynamics in the spanwise-minimal plane Couette flow with pressure gradients, so-called, Couette-Poiseuille (C-P) flow, are investigated using direct numerical simulation. The large-scale motion is limited in the spanwise box dimension as in the minimal-span channel turbulence of Flores & Jiménez (*Phys. Fluids*, vol. 22, 2010, 071704). The effect of the top wall, where normal pressure-driven Poiseuille flow is realised, is distinguished from the events on the bottom wall, where the pressure gradient results in mild or almost-zero wall-shear stress. A proper scaling of turbulence statistics in minimal-span C-P flows is presented. Also the 'shear-less' wall-bounded turbulence, where the Corrsin shear parameter is very weak compared to normal wall-bounded turbulence, represents local separation, which is also observed as spanwise streaks of reversed flow in full-size plane C-P turbulence. The local separation is a multi-scale event, which grows up to the order of the channel height even in the minimal-span geometry.

## 1. Introduction

The adverse-pressure-gradient (APG) effect on the turbulent boundary layer (TBL) has been a challenging problem for a long time, since the scaling law for the mean velocity profile and turbulence statistics is not fully understood. Increasing the computational resources would pave the way to understand the problem and for the development of a better turbulence model in complex geometries. Many studies have dealt with the self-similar aspect of an APG-TBL, in the sense that each of the terms in the governing equation has the same proportionality with streamwise position [1, 2, 3]. However, since the boundary layer with a strong APG grows rapidly, direct numerical simulation (DNS) of APG-TBL requires a very large computational domain with artificial inlet and outlet boundary conditions, so that the realisation of a healthy self-similar region is still limited [4].

In order to simplify the canonical problem of self-similar APG-TBLs, the present study investigates plane Couette flow with a pressure gradient, the so-called Couette-Poiseuille (C-P) flow, to achieve a zero-wall-shear stress on the stationary wall [5, 6]. As the pressure gradient is increased, the APG-TBL flow approaches that of a free-shear flow, and most of the large-scale coherent structures tend to be detached from the wall [7]. However, recent studies on the coherent structures in plane channel flow turbulence [8] and in a statistically-stationary homogeneous shear turbulence (SS-HST) [9, 10] have revealed that the statistical structure of intense Reynolds stress regions is common whether eddies are attached or detached [11]. The effects of pressure gradients on the turbulence structures in plane channel flow can be similar to APG-TBLs.

Here, we consider a minimal-span geometry where the spanwise box dimension ($L_z$) determines the largest length scale [12]. Since the effects of large-scale motions of C-P flow in a full size simulation [6] are not fully understood, these effects are artificially filtered in the present study. Johnstone *al.* [5] have reported that the mean turbulence statistics in the minimal-span simulations are quite different from those in the full simulation case. The effect of the wall-surface manipulation has also been investigated recently by MacDonald *et al.* [13]. They have shown that the roughness functions can be computed from the minimal-span simulations. Therefore, the basic near-wall mechanism and turbulence statistics in a minimal-span C-P can also provide useful information towards understanding the scaling of the 'shear-less' wall-bounded turbulence.

The aim of the present study is to characterise the minimal-span C-P flows and analyse them. At first, the numerical scheme is briefly introduced and the effect of the pressure gradient on a 'pure' Couette flow is presented in §2. The minimal-span C-P flows are characterised in §3, where the box dependency and the pressure gradient effect on the statistics are also discussed. In §4, the instantaneous reversed flow region over 'shear-less' walls is investigated, and the conclusion is provided in §5.

## 2. Numerical methodology

### *2.1. Governing equations and numerical schemes*

We consider the fluid motion between two walls as in Fig. 1(a): the top wall is moving in the streamwise direction, while the bottom wall is stationary. The governing equations are the Navier–Stokes and continuity equations,

$$\frac{\partial \boldsymbol{u}}{\partial t} + \boldsymbol{u}\cdot\nabla\boldsymbol{u} = -\nabla p + \nabla\cdot(\nu\nabla\boldsymbol{u}) + \boldsymbol{G}, \quad \nabla\cdot\boldsymbol{u} = 0, \tag{1}$$

where $\boldsymbol{u} = (u, v, w)$ is the total instantaneous velocity in the streamwise ($x$), wall-normal ($y$) and spanwise ($z$) directions, $t$ is time, $p$ is the pressure divided by the fluid density, which is unity throughout the paper, and $\nu$ is the kinematic viscosity. The frame of reference is the velocity of the bottom wall, $u|_{y=0} = 0$, and the velocity of the top wall ($y = \delta$) is $u|_{y=\delta} = U_{\mathrm{w}}$. The boundary condition is no-slip for the walls ($y = 0$, and $y = \delta$) and periodic in the homogeneous directions ($x, z$). The time-dependent spatially-invariant forcing term, $\boldsymbol{G} = G_x(t)\boldsymbol{i}_x$, is introduced to fix the mean bulk velocity [14],

$$u_b \equiv \frac{1}{\delta}\int_0^\delta U(y)\mathrm{d}y = \text{const.}, \tag{2}$$

where $U(y) = \langle u\rangle$, and $\langle\cdot\rangle$ is the velocity averaged over time and the homogeneous directions. The plane averaged values are defined as $\langle\cdot\rangle_{\mathrm{xz}}$, and $(\cdot)'$ is used for the turbulence intensities, like the root-mean-square values, $\boldsymbol{u}'$. The corresponding instantaneous vorticities are $\boldsymbol{\omega} = (\omega_x, \omega_y, \omega_z)$.

Eqs. (1) are reduced to the evolution equations for $v$, $\omega_y$ and the mean velocities, $\langle u\rangle_{xz}$ and $\langle w\rangle_{xz}$ [15], and the time integration is performed using the third-order Runge–Kutta scheme of Sekimoto et al. [9] with the modification of the viscous term being semi-implicit. Fourier–Chebyshev spectral transformations are applied to the variables,

$$\boldsymbol{u} = \sum_{k=0}^{N_y}\sum_{l=-N_x/2}^{N_x/2-1}\sum_{m=-N_z/2}^{N_z/2-1} \widehat{\boldsymbol{u}}_{k,l,m}\exp(\mathrm{i}(\alpha l x + \beta m z))T_k(y),$$

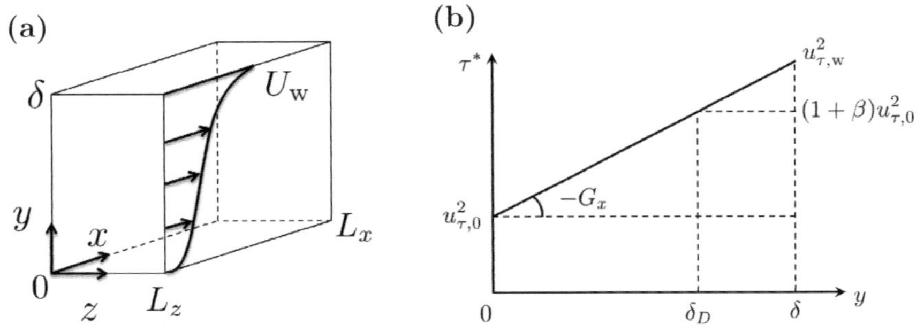


**Figure 1.** (a) The computational domain. (b) The schematic picture of the total Reynolds stress. See the text for the definition of variables.

where $N_x$ and $N_z$ are the number of Fourier modes in the homogeneous directions after 2/3 dealiasing, $k, l, m$ are integers, $\widehat{\boldsymbol{u}}_{k,l,m}$ are the Fourier–Chebyshev coefficients, $(\alpha, \beta) = (2\pi/L_x, 2\pi/L_z)$ are the fundamental wavenumbers, and $T_k$ are the Chebyshev polynomials. The corresponding 2nd- and 4th-order Helmholtz equations of the coefficients ($\widehat{\omega_y}$ and $\widehat{v}$) are solved using the Chebyshev–Legendre–Galerkin method [16], where the boundary condition in the wall-normal direction is imposed using basis functions based on Legendre polynomials. The basis functions which satisfy the boundary conditions for $\omega_y$ and $v$ are, respectively,

$$\{\Psi_\omega\}_k = L_k - L_{k+2} \quad (k = 0, 1, ..., N_y - 2), \tag{3}$$

and

$$\{\Psi_v\}_k = d_k^{-1}\left(L_k + a_k L_{k+2} + b_k L_{k+4}\right) \quad (k = 0, 1, ..., N_y - 4), \tag{4}$$

where

$$a_k = -\frac{2(2k+5)}{2k+7}, \quad b_k = \frac{2k+3}{2k+7}, \quad d_k = \sqrt{2(2k+3)^2(2k+5)}, \tag{5}$$

and $N_y$ is the number of collocation points in $y$. The spatial derivatives are computed using Chebyshev polynomials, and this scheme requires Chebyshev-Legendre transformations before solving the Helmholtz equations. After solving them, the coefficients of the basis functions are passed into the Legendre coefficients by the inverse Chebyshev-Legendre transformations. This scheme has the advantages of both the Legendre– and Chebyshev–Galerkin methods [16]. The snapshots are stored in the Fourier ($x$)–Fourier ($z$)–Chebyshev ($y$) space. The DNS data of the plane channel turbulence was used as an initial condition for the present DNSs.

### *2.2. Nondimensional parameters*

In the minimal-span simulations, the largest-scale motion is limited by the spanwise box dimension [12, 9], so that we explicitly determine the largest scale as $L_z$. The box aspect ratios are important parameters for turbulence statistics [12], and here they are $A_x = L_x/L_z$ and $A_y = \delta/L_z$ as in [9].

The global Reynolds number in a full box, $Re_\delta = U_{\rm w}\delta/\nu$ ($A_y < 1$), is modified as the minimal-span Reynolds number, $Re_z = U_{\rm w}L_z/\nu$, for a minimal-span box ($A_y > 1$). The friction Reynolds numbers are defined respectively for the bottom and top walls as $L^+_{z,0}$ and $L^+_{z,{\rm w}}$ based on the friction velocity, $u_{\tau,0} = {\rm sign}(\tau_0)|\tau_0|^{1/2}$ and $u_{\tau,{\rm w}} = |\tau_{\rm w}|^{1/2}$, where $\tau_0 = \nu\partial_y U|_{y=0}$ and $\tau_{\rm w} = \nu\partial_y U|_{y=\delta}$ are the wall-shear stress on the bottom and top wall. Throughout the paper,

$(\cdot)_0^+$ and $(\cdot)_{\mathrm{w}}^+$ are values normalised using $\nu$, $u_{\tau,0}$ and $u_{\tau,\mathrm{w}}$, respectively. When the mean flow has a reverse flow near the bottom wall, we use the negative values of $\tau_0$, $u_{\tau,0}$ and $L_{z,0}^+$. Note that $(\cdot)^+$ is the average of the top and bottom wall for 'pure' Couette flows.

The displacement thickness, $\delta_D$, is related to the mean bulk velocity in Eq. (2), and is defined in the conventional way analogous to that for a turbulent boundary layer,

$$\delta_D = \int_0^\delta \left(1 - \frac{U}{U_w}\right) \mathrm{d}y = \delta \left(1 - \frac{u_b}{U_w}\right). \tag{6}$$

By fixing the mean bulk velocity, the present DNS explicitly sets the displacement thickness. We only consider the negative pressure gradient, $G_x < 0$, then $0 < u_b/U_{\mathrm{w}} \leq 0.5$, then the flows with adverse-pressure gradients are realised over the bottom stationary wall.

### 2.3. Total Reynolds stress

The role of the pressure gradient $G_x$ is to redistribute the Reynolds stresses. By integrating the Reynolds stress equation, from $y = 0$ to $y = \delta$,

$$\frac{\partial \langle uv \rangle_{xz}}{\partial y} = \nu \frac{\partial^2 U}{\partial y^2} + G_x, \tag{7}$$

we have

$$G_x \delta + (\tau_{\mathrm{w}} - \tau_0) = 0. \tag{8}$$

The non-dimensional pressure gradient parameter, $\beta$, can also be calculated as in the turbulent boundary layers with adverse pressure gradients [1],

$$\beta = \frac{\delta_D}{\tau_0}(-G_x) = \frac{\delta_D}{\delta}\frac{\tau_{\mathrm{w}} - \tau_0}{\tau_0} = \left(1 - \frac{u_b}{U_w}\right)\frac{\tau_{\mathrm{w}} - \tau_0}{\tau_0}. \tag{9}$$

To simplify notation, we will use from now on $\sigma = -\langle uv \rangle_{xz}$ for the tangential Reynolds stress. As shown in Fig. 1(b), the local total shear stress, $\tau^* = \sigma + \nu \partial_y U = \tau_0 - G_x y$, can be written as

$$\tau^* = \tau_0 \left[1 + \left(\frac{y}{\delta_D}\right)\beta\right] = \tau_0 \left[1 + \left(\frac{y}{L_z}\right)\beta_z\right], \tag{10}$$

where $\beta_z = (L_z/\tau_0)(-G_x)$.

The numerical parameters are summarised in Table 1. We consider three cases of zero, mild and strong pressure gradients with different box aspect ratios. For example, 'CP43m' represents the Couette-Poiseuille turbulence with a mild pressure gradient in the box $(A_x, A_y)$=$(4, 3)$, and 'C20' is a 'pure' Couette turbulence in the full box ($A_x = 2$, $A_y = 0.637$). Note that the spanwise box dimension does not necessarily represent the characteristic outer scale in the full simulation, so that the nondimensional pressure gradient parameter, $\beta = \beta_z(\delta_D/L_z)$, is used to distinguish the intensity of the pressure gradient, i.e. 'CP30m' ($\beta_z = 17.7$) is defined as a full-box simulation with a mild pressure gradient ($\beta \approx 1.8$).

### 2.4. The effect of pressure gradients

The effect of the pressure gradient in the full-size simulation is briefly shown using cases C20, CP30m and CP30s. The control parameter $(u_b/U_{\mathrm{w}})$ is gradually decreased from 0.5 to 0.33, at which the wall-shear stress at the bottom wall is almost zero. By decreasing the mean bulk velocity, the mean velocity near the bottom wall behaves as in Fig. 2(a, b), similar to a turbulent boundary layer with an adverse pressure gradient [17]. Although, the wall-unit scaling cannot

IOP Conf. Series: Journal of Physics: Conf. Series **1001** (2018) 012020 doi:10.1088/1742-6596/1001/1/012020

**Table 1.** Numerical parameters for the Couette (C) and Couette-Poiseuille (CP) flows; the first and second numbers represent the streamwise and vertical aspect ratios, $A_x$ and $A_y$, respectively; and the last characters, 'm' and 's' stand for the 'mild' and 'strong' pressure gradients.

| case | $N_x$ | $N_y$ | $N_z$ | $Re_z$ | $A_x$ | $A_y$ | $u_b/U_w$ | $\delta_D/L_z$ | $\beta_z$ | $L^+_{z,0}$ | $L^+_{z,w}$ |
|---|---|---|---|---|---|---|---|---|---|---|---|
| C20 | 340 | 257 | 254 | 70700 | 2.0 | 0.637 | 0.5 | 0.318 | 0 | 1734 | 1735 |
| C21 | 170 | 257 | 126 | 39300 | 2.0 | 1.27 | 0.5 | 0.637 | 0 | 882.8 | 887.5 |
| C22 | 84 | 257 | 62 | 22000 | 2.0 | 2.55 | 0.5 | 1.27 | 0 | 421.6 | 420.3 |
| CP30m | 512 | 193 | 338 | 90500 | 3.0 | 0.159 | 0.42 | 0.093 | 17.7 | 1678 | 3280 |
| CP30s | 512 | 193 | 338 | 70900 | 3.0 | 0.159 | 0.325 | 0.107 | 8300 | 90.9 | 3303 |
| CP21s | 42 | 97 | 40 | 12600 | 2.0 | 1.27 | 0.333 | 0.85 | 1560 | 12.4 | 553.4 |
| CP22s | 42 | 129 | 40 | 12600 | 2.0 | 1.91 | 0.333 | 1.27 | -33.8 | -60.9 | 493.6 |
| CP23s | 42 | 193 | 40 | 12600 | 2.0 | 3.0 | 0.333 | 2.0 | -7.0 | -90.1 | 424.1 |
| CP24s | 42 | 257 | 40 | 12600 | 2.0 | 4.0 | 0.333 | 2.67 | -4.2 | -89.0 | 375.3 |
| CP43m | 128 | 257 | 40 | 12600 | 4.0 | 3.0 | 0.417 | 1.75 | 1.78 | 132.9 | 334.8 |
| CP53m | 170 | 257 | 40 | 12600 | 5.0 | 3.0 | 0.417 | 1.75 | 1.81 | 131.7 | 333.7 |
| CP63m | 170 | 257 | 40 | 12600 | 6.0 | 3.0 | 0.417 | 1.75 | 1.74 | 133.7 | 333.9 |
| CP83m | 256 | 257 | 40 | 12600 | 8.0 | 3.0 | 0.417 | 1.75 | 1.85 | 131.2 | 335.7 |
| CP33s | 84 | 257 | 40 | 12600 | 3.0 | 3.0 | 0.37 | 1.89 | 24.6 | 44.2 | 382.3 |
| CP43s | 128 | 257 | 40 | 12600 | 4.0 | 3.0 | 0.37 | 1.89 | 20.5 | 48.9 | 386.8 |
| CP53s | 170 | 257 | 40 | 12600 | 5.0 | 3.0 | 0.37 | 1.89 | 21.3 | 47.6 | 383.5 |
| CP63s | 170 | 257 | 40 | 12600 | 6.0 | 3.0 | 0.37 | 1.89 | 22.7 | 46.0 | 382.0 |
| CP83s | 256 | 257 | 40 | 12600 | 8.0 | 3.0 | 0.37 | 1.89 | 31.9 | 38.6 | 379.7 |
| CP24m | 42 | 257 | 40 | 12600 | 2.0 | 4.0 | 0.417 | 2.33 | 1.70 | 108.9 | 304.3 |
| CP44m | 84 | 257 | 40 | 12600 | 4.0 | 4.0 | 0.417 | 2.33 | 1.61 | 108.8 | 296.8 |
| CP64m | 128 | 257 | 40 | 12600 | 6.0 | 4.0 | 0.417 | 2.33 | 1.73 | 106.2 | 298.8 |
| CP84m | 170 | 257 | 40 | 12600 | 8.0 | 4.0 | 0.417 | 2.33 | 1.67 | 108.0 | 299.5 |
| CP24s | 84 | 257 | 40 | 12600 | 2.0 | 4.0 | 0.37 | 2.52 | -42.9 | -25.8 | 348.6 |
| CP44s | 170 | 257 | 40 | 12600 | 4.0 | 4.0 | 0.37 | 2.52 | -25.5 | -31.1 | 343.6 |
| CP64s | 256 | 257 | 40 | 12600 | 6.0 | 4.0 | 0.37 | 2.52 | -35.8 | -28.8 | 342.1 |
| CP84s | 340 | 257 | 40 | 12600 | 8.0 | 4.0 | 0.37 | 2.52 | -40.3 | -26.2 | 341.3 |

be defined for the 'shear-less' case (CP30s), the C-P turbulence with a mild pressure gradient exhibits a logarithmic-like profile as in Fig. 2(b). The logarithmic profile is, however, shifted slightly from the zero-pressure gradient case (C20). This behaviour is also observed in strong APG-TBLs [18]. As shown in Fig. 2(c), the pressure redistributes the tangential Reynolds stress as discussed in the previous section. The inclined profiles of $\sigma$ for CP30m and CP30s happen to cross with the zero-pressure gradient case (C20) at the same wall-normal position, $y/\delta = 0.27$. There is, however, no reason for $\sigma$ to cross at this point, and it is likely to be dependent on the Reynolds number.

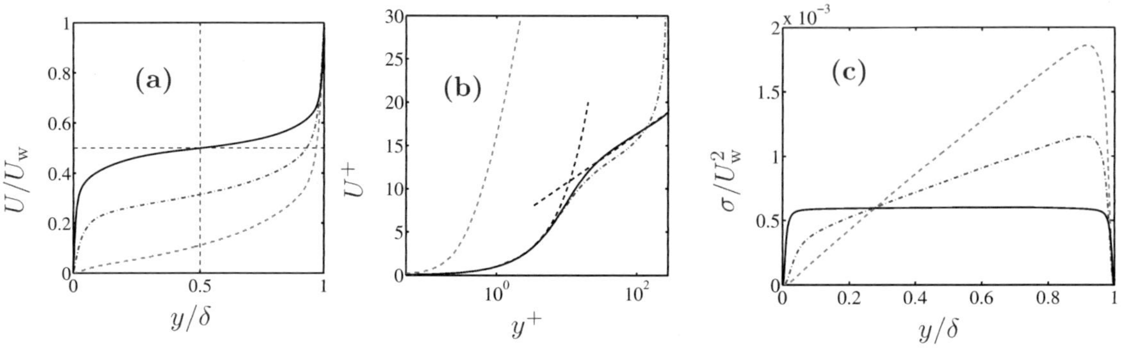


**Figure 2.** (a) Mean velocity profile. (b) The same as (a), but in wall units based on the stationary wall. (c) Tangential Reynolds stress. ——, 'pure' Couette turbulence (C20); —·— (in blue), C-P turbulence with mild pressure gradient (CP30m); - - - -(in red), C-P turbulence with strong pressure gradient (CP30s).

## 3. Minimal-span simulations

### 3.1. 'Pure' Couette turbulence

Here, DNS results of plane Couette turbulence, i.e. without pressure gradient ($u_b/U_w = 0.5$), are presented. For the cases C20, C21, and C22 (see Table 1), the streamwise aspect ratio, $A_x$, is fixed, and the friction Reynolds number, $\delta^+ \approx 1100$, is also adjusted to the same value as Avsarkisov *et al.* [19] ($A_x = 3.33$, $A_y = 0.106$). By increasing the vertical box aspect ratio $A_y$, the large-scale motion is constrained by the narrow computational domain, and the turbulence statistics are affected. The effective large length scale is limited by $L_z$, which is the same as the minimal-span plane Poiseuille channel by Flores & Jiménez [12] and also the statistically-stationary homogeneous shear turbulence [9]. The mean velocity profile of C20 does not differ from that of Avsarkisov *et al.* (not shown). As $A_y$ is increased, the mean velocity profile departs from the logarithmic profile (see Fig. 3(b)), indicating that the net Reynolds number decreases as $A_y$ increases. The departing point is roughly $y_d = L_z/3$ from the present minimal-span Couette turbulence with good agreement with 'pure' Poiseuille turbulence [12]. The mean velocity fluctuations are shown in Fig. 3(c) for the full simulation [19] and the present minimal-span cases (C20, C22). There are differences from the buffer to the outer layer. As the net Reynolds number $L_z^+$ is decreased, the buffer layer peaks of $u'^+$ and $w'^+$ become less marked.

### 3.2. Box dependency

The effect of the aspect ratios of the box, $A_x$ and $A_y$ is investigated in this section. In the full simulation, CP30s in Table 1, an almost zero wall-shear stress at the bottom wall has been realised by using $u_b/U_w = 0.33$. The cases of CP21s, CP22s, CP23s and CP24s in Table 1 are preliminary tests to investigate the $A_y$-dependency, and it appears that the net pressure gradient parameter, $\beta_z$, decreases as $A_y$ is increased. Reverse flows are observed for $A_y > 2$ near the bottom wall as shown in Fig. 4(a). Here, the mean velocity is scaled using $U_w/A_y$ for clarity. It turns out this reference velocity scales well the velocity fluctuations and the tangential Reynolds stress in the minimal-span simulations (see in Fig. 4 b). It is confirmed that the reversed flows are observed even at higher Reynolds numbers with the same aspect ratios and $u_b/U_w$ (not shown), indicating that $\beta_z$ (or $\beta$) is independent of the Reynolds numbers, at least when the turbulence is fully-developed. As it is apparent from Eq. (9), the nondimensional pressure gradient parameter is dependent only on the boundary condition ($u_b/U_w$) and the ratio of the wall shear stresses of the bottom and top walls. The tangential Reynolds stress is not significantly affected by the

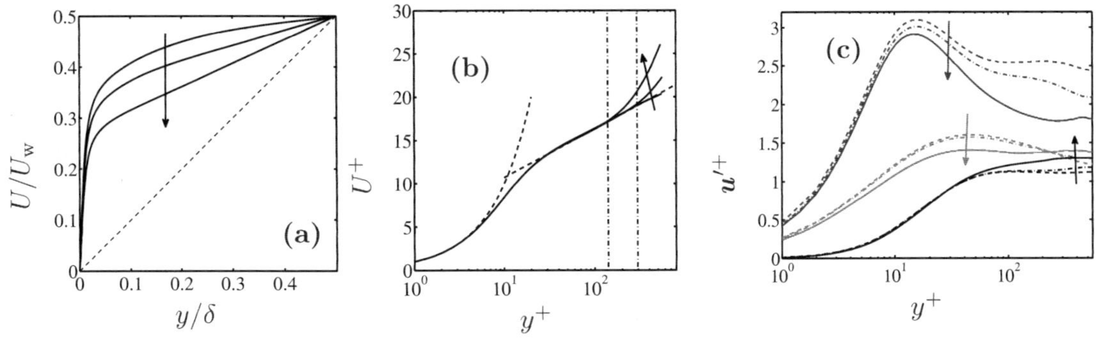


**Figure 3.** (a) Mean velocity for the 'pure' Couette turbulence, C20, C21 and C22. (b) The same as (a), but in wall units. The dashed lines are $u^+ = y^+$ and $u^+ = \kappa^{-1} \ln y^+ + 5.1$ ($\kappa = 0.41$). The vertical dash-dotted lines are $y = L_z/3$ for C21 and C22. (c) The velocity fluctuations for the reference [19] (- - - -), C20 (—·—), C22 (——): (blue lines) $u'^+$; (black lines) $v'^+$; (red lines) $w'^+$. The arrows are in the sense of increasing $A_y$.

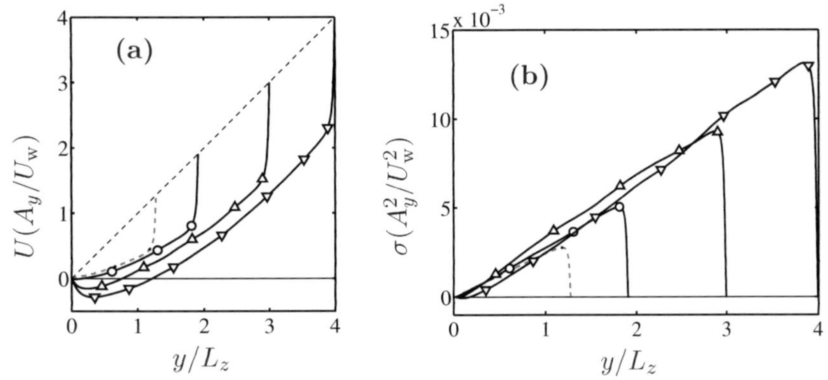


**Figure 4.** (a) Mean velocity. (b) Tangential Reynolds stress: - - - -(red), CP21; —○—, C22; —△—, C23; —▽—, C24.

reversed flow in the range of $|L_{z,0}^+| < 100$. It is probably because the minimal flow unit [20] can not fit in this spanwise length, and turbulence structures are not energetic. As $A_y$ increases, the profile of $\sigma$ changes roughly in a self-similar way. The effect of the streamwise box aspect ratio, $A_x$, does not produce significant differences for the minimal-span box simulations with $A_y = 3, 4$ (see Table 1). This is true both for the mild and strong pressure gradient cases.

The mean velocity profiles for relatively long-box simulations, CP83m, CP83s, CP84m and CP84s, are shown in Fig. 5(a–c). The wall-unit scaling on the bottom wall is not useful both for the mild and strong pressure gradient cases, while the mean velocity near the top wall is the same as in minimal-span channel turbulence. Note that CP84s has a slightly negative wall-shear stress on the bottom wall, so that the corresponding line is not shown in Fig. 5(b). Going back to the effect of $A_y$, the mild pressure-gradient cases with $A_y = 3, 4$ have the same mean velocity profile near the bottom wall, while the strong cases have slightly shifted profiles as shown in Fig. 5(a). It indicates that the logarithmic-like mean velocity profile is more robust when the minimal flow unit of turbulence is captured, $|L_{z,0}^+| > 100$, probably because it has some resilience to the sweeps from the outer layer. Such resilience disappears when $|L_{z,0}^+| < 100$, and the near-wall viscous layer is directly affected by the sweep.

IOP Conf. Series: Journal of Physics: Conf. Series **1001** (2018) 012020 doi:10.1088/1742-6596/1001/1/012020

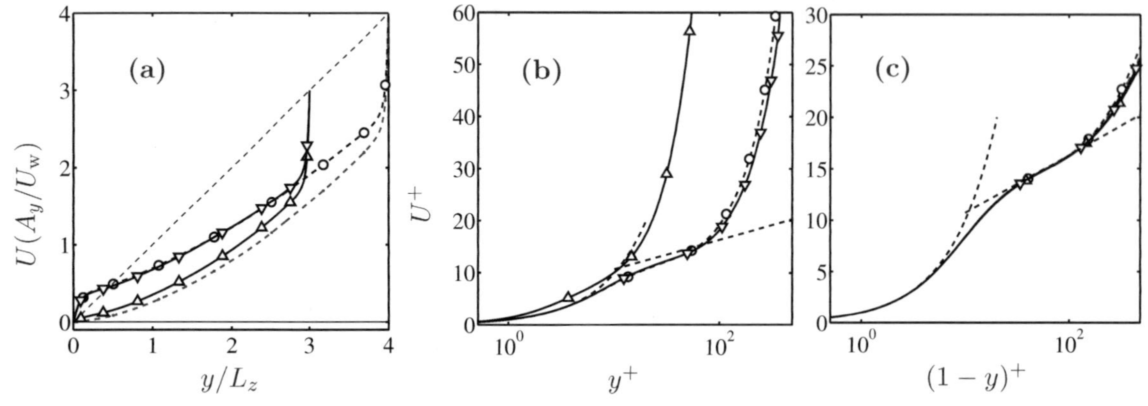

**Figure 5.** Mean velocity for CP84s (- - - -, in blue), CP84m (– –○– –), CP83s (—△—) and C83m (—▽—). (a) In outer units using $U_\mathrm{w}/A_y$ and $L_z$. (b,c) In wall units based on the bottom (b) and the top wall (c). The dashed lines are $u^+ = y^+$ and $u^+ = \kappa^{-1} \ln y^+ + 5.1$ ($\kappa = 0.41$).

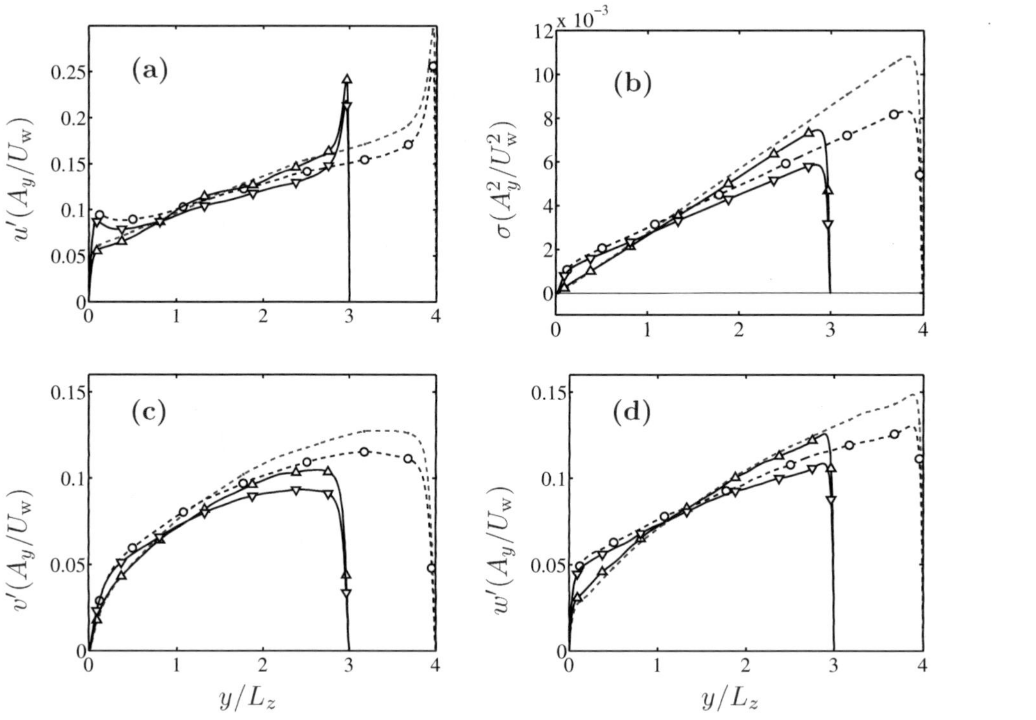

**Figure 6.** (a) $u'/(U_\mathrm{w}/A_y)$. (b) $\sigma/(U_\mathrm{w}/A_y)^2$. (c) $v'/(U_\mathrm{w}/A_y)$. (d) $u'/(U_\mathrm{w}/A_y)$: - - - -(in blue), CP84s; – –○– –, CP84m; —△—, CP83s; and —▽—, C83m.

The low-order turbulence statistics for relatively long-box simulations, CP83m, CP83s, CP84m and CP84s, are shown in Fig. 6(a–d). An interesting feature is that the velocity fluctuations do not decrease linearly near the wall, even when the wall-shear stress linearly approaches zero on the bottom wall (CP83s, CP84s). The origin of these velocity fluctuations is from above the outer layer. In a strong APG-TBL, more sweep events are observed near the

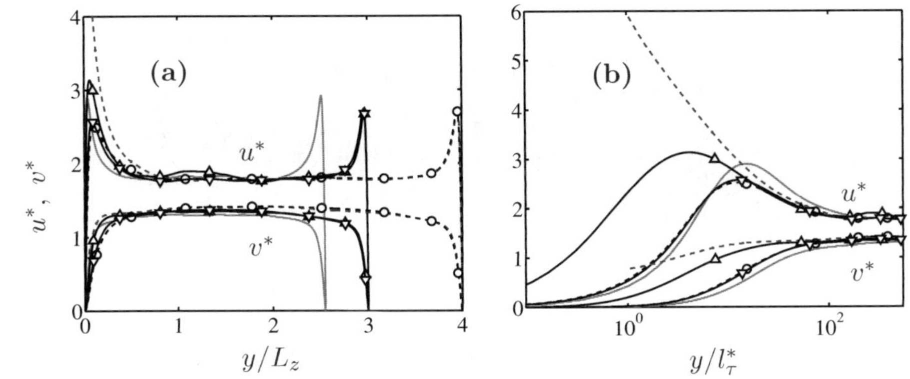

**Figure 7.** The streamwise and wall-normal velocity fluctuations scaled by the analytical local friction velocity, $u_\tau^* = \tau^{*1/2}$, as functions of (a) $y/L_z$ and (b) $y/l_\tau^*$: ——(in red), C22; - - - -(in blue), CP84s; --○--, CP84m; —△—, CP83s; and —▽—, C83m. Note that the near-wall profiles for CP84s with $\tau^*/\overline{\tau} < 0.01$, where $\overline{\tau} = (\tau_w - \tau_0)/2$ are truncated to eliminate the singular behaviour ($\tau^* = 0$).

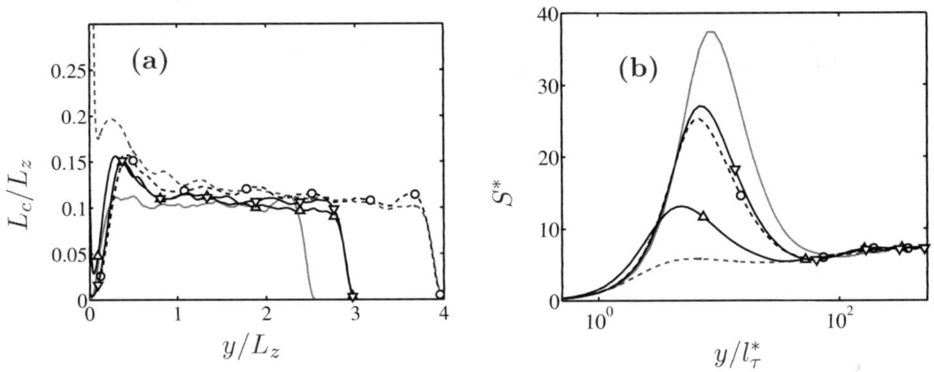

**Figure 8.** (a) The Corrsin length scale, $L_c$. (b) The nondimensional Corrsin shear parameter, $S^*$. Lines are as in Fig. 7.

wall than in zero-pressure gradient TBLs [21].

There is a region where $\sigma/q^2$ ($q^2 = u'^2 + v'^2 + w'^2$), is constant [22] for APG-TBLs, and it is true for the present minimal-span C-P turbulence. Fig. 7 shows $u'$ and $v'$ scaled by the local total shear stress, $\tau^*$, which is given in Eq.(10). The constant region for $y/L_z > 1$ indicates that there is a common mechanism which produces turbulence intensity as in the zero-pressure gradient case. The near-wall peaks increase as $\beta$ increases, eventually going to infinity when a reversed flow with negative $\tau^*$ is realised. In Fig. 7(b), the corresponding friction length scale, $l_\tau^* = \nu/(\tau^*)^{1/2}$, is used, rather than the conventional friction velocity and length. It appears that the peak positions approach the bottom wall asymptotically as $\beta \to \infty$, while $u^* \approx 1.8$, $v^* \approx w^* \approx 1.4$ for $y/l_\tau^* \gtrsim 100$. These values agree with those at around the centre plane of the zero-pressure gradient case, i.e. minimal-span 'pure' Couette flow (see also Fig. 3c), and also with statistically-stationary homogeneous shear turbulence [9].

In the end, the large length scale and local shear are investigated. Fig. 8(a) shows the Corrsin length scale, $L_c = (\varepsilon/(\partial_y U)^3)^{1/2}$ [23], where $\varepsilon$ is the dissipation rate. The effect of the pressure gradient is limited near the wall (roughly $y < L_z$), and $L_c/L_z$ is almost constant in the bulk

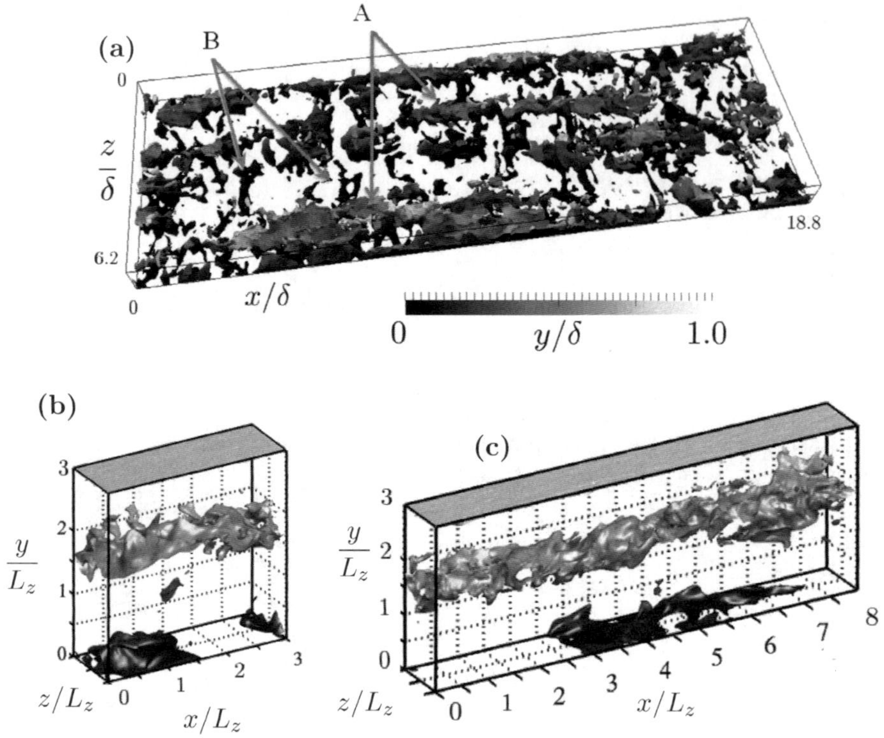


**Figure 9.** (a) The isosurface of $u/U_{\mathrm{w}} = 0.0$ for a full-box simulation (CP20s); the structures, 'A', represent large-scale reversed flow while 'B' structures represent spanwise streaks. The contour is gradually coloured by the distance from the bottom wall. (b, c) The isosurface of $u/U_{\mathrm{w}} = 0.0$ (dark red), $u_b/U_{\mathrm{w}}$ (orange) and 0.99 (yellow): (b) CP33s, (c) CP83s. The flow is from left to right.

uniformly sheared region in minimal-span Couette-Poiseuille flows. The 'pure' Couette flow has $L_c/L_z \approx 0.1$, which agrees well with that in SS-HST [9], while Couette-Poiseuille have a tendency to decrease roughly by 20–30% from $y/L_z = 1$ to the end of the uniformly sheared region ($1 < y/L_z \lesssim A_y - 1/3$). The mean velocities in the bulk region of C-P flows are not perfectly constant as already shown in Fig. 5, however it is not significant difference. The nondimensional Corrsin shear parameter, $S^* = \partial_y U q^2/\varepsilon$, is similar between Couette and C-P flows (Fig. 8). The statistically-homogeneous shear turbulence [9] also represents $S^* \approx 7$–8, which agrees with both 'pure' Couette and C-P flows for $y/l_\tau^* > 100$. As $\beta_z$ approaches $\infty$, the near-wall peak of $S^*$ drastically decreases, and the local shear would not produce active turbulent structures at the buffer layer ($y/l_\tau^* \approx 10$).

## 4. Locally reversed flows

Focusing on the 'shear-less' wall-bounded turbulence, the dynamics of the region of the reversed flow ($u/U_{\mathrm{w}} < 0$) is investigated. Figure 9 shows reversed flow regions, which are visualised by the isosurface of $u/U_{\mathrm{w}} = 0.0$. In the full C-P simulation (CP20s), there are two typical structures: streamwise elongated large-scale ejections ('A' in Fig. 9(a)); and near-wall spanwise reversed flows ('B' in Fig. 9(a)). The large-scale motions span the whole channel height and interact with both bottom- and top-wall events. In the minimal-span simulations, such large-scale events are artificially filtered, therefore only the spanwise reversed flow regions exist as

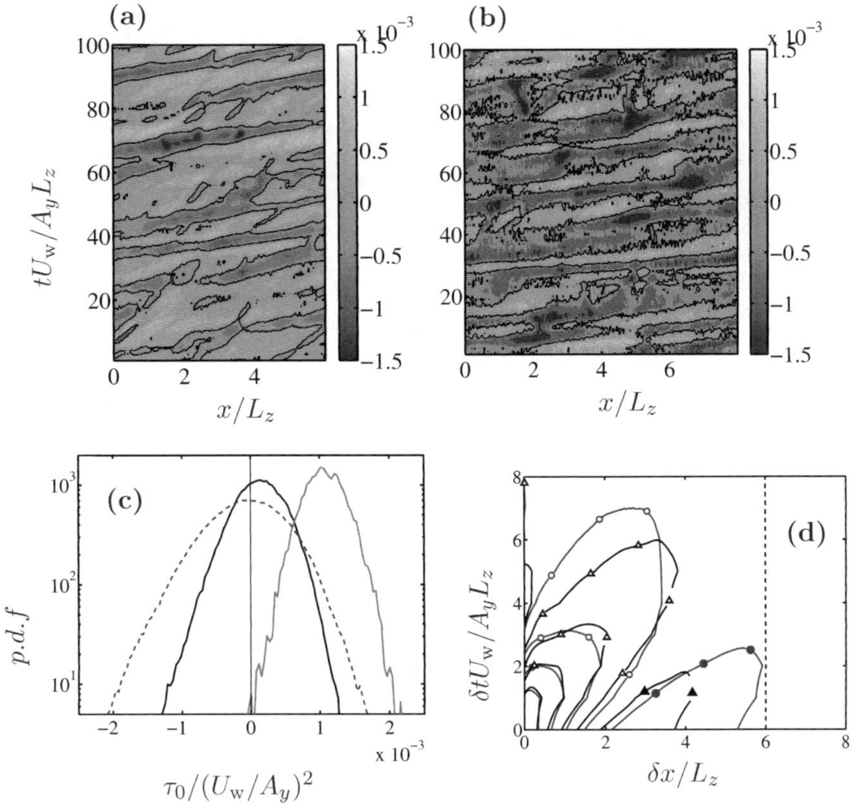


**Figure 10.** (a, b) The spanwise-averaged wall-shear stress, $\langle\tau_0\rangle_z/(U_w/A_y)^2$, as a function of time ($t$) and space ($x$) for (a) CP63s and (b) CP84s. (c) P.d.f.s of $\langle\tau_0\rangle_z(t,x)/(U_w/A_y)^2$ for CP43m (——, in red), CP63s (——, in black), and CP84s (- - - -, in blue). (d) Time–space correlations of $\langle\tau_0\rangle_z$ for local $\tau_0/(U_w/A_y)^2$; —○—, CP63s; —△—, CP84s. The contour levels are [-0.1 0.1 0.3 0.5 0.7] and the filled symbols represent the negative correlation.

shown in Fig 9(b,c). Here, the spanwise-averaged local wall-shear stress, $\langle\tau_0\rangle_z(t,x)$, reasonably detects such local separation events due to the narrow domains. Some examples of the time–space maps of $\langle\tau_0\rangle_z$ are shown in Fig. 10(a, b) for strong pressure-gradient cases, CP63s and CP84s. These two cases have almost-zero wall-shear stress but also quite different probability density function of $\tau_0$. Note that the mean $\langle\tau_0\rangle$ of CP84s is slightly negative. There are intermittent strong negative wall-shear-stress regions, which are observed as the dark blue spots in Fig. 10(a, b) and which make the tails of the p.d.f. wider for CP84s. Such intermittent wall-shear stress can be scaled by $U_w$ rather than $U_w/A_y$, indicating that local separations grows until the order of the box height, $\delta$. Figure 10(d) shows the time–space correlation of $\langle\tau_0\rangle_z$ for the 'shear-less' cases (CP63s, CP84s), exhibiting a good agreement for small-scale separation events of the size of $L_z$. The negative correlation at $\delta x/L_z = 0.5A_x$ also indicate the existence of large-scale separation (or sweep) events. The advection velocity is roughly $0.2U_w$, which is estimated from the slope of the time–space correlation.

## 5. Conclusion

The minimal-span Couette-Poiseuille (C-P) turbulence has been investigated using newly developed direct numerical simulations (DNSs), targeting 'shear-less' wall-bounded turbulence, as found in adverse-pressure gradient turbulent boundary layer (APG-TBL) flow at the verge of separation. A minimal-span simulation eliminates the upstream history effects, which complicates the analysis in TBLs, and drastically reduces the computational cost. It can be used as a fast method to characterise zero wall-shear ('shear-less') wall-bounded turbulence.

Since the large-scale motions are artificially limited by the narrow computational domain, the box-dependency on the Couette-Poiseuille flow has been revealed. The nondimensional numerical parameters are four: the Reynolds number, $Re_z$, box aspect ratios, ($A_x$, $A_y$) and the nondimensional pressure gradient parameter $\beta$, which is defined as for APG-TBL flows. The minimal-span 'pure' Couette turbulence features a limited logarithmic layer until $y \approx L_z/3$, which agrees with that in the minimal-span channel turbulence [12]. In this minimal Couette flow, the homogeneously sheared region which exists above the limited logarithmic layer is similar to the statistically-stationary homogeneous shear turbulence (SS-HST) [9].

The effects of mild and strong pressure gradient on the turbulence statistics have been investigated. The mean velocity in the minimal-span C-P flow also results in a roughly uniform shear region. The latter is considered to be relevant to the outer layer in the APG-TBLs. As expected, the standard pressure-driven channel flow is realised over the top wall. It has been shown that the almost 'shear-less' case still has strong turbulent intensity near the wall, although the tangential Reynolds stress is linearly approaching zero. Such region can be considered as 'inactive', since the Corrsin's nondimensional shear parameter is very small.

It is considered that the sweeps (or bursts when we consider the rotational symmetry) from the upper layer are the primary ingredient to generate such local separation and velocity fluctuations in the 'inactive' layer. The dynamics of the local separation events in the minimal-span domain have also been investigated. Although the effect of large-scale motions in a full simulation is eliminated from the dynamics of local separation events, it appears that the reverse flow (separating region) is not limited by the spanwise box dimension, $L_z$. The local reversed flow is mostly attached to the wall and occasionally grows up to the order of the channel height, $\delta$, being advected with the local advection velocity even in the minimal-span simulations. The wall-attached regions of the large-scale reversed flow advect with roughly $0.2U_{\rm w}$, independently of the aspect ratios, $A_x$ and $A_y$. The local separation events are found to be multi-scale phenomena even in minimal-span simulations, with the effect of large-scale motions in a full simulation also requiring further investigation.

## Acknowledgements
This work was funded in part by the Coturb program of the European Research Council (ERC). The authors would like to acknowledge the support from the Australian Research Council (ARC) for this work which is gratefully acknowledged. Dr Atkinson was supported by an ARC Discovery Early Career Researcher Award (DECRA) fellowship. The computational resources were provided by the Australian National Computational Infrastructure (NCI) through NCMAS which is supported by the Australian government. The authors are grateful to Dr. J. Cardesa for his helpful comments on the manuscript.

## References
[1] F. H. Clauser. Turbulent boundary layers in adverse pressure gradients. *J. Aero. Sci.*, 21:91–108, 1954.
[2] A. A. Townsend. The development of turbulent boundary layers with negligible wall stress. *J. Fluid Mech.*, 8:143–155, 1960.
[3] G. L. Mellor and D. M. Gibson. Equilibrium turbulent boundary layers. *J. Fluid Mech.*, 24:225–253, 1966.
[4] V. Kitsios, A. Sekimoto, C. Atkinson, J.A. Sillero, G. Borrell, A. G. Gungor, J. Jiménez, and J. Soria. Direct numerical simulation of a self-similar adverse pressure gradient turbulent boundary layer at the verge of separation. *J. Fluid Mech.*, 829:392–419, 2017.
[5] R. Johnstone, G. N. Coleman, and P. R. Spalart. The resilience of the logarithmic law to pressure gradients: evidence from direct numerical simulation. *J. Fluid Mech.*, 643:163–175, 2010.

[6] S. Pirozzoli, M. Bernardini, and P. Orlandi. Large-scale motions and innner/outer layer interactions in turbulent Couette–Poiseuille flows. *J. Fluid Mech.*, 680:534–563, 2011.

[7] Y. Maciel, M. P. Simens, and A. G. Gungor. Coherent structures in a non-equilibrium large-velocity-defect turbulent boundary layer. *Flow Turbul. Combust.*, 98:1–20, 2016.

[8] A. Lozano-Durán, O. Flores, and J. Jiménez. The three-dimensional structure of momentum transfer in turbulent channels. *J. Fluid Mech.*, 694:100–130, 2012.

[9] A. Sekimoto, S. Dong, and J. Jiménez. Direct numerical simulation of statistically stationary and homogeneous shear turbulence and its relation to other shear flows. *Phys. Fluids*, 28:035101, 2016.

[10] S. Dong, A. Lozano-Durán, A. Sekimoto, and J. Jiménez. Coherent structures in statistically stationary homogeneous shear turbulence. *J. Fluid Mech.*, 816:167208, 2017.

[11] A. A. Townsend. *The Structure of Turbulent Shear Flows, 2nd ed.* Cambridge U. Press, 1976.

[12] O. Flores and J. Jiménez. Hierarchy of minimal flow units in the logarithmic layer. *Phys. Fluids*, 22:071704, 2010.

[13] M. MacDonald, D. Chung, N. Hutchins, L. Chan, A. Ooi, and R. Garcia-Mayoral. The minimal-span channel for rough-wall turbulent flows. *J. Fluid Mech.*, 816:5–42, 2017.

[14] P. Orlandi, M. Bernardini, and S. Pirozzoli. Poiseuille and Couette flows in the transitional and fully turbulent regime. *J. Fluid Mech.*, 770:424441, 2015.

[15] J. Kim, P. Moin, and R. D. Moser. Turbulent statistics in fully developed channel flow at low Reynolds number. *J. Fluid Mech.*, 177:133–166, 1987.

[16] J. Shen. Efficient Chebyshev–Legendre Galerkin methods for elliptic problems. In *Int. Conf. spectral and high-order methods*, pages 233–239, 1996.

[17] V. Kitsios, C. Atkinson, J. A. Sillero, G. Borrell, A. G. Gungor, J. Jiménez, and J. Soria. Direct numerical simulation of a self-similar adverse pressure gradient turbulent boundary layer. *Int. J. Heat Fluid Fl.*, 61:129–136, 2016.

[18] M. Skote and D.S. Henningson. Direct numerical simulation of a separated turbulent boundary layer. *J. Fluid Mech.*, 471:107–136, 2002.

[19] V. Avsarkisov, S. Hoyas, M. Oberlack, and J. P. Garcia-Galache. Turbulent plane Couette flow at moderately high Reynolds number. *J. Fluid Mech.*, 751, 2014.

[20] J. Jiménez. Near-wall turbulence. *Phys. Fluids*, 25:101302, 2013.

[21] P.-Å. Krogstad and P. E. Skåre. Influence of a strong adverse pressure gradient on the turbulent structure in a boundary layer. *Phys. Fluids*, 7:2014–2024, 1995.

[22] P. R. Spalart and J. H. Watmuff. Experimental and numerical study of a turbulent boundary layer with pressure gradients. *J. Fluid Mech.*, 249:337–371, 1993.

[23] S. Corrsin. Local isotropy in turbulent shear flow. *Res. Memo. NACA.*, 58B11, 1958.

**Institute of Physics**
Dirac House, Temple Back
Bristol BS1 6BE UK

ISSN: 1742-6588
ISBN 978-1-5108-6168-8

CURRAN ASSOCIATES INC.
proceedings.com